Recent Advances of Spectrometric and Spectroscopic Techniques in Food Quality and Safety

Recent Advances of Spectrometric and Spectroscopic Techniques in Food Quality and Safety

Editors

Weiying Lu
Yanping Chen

MDPI • Basel • Beijing • Wuhan • Barcelona • Belgrade • Manchester • Tokyo • Cluj • Tianjin

Editors

Weiying Lu
Department of Food Science
Engineering, School of
Agriculture and Biology
Shanghai Jiao Tong
University
Shanghai
China

Yanping Chen
Department of Food Science
Engineering, School of
Agriculture and Biology
Shanghai Jiao Tong
University
Shanghai
China

Editorial Office
MDPI
St. Alban-Anlage 66
4052 Basel, Switzerland

This is a reprint of articles from the Special Issue published online in the open access journal *Molecules* (ISSN 1420-3049) (available at: www.mdpi.com/journal/molecules/special_issues/food_quality_and_safety).

For citation purposes, cite each article independently as indicated on the article page online and as indicated below:

LastName, A.A.; LastName, B.B.; LastName, C.C. Article Title. *Journal Name* **Year**, *Volume Number*, Page Range.

ISBN 978-3-0365-8231-3 (Hbk)
ISBN 978-3-0365-8230-6 (PDF)

Contents

About the Editors

Weiying Lu

Dr. Weiying Lu is an Associate Professor at Shanghai Jiao Tong University. He earned his Ph.D. degree in analytical chemistry from Ohio University. He specializes in the research fields of food analysis chemistry, food safety, and quality control. His research interests encompass theoretical and applied aspects of combining fingerprinting techniques such as high-performance liquid chromatography (HPLC), ultra-high-performance liquid chromatography–time-of-flight mass spectrometry (UHPLC-TOF-MS), gas chromatography (GC), and chemometrics in food fingerprinting analysis. These techniques are focused on food products such as olive oil and dairy products. He also investigates the traceability and quality identification of functional health food products such as goji berries and Plantago seeds and applications of deep learning techniques in food quality assessment.

Dr. Lu has led various research projects, including grants from the National Natural Science Foundation of China and industry-sponsored collaborations. His scholarly contributions include over 30 publications in prestigious domestic and international journals. Additionally, he has obtained multiple patents and software copyrights. Dr. Lu's outstanding work has garnered recognition, including the American Chemical Society Member Award and the Journal of Agricultural and Food Chemistry Excellence in Review Award.

Yanping Chen

Dr. Yanping Chen is an Assistant Professor in the Department of Food Science and Technology at Shanghai Jiao Tong University. Her research interests include food flavor chemistry, sensory evaluation, and flavor perception. She obtained a Ph.D. degree from the Chinese University of Hong Kong in 2016 and obtained master's and bachelor's degrees from Jiangnan University in 2013 and 2010. She has published 36 academic papers, including 2 highly cited papers and 2 papers with IF>15. She has been involved in more than 20 research projects funded by the Chinese national natural science foundation and the food industry as the principal investigator or a co-investigator. She is a member of the Youth Working Committee of Shanghai Food Society, a member of the China Animal Products Processing Research Association, and a member of the American Chemical Society. She has served as Guest Editor for the *Journal of Food Biochemistry*, *Biosensors*, *Foods*, and other journals. She is also a reviewer for food-related journals. She has guided many students to win awards in various food innovation competitions and receive other honorary titles such as Shanghai Jiao Tong University Outstanding Graduate and Shanghai Outstanding Graduate.

molecules

MDPI

Article

In Vivo Detection of Tetrodotoxin in *Takifugu obscurus* Based on Solid-Phase Microextraction Coupled with Ultrahigh-Performance Liquid Chromatography–Tandem Mass Spectrometry

Hengli Meng [1,2], Shui Jiang [1,2,*], Yin Zhang [3], Yun Hu [4,*] and Yuan Liu [1,2]

[1] Department of Food Science & Technology, School of Agriculture & Biology, Shanghai Jiao Tong University, Shanghai 200240, China
[2] Shanghai Engineering Research Center of Food Safety, Shanghai 200240, China
[3] Key Laboratory of Meat Processing of Sichuan, Chengdu University, Chengdu 610106, China
[4] Yangzhou Center for Food and Drug Control, Yangzhou 225000, China
* Correspondence: jiangshui@sjtu.edu.cn (S.J.); hy@yzzjs.com (Y.H.)

Abstract: Pufferfish is nutritious and delicious, but the tetrodotoxin (TTX) that may exist in its body poses a serious safety hazard. It is important to use scientific and effective methods to detect the TTX in pufferfish, but most of the existing methods require complex pre-treatment steps and have sample lethality. The solid-phase microextraction (SPME) technology can be used for in vivo detection due to its advantages such as no solvent demand, simple operation, and fast detection speed. In this study, the GO-PAN@PNE SPME fibers were made via a dipping method, and their extraction effect was verified in the TTX aqueous and spiked fish. The established method has good reproducibility, and the limit of detection of TTX in pufferfish was 32 $ng{\cdot}g^{-1}$, and the limit of quantitation was 150 $ng{\cdot}g^{-1}$, which can meet the detection needs of pufferfish for safe consumption. This method was used to in vivo detect the *Takifugu obscurus* exposed to the TTX, to determine the content of TTX in the pufferfish muscle. The detection method established in this study can relatively quickly and easily realize the in vivo detection of TTX in the pufferfish, which can provide theoretical support for improvement in the food safety level of the pufferfish.

Keywords: tetrodotoxin; *Takifugu obscurus*; solid-phase microextraction; ultrahigh-performance liquid chromatography–tandem mass spectrometry

Citation: Meng, H.; Jiang, S.; Zhang, Y.; Hu, Y.; Liu, Y. In Vivo Detection of Tetrodotoxin in *Takifugu obscurus* Based on Solid-Phase Microextraction Coupled with Ultrahigh-Performance Liquid Chromatography–Tandem Mass Spectrometry. *Molecules* **2022**, 27, 6122. https://doi.org/10.3390/molecules27186122

Academic Editor: Francesco Cacciola

Received: 19 August 2022
Accepted: 15 September 2022
Published: 19 September 2022

1. Introduction

Pufferfish is a general name for a group of poisonous economic fish species of Pelecypoda, containing more than 200 species, distributed in temperate, tropical, and subtropical regions. Pufferfish is rich in flavoring amino acids, nucleotides, and some flavor peptides, which form the unique flavor of pufferfish [1], so it is welcomed by consumers all over the world, especially in China, Japan and Korea. However, the possible presence of tetrodotoxin (TTX) in pufferfish poses a great threat to the safety of pufferfish consumption. Pufferfish poisoning incidents frequently occur all over the world, especially in Asian countries [2,3].

TTX is generally believed to be produced by marine bacteria and enriched in higher organisms such as pufferfish through the food chain and mostly accumulates in the liver and ovaries of pufferfish [4]. Tetrodotoxin is a neurotoxin that can selectively block sodium channels on the surface of muscle and nerve cell membranes when ingested, thereby blocking action potentials and causing nerve and muscle paralysis, which can lead to death in severe cases [5]. Furthermore, tetrodotoxin has high stability, which is difficult to be destroyed by normal cooking methods, and there is no targeted antidote or antitoxin after poisoning [6]. Therefore, it is very important to detect tetrodotoxin in pufferfish using scientific methods to avoid the occurrence of tetrodotoxin poisoning.

Traditionally, the detection of TTX relies on the mouse bioassay, immunoassay, and instrumental assay. The mouse bioassay is a traditional assay method that determines the TTX content based on the time of death of mice after the injection of sample extracts. Different mouse strains have different susceptibility to TTX [7]. The limit of detection of TTX using Kunming strain mice is 0.56 $\mu g \cdot g^{-1}$ [8]. The mouse bioassay is intuitive and has low costs, but it has a high detection limit, poor reproducibility, and no specificity. The principle of an immunoassay is the qualitative or quantitative analysis of the substance to be measured in the sample by the specific binding of antigen and antibody. The commonly used immunoassay for the detection of TTX is the enzyme-linked immunosorbent assay (ELISA) and the colloidal gold strip test [9]. The immunoassay method is highly sensitive, specific, and easy to operate. In instrumental assays, the level of TTX in the sample is obtained by extracting the TTX from the specimen with a specific solution and then measuring the TTX content with an instrument. The LC-MS method is a commonly used instrumental method for TTX detection, which has a low limit of detection and a good linear range [10,11]. All of the above methods can achieve the detection of TTX, and each has its advantages and disadvantages. However, most of the above methods require complex pre-treatment steps and have sample lethality, which destroys the higher commercial value of pufferfish. Therefore, it makes sense to find a rapid in vivo method for detecting TTX. The pre-treatment of samples often requires the biological sample to be killed and the corresponding solvent to be used for the extraction of the substances to be tested, which is not only a complex and cumbersome procedure but also unethical.

The solid-phase microextraction (SPME) technique, first proposed by Professor Janusz Pawliszyn in Canada in 1990 [12], is a solvent-free extraction method. It is a non-exhaustive extraction method, based on the partitioning equilibrium of the substance to be measured in the sample and the extraction coating, a phase equilibrium process. This technique was selected as one of the six greatest ideas in the field of analytical science in the 1990s, together with electrospray ionization and capillary electrophoresis [13]. Since the sampling process of SPME does not require the use of solvents and is less damaging to the analytical system, it can be applied to in vivo testing (in vivo SPME). In vivo SPME does not require extraction, centrifugation, purification, and other processes; rather, it simplifies these steps into one step, which greatly shortens its detection time and significantly improves the sampling efficiency; moreover, in vivo SPME technology is less damaging to the analyte system and is more suitable for the analysis of high-value or rare plants and animals such as pufferfish. In addition, in vivo SPME can more accurately reflect the contents and changes in analytes in living organisms, which is conducive to the tracking study of individual organisms [14]. Given these advantages of in vivo SPME technology, in vivo SPME has been used in studies about plants [15], animals [16,17], insects [18], and even humans [19].

In this work, we applied the SPME technique to the detection of TTX, constructed a rapid analysis method for TTX with a self-made SPME fiber, and attempted to use the fiber for the in vivo detection of TTX, to realize the hazard identification and risk assessment of TTX and provide theoretical support for improving the food safety of pufferfish.

2. Results and Discussion

2.1. Preparation and Characterization of SPME Fibers

GO-PAN@PNE fibers with brown color and uniform coating thickness were successfully prepared via a dipping method (Figure 1). In recent years, many new materials are used as novel SPME fiber coatings, including some carbon materials such as graphene oxide (GO) [20]. GO is an oxide of graphene and has been used for the SPME of various polar substances [21–23]. After oxidation treatment, GO still maintains the layer structure of graphite, and oxygen-containing groups that are highly hydrophilic and have a large specific surface area, such as carboxyl, hydroxyl, and epoxy groups, are introduced on each layer of graphene monoliths [24]. The aforementioned advantages make the GO a potential material with good adsorption capacity to TTX. The PAN is used as a biocompatible polymer in biomedical applications, such as dialysis and ultrafiltration. Its high levels of chemical and mechanical stability allow PAN to be used as an ideal stationary phase fixative [15]. The NE, formed when N-methyl is removed from epinephrine, is a catecholamine neurotransmitter that shares a similar oxidative polymerization process with the common neurotransmitter dopamine [25]. Polynoradrenaline (PNE) is the product of a polymerization reaction, and it is capable of sheath wrapping on its own and possesses good biointerfacial properties. In this study, PNE was used to improve the extraction performance and corrosion resistance of the SPME fibers.

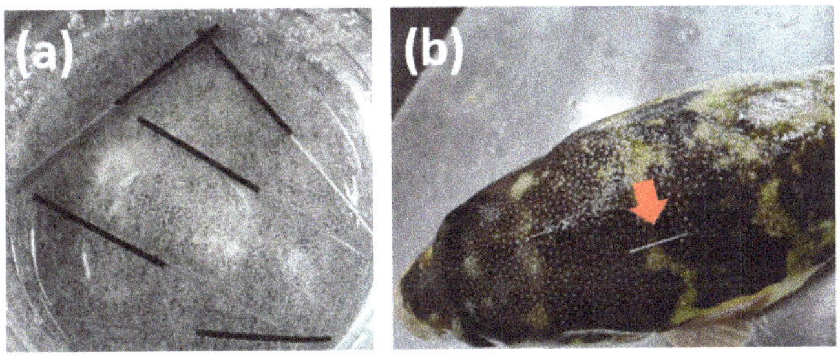

Figure 1. The GO-PAN@PNE fibers made via dipping method (**a**) and the in vivo sampling in the dorsal–epaxial muscle of living pufferfish using the SPME fiber (**b**).

During heat treatment, the carboxyl group on GO sheets nucleophilically attacks the carbon of the cyano group to initiate the anionic cyclization of the PAN molecular chain, which in turn connects GO and PAN to form a more stable membrane structure [26]. Graphene oxide–polyacrylonitrile composites have applications in the fields of electrochemistry, dye removal, and VOC adsorption [27–30] but have been rarely used in direct immersion SPME.

A Nova Nano-SEM 450 field-emission scanning electron microscope (FESEM, FEI, USA) was used to characterize the fibers. The fiber surfaces were observed at magnifications of 400, 800, and 3000 (at higher magnifications, the coating was damaged by the high-energy electron beam). As shown in Figure 2, at lower magnifications, the overall surface of the fiber coating showed a smooth and uniform morphology without large bumps, depressions, and agglomerates, and this uniform distribution facilitated the reproducibility of the fibers. At higher magnification, it can be observed that the surface of the fiber gradually became rougher and appeared folded and porous. The DMF was evaporated from the coating solution, and the remaining PAN and GO were fixed on the surface of the fiber, which formed a stable structure. Such a structure was beneficial to increasing the contact area and improving the loading capacity of the fiber, which enhanced the extraction performance.

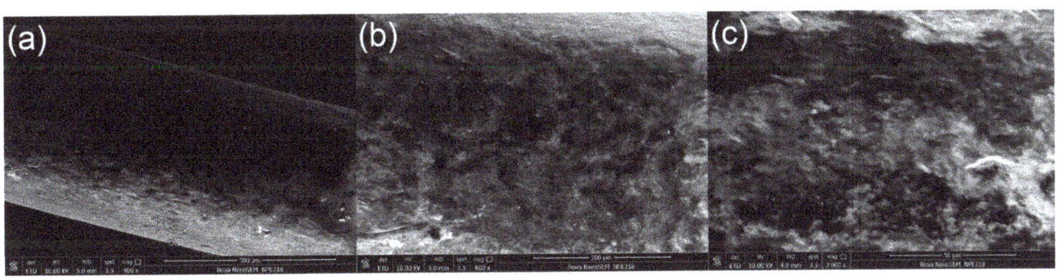

Figure 2. SEM images of the GO-PAN@PNE fiber in different magnifications: (**a**) 400×; (**b**) 800×; (**c**) 3000×).

2.2. In Vitro Evaluation of SPME Fibers

The extraction of a 1 $\mu g \cdot mL^{-1}$ aqueous solution of TTX and 1 $\mu g \cdot g^{-1}$ spiked fish was performed using blank PAN fibers, GO-PAN fibers, and GO-PAN@PNE fibers, respectively, to evaluate the extraction performance of the self-made fibers. As shown in Figure 3, the extraction performance of the SPME fibers with the addition of GO was significantly improved, compared with the blank PAN fibers, because GO has a large specific surface area and strong adsorption to small molecules. Furthermore, the surface of GO contains oxygen-containing groups that are highly hydrophilic and can produce hydrogen bonding and induce forces on tetrodotoxin molecules during the extraction process. It can also be observed that the extraction amount of TTX in the fish meat was lower than that in the aqueous solution when the fibers were not wrapped with the PNE bionic sheath, while the GO-PAN@PNE fibers with the addition of the PNE bionic sheath had better extraction performance in the fish meat. This is because the fish meat samples not only contained TTX but also biomolecules such as proteins, which may reduce the loading capacity of the fibers by binding to the surface and competing with TTX for the adsorption sites during the extraction process. At the same time, the biomolecules in the elution solvent reduced the ionization efficiency of the sample during the LC-MS/MS analysis. These biomolecules that adhere to the fiber surface are not easily cleaned off, thus reducing the loading capacity of the fibers for the next use. PNE with good resistance to biocorrosion can inhibit the attachment of biomacromolecules such as proteins to the surface and enhance the extraction performance of the fibers in fish. Meanwhile, the interface of the PNE sheath layer is hydrophilic and does not hinder the binding of TTX to the coating.

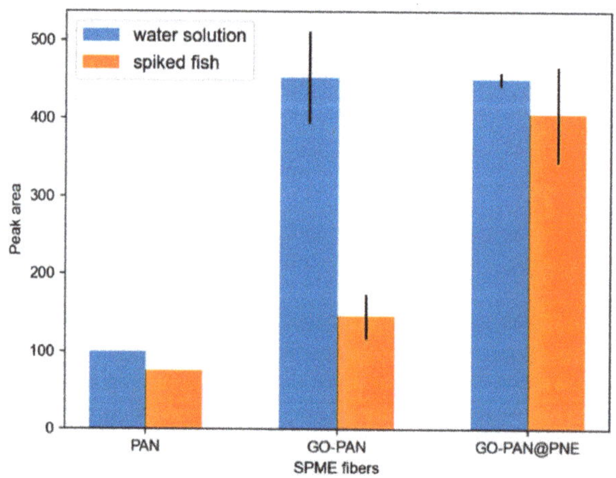

Figure 3. In vitro evaluation of extraction performance of PAN, GO-PAN, and GO-PAN@PNE fibers.

2.3. *Method Optimization*

The results of SPME were related to various factors. The extraction time, elution solvent, and elution time were used as variables in turn to investigate the effects of these factors on the extraction amount and to select the best experimental method for subsequent evaluation and in vivo SPME extraction.

2.3.1. Extraction Time

The SPME process is essentially an equilibrium of the analytes between the matrix and the extraction phase, which requires a certain amount of time to reach equilibrium. Extraction was carried out in the fish meat samples with 1 $\mu g \cdot g^{-1}$ TTX in the time range of 5–60 min. As shown in Figure 4a, the extraction efficiency increased with the extraction time and had little change after 30 min; thus, the extraction time was set as 30 min for subsequent experiments. This extraction time is close to some previous in vivo SPME studies [15,17,31].

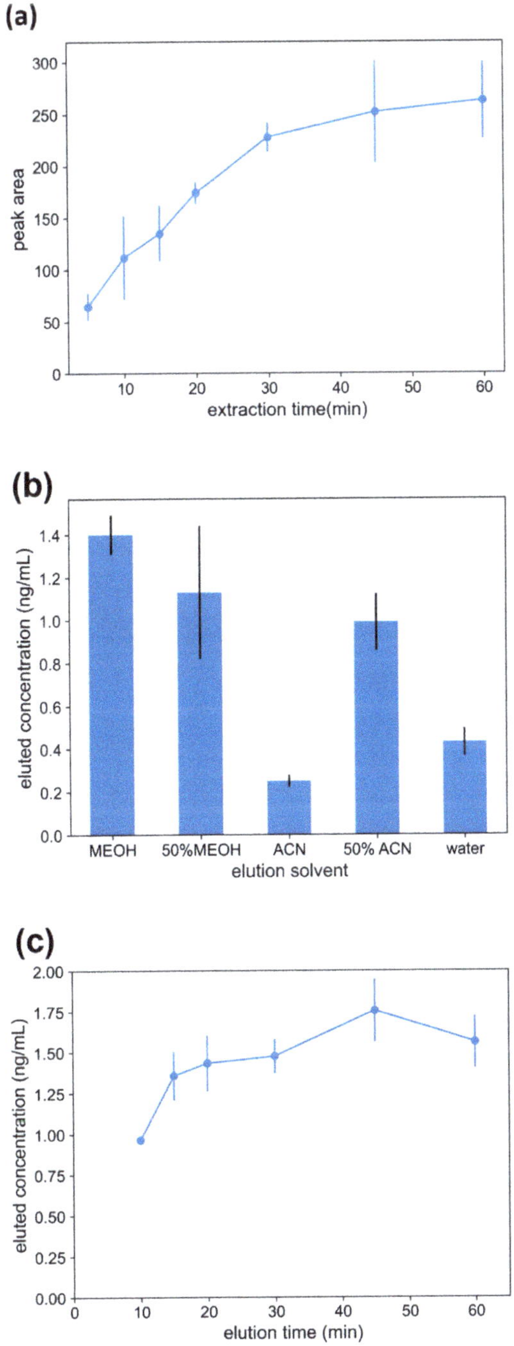

Figure 4. Optimization of SPME method using GO-PAN@PNE fibers in spiked fish samples: (**a**) extraction time; (**b**) elution solvent; (**c**) elution time.

2.3.2. Elution Solvent

The elution solvent greatly affects the efficiency of analyte removal from the fibers. Pure water, 50% (*v*/*v*) methanol/water, pure methanol, 50% (*v*/*v*) acetonitrile/water, and pure acetonitrile were selected as elution solvents for the experiments to select the most suitable solvent for elution. It could be observed that pure methanol had the best effect on the elution of TTX, as shown in Figure 4b, so methanol was selected as the best elution solvent, which is consistent with the previous research [31].

2.3.3. Elution Time

Similar to the extraction process, the elution process also needs a certain time to reach equilibrium. As shown in Figure 4c, the best extraction performance was obtained when the elution time was 30 min.

2.4. Method Evaluation

The proposed in vivo SPME extraction method was further evaluated with the linear range, the limit of detection (LOD), the limit of quantitation (LOQ), and reproducibility. Eight concentration gradients of 5, 10, 50, 100, 250, 500, 750, and 1000 ng·mL^{-1} (or ng·g^{-1}) were determined in the TTX aqueous and TTX spiked fish meat, respectively. As shown in Table 1, the linear range was good in the concentration range of 150–1000 ng·g^{-1}, and the R^2 values were 0.9955 in the aqueous and 0.9895 in the spiked fish. The LOD and LOQ of the method were calculated based on the signal-to-noise ratios of 3 and 10, respectively. The LOD and LOQ of the fibers in the TTX aqueous solution were lower than those in the TTX spiked fish meat, which is probably because the biomolecules in the fish samples bound to the fiber surface and competed with TTX for the adsorption sites during extraction, leading to a decrease in the specificity of the fibers for TTX.

Table 1. Linear range, limit of detection (LOD), limit of quantitation (LOQ), and reproducibility.

Samples	LOD	LOQ	Linear Range	R^2	Reproducibility (RSD, %, *n* = 6)	
					Intra-Fiber	Inter-Fiber
TTX aqueous	11.8 ng·mL^{-1}	81.3 ng·mL^{-1}	100–1000 ng·mL^{-1}	0.9955	13.0	4.63
TTX spiked fish	32 ng·g^{-1}	150 ng·g^{-1}	150–1000 ng·g^{-1}	0.9895	36.1	8.82

The LOD for the fish analyzed with this method is far lower than that of the traditional mouse biological method (0.56 μg·g^{-1}) [8]. However, the LOD of this method is higher than that of the instrumental analysis using the liquid-phase extraction method [10,11]. Liquid-phase extraction could completely extract the TTX in the sample by grinding the sample to minced status and using the solvent. The SPME technology is an in vivo extraction method that belongs to non-exhaustive extraction, and only a small part of the TTX in the sample was extracted by using the SPME fibers. The lethal dose of TTX for humans is about 10,000 MU (about 2 mg) [3]. A concentration of 2 μg·g^{-1} of TTX in a 1 kg pufferfish is required to reach lethal levels, so the LOQ of 150 ng·g^{-1} of this method can fully meet the needs of food safety testing for pufferfish.

The reproducibility of the proposed method was determined in a 1000 ng·mL^{-1} TTX aqueous solution and 1000 ng·g^{-1} spiked fish. As shown in Table 1, the inter-fiber RSD values of the SPME fibers in the aqueous solution and spiked fish meat were 4.63 and 8.82, respectively, indicating that the uniformity of the SPME fibers' preparation was good. However, the intra-fiber reproducibility was not very ideal, especially in the spiked fish. Although PNE was used as the biological sheath in the fiber preparation process to improve the anti-biological contamination ability of the fibers, the extraction stability of a single fiber in spiked fish meat still decreased with the increase in extraction time, possibly because the PNE sheath was scraped off in the fish flesh. The biocorrosion resistance and service life of the fibers need to be improved.

2.5. In Vivo SPME Extraction

The results of in vivo SPME are shown in Table 2. On the day of the start of the exposure experiment, one of the pufferfish (No. 1) was detected using SPME extraction, and no TTX was detected, which indicated that the pufferfish was suitable for the exposure experiment. After the pufferfish were exposed to TTX for 7 days, the dorsal muscles of the pufferfish surviving in good condition were extracted in vivo using the self-made SPME fibers to obtain the level of TTX. The results of fish No. 4 and No. 6 were higher than the LOD (32 ng·g^{-1}) but lower than the LOQ (150 ng·g^{-1}), which can indicate the presence of TTX in the fish muscle but cannot be precisely quantified. While the TTX concentration of fish No. 5 was below the LOD, it could not be determined that fish No. 5 contained TTX. During the feeding process of the pufferfish, the phenomenon of fish death occurred due to long transportation time, the replacement of the feeding environment, and changes in the feeding water quality. There were several fish that died 2 days after exposure (No. 2), and 6 days after exposure (No. 3). In order to explore the content of TTX in the body of pufferfish after death, we also performed in vivo SPME extraction on the dead pufferfish, using the method described in Section 3.6 with the anesthesia step omitted.

Table 2. In vivo SPME results of TTX.

Pufferfish Number	Exposure Days	Calculated TTX Content (ng·g^{-1})
1	0	0
2	2	141
3	6	480
4	7	91
5	7	25
6	7	53

Pufferfish No. 1, 2, and 3 died before testing, and pufferfish No. 1 had not been exposed to TTX.

The result of the exposure experiment demonstrated that pufferfish could accumulate TTX in the muscles when eating TTX-contained food, and the TTX contents of the different pufferfish individuals were different. Furthermore, TTX could not be detected in the muscle of those pufferfish that were not fed with TTX-contained food. These results of the exposure experiment could be explained by the exogenous theory in the TTX source theory [4]. Generally speaking, when the TTX concentration of pufferfish is lower than 100 MU·g^{-1} (about 20 μg·g^{-1}), it can be considered that pufferfish are non-toxic or low-toxic to the human body [17], so all the pufferfish in this research were considered non-toxic to humans.

At the same time, we also found that after the fish died, the level of TTX in the muscles significantly increased, and this increase might even be close to 10 times. As shown in Table 2, the TTX contents in the dorsal muscles of two dead pufferfish (No. 2 and No. 3) were higher than that of the living fish. Even the fish that had been fed TTX for only 1 day (No. 2) had a higher TTX content in the muscle after death than the living pufferfish (No. 4, 5, and 6) that had been fed for 7 days. The reason for this phenomenon may be that, after the fish died, the TTX in the blood, liver, gonads, and other parts with high content of TTX diffused into the muscle, resulting in a substantial increase in the content of TTX in the muscle. Therefore, when cooking pufferfish, it should be handled in time after the death of the fish to avoid the increase in TTX in the muscles.

3. Materials and Methods

3.1. Chemicals and Materials

The species of the pufferfish used in this experiment is *Takifugu obscurus*, which was purchased from Jiangsu Zhongyang Company (Nantong, China). The weight of the pufferfish was about 250g. Medical stainless steel wires (0.5 × 2.5 cm) were purchased from Guangdong Longxin Co. (Zhongshan, China). Graphene oxide (GO) powder (>98 wt%) was purchased from 3A Chemicals Co. (Shanghai, China). Polyacrylonitrile (PAN, average

Mw 85000), norepinephrine (NE, 98%), and glacial acetic acid (analytical purity) were purchased from Macklin Co. (Shanghai, China). N,N-dimethylformamide (DMF, analytical purity), tetrodotoxin (≥99%), and methanol (chromatographic grade) were purchased from Aladdin Reagent Co. (Shanghai, China). Tris(hydroxymethyl)amine (TRIS, ≥99%) was purchased from Solarbio Co. (Beijing, China). And ultrapure water was prepared by the water purifier purchased from Chengdu Haokang Technology Co. (Chengdu, China).

3.2. *Preparation of SPME Fibers*

The GO-PAN@PNE fibers used in this study were prepared via a simple dipping method, with the stainless steel wires as the substrates.

3.2.1. Preparation of Coating Solution

Firstly, 0.1 g of PAN solid was added to 0.8 g of DMF in a 1.5 mL centrifuge tube, and the suspension was heated at 90 °C for 1 h to fully dissolve the PAN solid and form PAN glue. Then, 20 mg of GO powder was added to 250 μL of DMF, and the mixture was ultrasonicated for 1 h to make the GO uniformly distributed; then, it was added to the PAN glue and stirred well to form a coating solution.

3.2.2. Dipping Process of Fibers

The medical stainless steel wires were cut into small segments with a length of 2.5 cm and ultrasonicated in concentrated hydrochloric acid for 10 min to activate the steel wire surface. Additionally, the GO-PAN@PNE fiber was prepared via a dipping method: One end of the treated stainless steel wire was dipped vertically into the coating solution and taken out slowly; then, the fiber was heated at 90 °C for 3 min to volatilize DMF and fix the coating. The dipping and heating steps were repeated several times so that the coating was wrapped as evenly as possible on the steel wire, and the final thickness of the coating was about 50 μm.

3.2.3. Addition of PNE Bionic Sheath

A certain amount of NE powder was dissolved in a mixed solvent of Tris buffer (10 mM, pH = 8.5) and methanol (1:1, v/v) to prepare the NE solution (2.0 mg·mL^{-1}). Afterward, the prepared GO-PAN fibers were thoroughly submerged in the NE solution for 24 h.

3.3. *Characterization of SPME Fibers*

The self-made GO-PAN@PNE SPME fibers were cut to the appropriate size (about 1 cm) and attached to the sample stage with a conductive adhesive. A Nova Nano-SEM 450 field-emission scanning electron microscope (FESEM, FEI, Hillsboro, OR, USA) was used to observe the surface morphology of the fibers.

3.4. *Extraction Performance Evaluation*

In addition to the final GO-PAN@PNE fibers, blank PAN fibers without GO addition and the GO-PAN fiber without the PNE sheath layer were prepared as references. The TTX aqueous (1 μg·mL^{-1}) and TTX spiked fish (1 μg·g^{-1}) were used as samples to conduct the in vitro SPME extraction. The extracted TTX amount was used to evaluate the performances of these three fibers and to confirm the role of GO and PNE in the extraction. The sample preparation and extraction method for the spiked fish were as follows: A few *Takifugu obscurus* purchased from Jiangsu Zhongyang Company (Nantong, China) were dissected, and 5 g of their dorsal muscles were ground into minced meat and then added into 10 mL glass sample bottles. Then, 5 mL of the aqueous solution of TTX was added to prepare a spiked fish sample of TTX, which was used for the in vitro SPME extraction. The pretreated fibers were inserted vertically into the glass bottles containing the spiked fish sample to ensure that the coating was completely immersed in the samples. The extraction process was controlled for a certain period. After extraction, the fiber was rinsed with ultrapure water for 3 s and then dried with a Kimwipe paper towel. The extracted TTX was eluted from the fibers with 250 μL of elution solvent for a certain time. The elution solution was analyzed via the LC-MS/MS analysis. The in vitro extraction was repeated to optimize the extraction time, elution time, and elution solvent.

3.5. Exposure Experiment

The pufferfish were raised in a fish tank containing 40 L of oxygenated dechlorinated tap water and fed with eel meal containing TTX. The exposure dose of TTX was 2 MU·g^{-1} body mass·day-1 (each pufferfish was fed with about 100 μg of TTX every day; MU: mouse unit, 1 MU represents the average amount of toxin that kills a male mouse weighing 20 g within 30 min after intraperitoneal administration, and 1 MU is equivalent to about 0.2 mg). The pufferfish exposure experiment lasted for 7 days. The feeding condition was controlled during the exposure period. The water temperature was controlled at about 25 °C with an air conditioner. The water was continuously oxygenated with an air pump to ensure the oxygen content of the water body. A filter device was used to filter the residual bait and feces. Nitrifying bacteria were added to establish a nitrification system. Additionally, 0.5% of NaCl was added to maintain water osmotic pressure. During the exposure experiment, the in vivo SPME extraction was performed to determine the TTX content in the pufferfish.

3.6. In Vivo SPME Extraction

The procedure of in vivo SPME extraction in the fish muscle was similar to previous research [17,31]. The pufferfish was picked up from the water and submerged in a 0.1% eugenol (*v*/*v*) aqueous solution until its body was unbalanced. A medical syringe needle was inserted into the side of the dorsal muscle near the fin of the pufferfish and then was withdrawn to obtain an approximately 1.5 cm deep pinhole. Afterward, the SPME fiber was inserted into the pinhole to extract the TTX in the muscle, as shown in Figure 1b. During the extraction process, the pufferfish was placed back into the clean water to make it free to swim. After the extraction, the pufferfish was picked up from the water again, and the SPME fiber was withdrawn from the muscle. Finally, the TTX was eluted by the solvent from the SPME fiber for the instrumental analysis to obtain the TTX content in the pufferfish muscle.

3.7. Instrumentation (UPLC-MS/MS)

The instrumental analysis was based on a UPLC system coupled to a triple-quadrupole tandem mass spectrometer (Nexera LC30AD&SCIEX SelexION Triple Quad 5500 System). An ACQUITY BEH HILIC column (2.1 × 100 mm, 1.7 μm) was used for separation, and the column temperature was 40 °C, and the injection volume is 2 μL. Gradient elution was applied by using water with 0.1% formic acid as mobile phase A and ACN with 0.1% formic acid as mobile phase B. The initial gradient of 5% A was kept unchanged for 0.5 min, then ramped to 50% A in 2.5 min and ramped to 95% A in 1 min, and kept for 2 min, then decreased to 5% A in 1 min, and kept for 1.9 min. The flow rate was set at 0.4 mL·min^{-1}, and the total analysis time for each sample was 8 min.

The mass spectrum parameters were as follows: electrospray ionization (ESI); the positive ionization mode; curtain gas 35 psi; ion spray voltage 5500 V; desolvation temperature 600 °C; nebulizing gas 55 psi; auxiliary heating gas 55 psi; analysis of the samples via multiple reaction monitoring (MRM). The transition was monitored with the positive ion mode at m/z 320.1/162.0, and the detailed mass spectrometry parameters are shown in Table 3.

Table 3. Mass spectral parameters of TTX.

Precursor Ion (*m*/*z*)	Product Ion (*m*/*z*)	Declustering Potential (V)	Collision Energy (eV)
320	302	80	40
	162	80	35

In the UPLC-MS/MS results, the characteristic ion and retention time (around 2.43 min) were used for qualitative analysis, and the characteristic peak areas for quantitative analysis. The UPLC-MS/MS chromatograms of the TTX standard are shown in Figure 5.

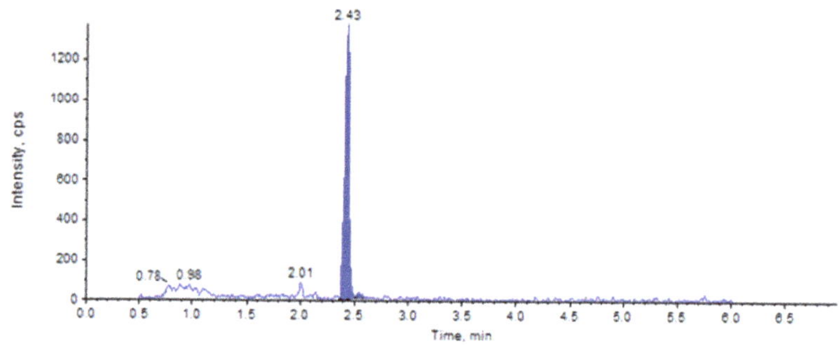

Figure 5. UPLC-MS/MS chromatogram of TTX standard.

4. Conclusions

In this study, a detection method of TTX in pufferfish was established based on in vivo SPME extraction and UPLC-MS/MS. The GO-PAN fibers were fabricated through a dipping method, followed by biomimetic PNE sheath modification. The self-made SPME fibers could successfully extract TTX in both the spiked fish samples and living pufferfish, and the LOD of the proposed method was 32 ng·g^{-1}, which can meet the requirement for safe consumption of pufferfish. The exposure experiment demonstrated that the pufferfish could accumulate TTX in the muscles when eating TTX-contained food, and the TTX level in the muscles of the dead pufferfish significantly increased up to 10 times that of the living pufferfish. The established method could quickly and easily detect the TTX content in pufferfish in vivo without damage to the living pufferfish. This study provides theoretical support for improvement in the food safety level of pufferfish.

Author Contributions: Conceptualization, H.M., S.J. and Y.H.; methodology, H.M. and S.J.; validation, S.J. and Y.L.; formal analysis, H.M.; investigation, H.M.; writing—original draft preparation, H.M.; writing—review and editing, S.J., Y.Z. and Y.L.; supervision, S.J., Y.Z., Y.H. and Y.L.; project administration, Y.H. and Y.L.; funding acquisition, Y.H. All authors have read and agreed to the published version of the manuscript.

Funding: This research was funded by the Key Research and Development Project of Yangzhou in 2020 (YZ2020047), the Open Project Program of State Key Laboratory of Food Nutrition and Safety, Tianjin University of Science and Technology (No. SKLFNS-KF-201801), and Shanghai Science and Technology Commission (Grant No. 19DZ2284200).

Institutional Review Board Statement: Ethical review and approval were waived for this study due to the fact that the fish used in this study were purchased from the general market and were raised in a normal environment, and the experiment caused little damage to them.

Informed Consent Statement: Not applicable.

Data Availability Statement: Not applicable.

Conflicts of Interest: The authors declare no conflict of interest.

Sample Availability: Samples of the compounds are not available from the authors.

References

1. Zhang, N.L.; Ayed, C.; Wang, W.L.; Liu, Y. Sensory-Guided Analysis of Key Taste-Active Compounds in Pufferfish (*Takifugu obscurus*). *J. Agric. Food Chem.* **2019**, *67*, 13809–13816. [CrossRef] [PubMed]
2. Lago, J.; Rodríguez, L.; Blanco, L.; Vieites, J.; Cabado, A. Tetrodotoxin, an Extremely Potent Marine Neurotoxin: Distribution, Toxicity, Origin and Therapeutical Uses. *Mar. Drugs* **2015**, *13*, 6384–6406. [CrossRef] [PubMed]

3. Noguchi, T.; Arakawa, O. Tetrodotoxin—Distribution and accumulation in aquatic organisms, and cases of human intoxication. *Mar. Drugs* **2008**, *6*, 220–242. [CrossRef] [PubMed]
4. Noguchi, T.; Arakawa, O.; Takatani, T. TTX accumulation in pufferfish. *Comp. Biochem. Physiol. Part D Genom. Proteom.* **2006**, *1*, 145–152. [CrossRef] [PubMed]
5. Cui, Z. Study on Processing of Tetrodotoxin Mechanism of Poisoning and Preparation. *Food Sci.* **2003**, *24*, 179–182.
6. Noguchi, T.; Ebesu, J.S.M. Puffer Poisoning: Epidemiology and Treatment. *J. Toxicol. Toxin Rev.* **2001**, *20*, 1–10. [CrossRef]
7. Suzuki, H. Differences in susceptibility of mouse strains to tetrodotoxin. *Toxicon* **2016**, *119*, 168–170. [CrossRef]
8. Jing, W.; Lijun, Y.; Zhaojie, L.I.; Meng, Z.; Xiaohua, S.; Yumin, L.I.U. Quantitative Detection of Tetrodotoxin Using Kunming Strain Mice. *Food Sci.* **2011**, *32*, 181–184.
9. Ling, S.; Chen, Q.A.; Zhang, Y.; Wang, R.; Jin, N.; Pang, J.; Wang, S. Development of ELISA and colloidal gold immunoassay for tetrodotoxin detetcion based on monoclonal antibody. *Biosens. Bioelectron.* **2015**, *71*, 256–260. [CrossRef]
10. Chen, X.W.; Liu, H.X.; Jin, Y.B.; Li, S.F.; Bi, X.; Chung, S.; Zhang, S.S.; Jiang, Y.Y. Separation, identification and quantification of tetrodotoxin and its analogs by LC-MS without calibration of individual analogs. *Toxicon* **2011**, *57*, 938–943. [CrossRef]
11. Nzoughet, J.K.; Campbell, K.; Barnes, P.; Cooper, K.M.; Chevallier, O.P.; Elliott, C.T. Comparison of sample preparation methods, validation of an UPLC-MS/MS procedure for the quantification of tetrodotoxin present in marine gastropods and analysis of pufferfish. *Food Chem.* **2013**, *136*, 1584–1589. [CrossRef] [PubMed]
12. Arthur, C.L.; Pawliszyn, J. Solid phase microextraction with thermal elution using fused silica optical fibers. *Anal. Chem.* **1990**, *62*, 2145–2148. [CrossRef]
13. Handley, J.; Harris, C.M. Great Ideas of a Decade. *Anal. Chem.* **2001**, *73*, 23. [CrossRef] [PubMed]
14. Ouyang, G.F.; Vuckovic, D.; Pawliszyn, J. Nondestructive Sampling of Living Systems Using in Vivo Solid-Phase Microextraction. *Chem. Rev.* **2011**, *111*, 2784–2814. [CrossRef] [PubMed]
15. Fang, X.A.; Chen, G.S.; Qiu, J.L.; Xu, J.Q.; Wang, J.H.; Zhu, F.; Ouyang, G.F. Determination of four salicylic acids in aloe by in vivo solid phase microextraction coupling with liquid chromatography-photodiode array detection. *Talanta* **2018**, *184*, 520–526. [CrossRef]
16. Qiu, J.L.; Chen, G.S.; Liu, S.Q.; Zhang, T.L.; Wu, J.Y.; Wang, F.X.; Xu, J.Q.; Liu, Y.; Zhu, F.; Ouyang, G.F. Bioinspired Polyelectrolyte-Assembled Graphene-Oxide-Coated C18 Composite Solid-Phase Microextraction Fibers for In Vivo Monitoring of Acidic Pharmaceuticals in Fish. *Anal. Chem.* **2016**, *88*, 5841–5848. [CrossRef]
17. Tang, Y.; Huang, S.; Xu, J.; Ouyang, G.; Liu, Y. PLGA-based nanofibers with biomimetic polynoradrenaline sheath for rapid in vivo sampling of tetrodotoxin and sulfonamides in pufferfish. *J. Mater. Chem. B* **2018**, *6*, 3655–3664. [CrossRef]
18. Rodstein, J.; McElfresh, J.S.; Barbour, J.D.; Ray, A.M.; Hanks, L.M.; Millar, J.G. Identification and Synthesis of a Female-Produced Sex Pheromone for the Cerambycid Beetle Prionus Californicus. *J. Chem. Ecol.* **2009**, *35*, 590–600. [CrossRef]
19. Poli, D.; Goldoni, M.; Corradi, M.; Acampa, O.; Carbognani, P.; Internullo, E.; Casalini, A.; Mutti, A. Determination of aldehydes in exhaled breath of patients with lung cancer by means of on-fiber-derivatisation SPME-GC/MS. *J. Chromatogr. B Anal. Technol. Biomed. Life Sci.* **2010**, *878*, 2643–2651. [CrossRef]
20. Zheng, J.; Huang, J.; Yang, Q.; Ni, C.; Xie, X.; Shi, Y.; Sun, J.; Zhu, F.; Ouyang, G. Fabrications of novel solid phase microextraction fiber coatings based on new materials for high enrichment capability. *Trends Anal. Chem.* **2018**, *108*, 135–153. [CrossRef]
21. Grajek, H.; Jonik, J.; Witkiewicz, Z.; Wawer, T.; Purchala, M. Applications of Graphene and Its Derivatives in Chemical Analysis. *Crit. Rev. Anal. Chem.* **2020**, *50*, 445–471. [CrossRef] [PubMed]
22. Jian, Y.; Deng, J.; Zhou, H.; Cheng, J. Fabrication of graphene oxide incorporated polymer monolithic fiber as solid phase microextraction device for determination of organophosphate esters in soil samples. *J. Chromatogr. A* **2019**, *1588*, 17–24. [CrossRef] [PubMed]
23. Tashakkori, P.; Erdem, P.; Merdivan, M.; Bozkurt, S.S. Determination of Phthalate Esters in Water and Coffee by Solid-Phase Microextraction Using Vinyl Terminated Imidazolium Based Ionic Liquid Grafted on Graphene Oxide Coatings. *ChemistrySelect* **2019**, *4*, 2307–2313. [CrossRef]
24. Zhu, Y.; Murali, S.; Cai, W.; Li, X.; Suk, J.W.; Potts, J.R.; Ruoff, R.S. Graphene and graphene oxide: Synthesis, properties, and applications. *Adv. Mater.* **2010**, *22*, 3906–3924. [CrossRef] [PubMed]
25. Taskin, M.B.; Xu, R.; Zhao, H.; Wang, X.; Dong, M.; Besenbacher, F.; Chen, M. Poly(norepinephrine) as a functional bio-interface for neuronal differentiation on electrospun fibers. *Phys. Chem. Chem. Phys.* **2015**, *17*, 9446–9453. [CrossRef] [PubMed]
26. Gao, A.; Xu, D.; Li, C.; Wang, Y.; Xu, L. Interfacial reactions in graphene oxide/polyacrylonitrile composite films. *Compos. Interfaces* **2020**, *28*, 159–173. [CrossRef]
27. Guo, Z.Y.; Huang, J.T.; Xue, Z.H.; Wang, X.M. Electrospun graphene oxide/carbon composite nanofibers with well-developed mesoporous structure and their adsorption performance for benzene and butanone. *Chem. Eng. J.* **2016**, *306*, 99–106. [CrossRef]
28. Jang, W.; Yun, J.; Seo, Y.; Byun, H.; Hou, J.; Kim, J.H. Mixed Dye Removal Efficiency of Electrospun Polyacrylonitrile-Graphene Oxide Composite Membranes. *Polymers* **2020**, *12*, 2009. [CrossRef]
29. Lu, Y.R.; Zhang, W.; Wang, M.; Zhang, H.; Li, J.H.; Luo, W.H. Fabrication of GO/PAN Nanofiber Membrane Grafted with Chitosan as Efficient Adsorbent for Dye Removal. *J. Polym. Environ.* **2022**, *30*, 2943–2954. [CrossRef]

30. Zhao, H.; Wu, X.W.; Liu, Y.G.; Cheng, B.H.; Huang, Z.H.; Fang, M.H.; Min, X. Processing, microstructure and electrochemical properties of reduced graphene oxide reinforced carbon nanofiber formed by gyration. *Chem. Phys. Lett.* **2021**, *767*, 138393. [CrossRef]
31. Chen, L.; Qiu, J.; Tang, Y.; Xu, J.; Huang, S.; Liu, Y.; Ouyang, G. Rapid in vivo determination of tetrodotoxin in pufferfish (Fugu) muscle by solid-phase microextraction coupled to high-performance liquid chromatography tandem mass spectrometry. *Talanta* **2017**, *171*, 179–184. [CrossRef] [PubMed]

molecules MDPI

Article

An Improved POD Model for Fast Semi-Quantitative Analysis of Carbendazim in Fruit by Surface Enhanced Raman Spectroscopy

Qiaoling Yang [1,2], Hong Lin [3], Jinge Ma [4], Niannian Chen [4], Chaomin Zhao [4], Dehua Guo [4], Bing Niu [2], Zhihui Zhao [5], Xiaojun Deng [4,6,*] and Qin Chen [2,*]

1 School of Environmental and Chemical Engineering, Shanghai University, Shanghai 200444, China; yangqiaoling@shu.edu.cn
2 School of Life Sciences, Shanghai University, Shanghai 200444, China; bingniu@shu.edu.cn
3 Department of Endocrine and Metabolic Diseases, Shanghai Institute of Endocrine and Metabolic Diseases, Ruijin Hospital, Shanghai Jiao Tong University School of Medicine, Shanghai 200025, China; lh12263@rjh.com.cn
4 Tech Ctr Anim Plant & Food Inspect & Quarantine, Shanghai Customs, Shanghai, 200135, China; 15678092339@163.com (J.M.); n_smily@163.com (N.C.); chaominzhao@126.com (C.Z.); guodehua@customs.gov.cn (D.G.)
5 Shanghai Oceanhood Instrument Equipment Co., Ltd., Shanghai 201608, China; zhihui.zhao@oceanhood.com
6 School of Kinesiology, Shanghai University of Sport, Shanghai 200438, China
* Correspondence: xjdeng@yeah.net (X.D.); chenqincc@shu.edu.cn (Q.C.); Tel.: +86-13917741123 (X.D.); +86-13611865576 (Q.C.)

Citation: Yang, Q.; Lin, H.; Ma, J.; Chen, N.; Zhao, C.; Guo, D.; Niu, B.; Zhao, Z.; Deng, X.; Chen, Q. An Improved POD Model for Fast Semi-Quantitative Analysis of Carbendazim in Fruit by Surface Enhanced Raman Spectroscopy. *Molecules* **2022**, 27, 4230. https://doi.org/10.3390/molecules27134230

Academic Editor: Daniel Cozzolino

Received: 8 June 2022
Accepted: 28 June 2022
Published: 30 June 2022

Abstract: The current detection method of carbendazim suffers from the disadvantages of complicated preprocessing and long cycle time. In order to solve the problem of rapid quantitative screening of finite contaminants, this article proposed a qualitative method based on characteristic peaks and a semi-quantitative method based on threshold to detect carbendazim in apple, and finally the method is evaluated by a validation system based on binary output. The results showed that the detection limit for carbendazim was 0.5 mg/kg, and the detection probability was 100% when the concentration was no less than 1 mg/kg. The semi-quantitative analysis method had a false positive rate of 0% and 5% at 0.5 mg/kg and 2.5 mg/kg, respectively. The results of method evaluation showed that when the added concentration was greater than 2.5 mg/kg, the qualitative detection method was consistent with the reference method. When the concentration was no less than 5 mg/kg, the semi-quantitative method is consistent between different labs. The semi-quantitative method proposed in this study can achieve the screening of finite contaminants in blind samples and simplify the test validation process through the detection probability model, which can meet the needs of rapid on-site detection and has a good application prospect.

Keywords: surface-enhanced Raman spectroscopy; carbendazim; probability of detection model; semi-quantitative analysis; rapid detection method evaluation

1. Introduction

Carbendazim (CBZ) is a broad-spectrum fungicide that is effective against diseases caused by fungi (e.g., Demodex, Polychaeta) in various crops. However, its residues are toxic to mammals and can cause liver disease and chromosomal aberrations [1,2]. Prashantkumar et al. found that exposure to CBZ in male goats caused testicular damage and impaired liver, kidney and blood function [3]. Meanwhile, the study found that CBZ can change the antioxidant defense system [4]. Hence, in the field of food safety, CBZ is restricted in different countries and foods. China stipulates that the maximum residue limit (MRL; The maximum legal allowable residue concentration of pesticides in an agricultural product, food and feed.) of CBZ in apple is <5 mg/kg [5], and the EU stipulates that the

MRL is <2 mg/kg [6]. The existing detection methods of CBZ include high performance liquid chromatography (HPLC) [7], liquid chromatography-tandem mass spectrometry (LC-MS/MS) [8], electrochemistry [9], immunosensor [10] and so on. Lee et al. used QuEChERS and LC-MS/MS to simultaneously detect the concentrations of thiophanate-methyl and CBZ in pears. The results showed that the detection limit of CBZ was as low as 0.0012 mg/kg [11]. Liu et al. applied HPLC with fluorescence detection to determinate CBZ and thiabendazole in apple juice. The limit of detection was 0.8 μg/kg for CBZ [12]. Although these detection techniques are highly sensitive and reliable, their preprocessing is complex, time-consuming and costly. Furthermore, these detection techniques cannot perform rapid detection for a large number of samples. Nowadays, rapid detection techniques including near-infrared spectroscopy, Raman spectroscopy, hyperspectral, etc. have been widely applied in the field of food quality and safety monitoring. Among them, Surface-enhanced Raman spectroscopy (SERS) [13,14] developed rapidly due to its outstanding sensitivity, specificity, real-time response and "fingerprint" identification characteristics, and is widely performed in food safety assessment [15], especially the qualitative analysis of finite contaminants food [16,17], including antibiotics [18], metal ions [19], microorganisms [20], pesticides [21], etc. For example, Chen et al. detected CBZ in oolong tea by SERS, and used the partial least squares method to quantitatively analyze the content of CBZ. The results showed a good linear relationship between the spiked and predicted carbendazim in methanol–water solution (R = 0.972; slope = 0.975; RMSEP = 0.819 mg/L) [22].

Quantitative analysis of limited finite contaminants has been widely applied in the field of food finite contaminants analysis [23]. The quantitative analysis of Raman spectroscopy is based on a linear proportional relationship between the Raman characteristic peak intensity and the concentration of the analyte. Although SERS technology has high sensitivity and advantages in the field of trace detection, the Raman peak intensity detected is affected by some factors such as the randomness of surface-enhanced particle aggregation, matrix, sample transparency and so on [24]. Hence, Raman peak intensities collected by SERS cannot form a clear linear proportional relationship with the concentration. To resolve this problem, existing standard curve-based quantification methods need to be improved according to the requirements of rapid testing. Recent studies suggested that the Raman peak intensity detected by SERS conforms to a Gaussian distribution [25], which inspired us to use Raman characteristic peak intensity of CBZ for semi-quantitative modeling to achieve rapid screening of limited finite contaminants, thereby avoiding complex quantitative analysis of SERS and improving detection efficiency.

As one of the rapid detection methods for food, SERS technology requires implemental technical specifications to validate its applicability in the screening of food finite contaminants. Both qualitative and semi-quantitative SERS assays are binary outputs, which can be evaluated by a validation system based on binary outputs. At present, there are many methods that can effectively evaluate the accuracy of binary results, such as Wilrich based on ISO 5725 [26], Fleiss's Kappa statistics [27], probability of detection (POD) model [28], Wieringen [29] and so on. POD is a model that reflects the change of detection probability with concentration [30] which can characterize metrics such as sensitivity, specificity, false positive and false negative rates. The POD model can plot qualitative data as a centralized function, and coordinate with the statistical parameters of quantitative method validation simultaneously, providing a unified statistical method for all method validation, solving the statistical problem of the unpaired test part. The evaluation process is concise, efficient, easy to understand, and implemented by non-statisticians, which has been successfully applied to the validation of chemical and microbiological methods [31,32] and to assess the reliability of non-destructive testing [33]. The evaluation method of SERS based on POD model not only satisfied the requirements of regulations, but also can make up for the poor quantitative ability of SERS and the lack of applicable evaluation methods.

Here, the intensity of Raman signature peak at the MRL of CBZ is used as the threshold, and the result is a quickly and effective method of determining whether the added concentration exceeds the MRL. Based on POD model, a standard SERS assay validation system was established to promote the commercial application of SERS in the field of rapid food detection.

2. Results

2.1. Establishment of a Rapid Detection Method for CBZ in Apple

2.1.1. Selection of Raman Characteristic Peaks

The SERS results of CBZ standard, spiked sample (5 mg/kg) and blank sample are shown in Figure 1. Six Raman shift peaks (630 cm^{-1}, 728 cm^{-1}, 1000 cm^{-1}, 1218 cm^{-1}, 1260 cm^{-1} and 1315 cm^{-1}) were observed for CBZ standard and spiked sample compared to the blank sample. The six Raman shift peaks can be used as the qualitative characteristic peaks of CBZ in apple due to the clear peak shapes and distinct intensities. According to the SERS spectrum of CBZ and referring to the relevant literature [34–36]., the assignment of Raman shift and Vibrational Description were calculated and shown in Table 1.

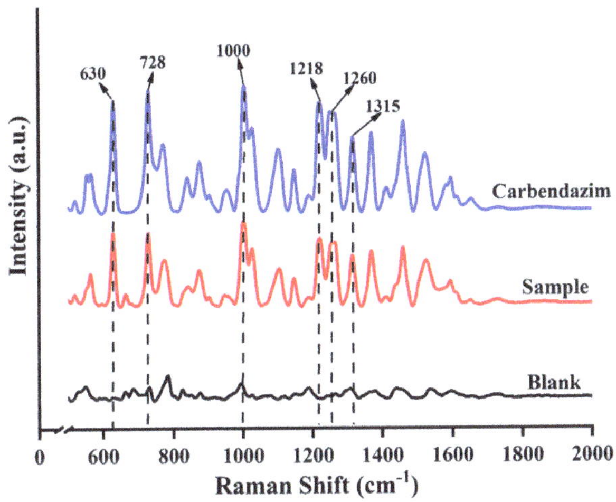

Figure 1. Raman spectra of different samples.

Table 1. Assignments and Raman shifts (cm^{-1}) for SERS spectra of CBZ.

Wavenumber (cm^{-1})	Vibrational Description	Wavenumber (cm^{-1})	Vibrational Description
630	ring stretching and C-C bending	1315	ring stretching
728	C-C bending and C-O-CH_3 bending	1370	C—N stretch
1000	C-N bending and C-C bending and C-O-CH3 stretching	1460	N-H bending and C-H bending
1218	C-C stretch, C-C bending and N-H bending	1523	N-H bending and C-N stretch
1260	C-H bending and N-H bending		

2.1.2. Establishment of POD Model

Based on the Raman characteristic peaks of CBZ, the detection of samples with different spiked concentrations was counted (Table 2). The results showed that the higher the added concentration, the greater the POD of CBZ in the spiked samples. When the added concentration was no less than 0.5 mg/kg, the POD range was (0.975, 1), which meets the requirement of the POD (>0.95). Hence, the LOD of this method was determined to be 0.5 mg/kg.

Table 2. The POD of CBZ in apple at different concentrations.

Concentration (mg/kg)	x	N	POD	LCL	UCL
0.000	0	10	0.000	0.000	0.280
0.050	0	50	0.000	0.000	0.070
0.100	40	50	0.800	0.670	0.890
0.500	39	40	0.975	0.871	1.000
1.000	40	40	1.000	0.910	1.000
2.500	40	40	1.000	0.910	1.000
5.000	40	40	1.000	0.910	1.000

Note: x is the number of positive samples with positive results detected; N is the total number of samples; POD is the probability of detection; LCL is the lower limit of the 95% confidence interval; UCL is the upper limit of the 95% confidence interval.

Based on the Table 2, a POD model was established with the additive concentration of CBZ in apple as the x-axis and the POD under different additive concentrations as the y-axis (Figure 2). According to 'Technology specification for the evaluation of food rapid detection products' (DB36/T 1334-2020) [37], when the sensitivity of the rapid detection method is greater than 95%, the additive concentration is LOD. When the added concentration is 0 mg/kg, this corresponds to a false positive rate for POD, where the sum of specificity and false positive rate is 100%. When added at concentration other than 0 mg/kg, sensitivity and POD were equal and the sum of the false negative rate and sensitivity is 100%.

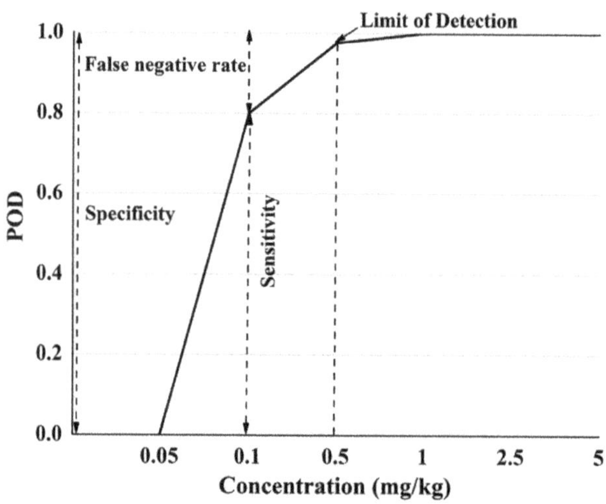

Figure 2. The POD model diagram of SERS detection method for CBZ in apple.

Figure 2 shows that when the sample does not contain CBZ, the POD is 0, indicating that the false positive rate of this method is 0. Since the sum of the false positive rate and specificity is 100%, so the specificity is 100%. When the additive concentration is not more than 0.05 mg/kg, the detection probability is 0, the sensitivity of the method at this concentration is 0, and the false negative rate is 100%. When the additive CBZ concentration is in the range of 0.05 mg/kg–1 mg/kg, the POD is different under different additive concentrations, the sensitivity and false negative rate are different, of which 0.5 mg/kg (sensitivity > 95%) is the LOD. When the added concentration is greater than 1 mg/kg, the POD is 100%. Based on the POD model plots, the specificity, sensitivity, false positive and false negative rates of SERS detection of CBZ in apples at different spiked concentrations can be visually analyzed.

2.1.3. Consistency Evaluation of Qualitative Methods Based on POD Model

(1) Consistency evaluation between qualitative method and reference method

At different additive concentrations, the qualitative POD of SERS method and HPLC [38] (GB/T 23380-2009) is shown in Table 3. It can be seen from Table 3 that when the concentration of CBZ is less than 0.01 mg/kg, the POD of the two methods is 0, and the dPOD is 0. The POD of the SERS method was lower than that of the HPLC method when the additive concentration was in the range of (0.01,2.5) mg/kg, and the POD of the HPLC was 100%. When the additive concentration was not lower than 2.5 mg/kg, the POD of both the SERS method and the reference method was 100%, and the dPOD is 0. This suggests that the SERS qualitative assay method has good reproducibility and is the same as the reference method.

Table 3. Evaluation of the consistency between the Raman qualitative method and the reference method for CBZ in apple.

Method	SERS					HPLC					Difference in POD (dPOD)
Concentration (mg/kg)	x	N	POD	LCL	UCL	x	N	POD	LCL	UCL	
0.00	0	10	0.000	0.000	0.280	0	10	0	0.000	0.280	0.000
0.01	0	50	0.000	0.000	0.070	0	50	0	0.000	0.070	0.000
0.10	40	50	0.800	0.670	0.890	50	50	1	0.930	1.000	0.200
0.50	39	40	0.975	0.870	1.000	40	40	1	0.910	1.000	0.025
2.50	40	40	1.000	0.910	1.000	40	40	1	0.910	1.000	0.000
5.00	40	40	1.000	0.910	1.00	40	40	1	0.910	1.000	0.000

Note: x is the number of positive samples with positive results detected; N is the total number of samples; POD is the probability of detection; LCL is the lower limit of the 95% confidence interval; UCL is the upper limit of the 95% confidence interval.

The POD curve and dPOD curve of Raman qualitative method and reference method were obtained (Figure 3) according to the consistency evaluation table of Raman qualitative method and reference method for CBZ in apple (Table 3). Based on the analysis of the POD (Figure 2), the false positive rate of SERS and HPLC in Figure 3A is 0, and the specificity is 100%. When the additive concentration was between 0.01 mg/kg and 2.5 mg/kg, the false negative rate of SERS was higher than that of HPLC. When the additive concentration was 5.0 mg/kg, the POD of SERS and HPLC are both 100%, the sensitivities are both 100% (Figure 3A). It can be seen from Figure 3B that when the added concentration is no less than 2.5 mg/kg, the dPOD of the two detection methods is 0, indicating that the SERS qualitative detection method and the reference method have the same detection results. This suggests that the SERS qualitative assay can meet the needs of limited CBZ detection at the MRL level of CBZ (5 mg/kg), while improving detection efficiency.

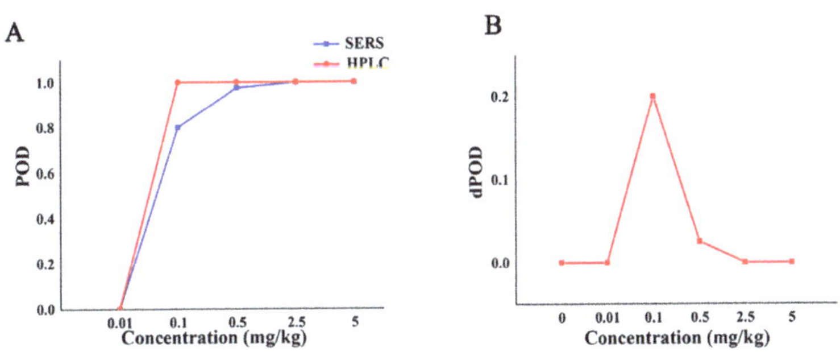

Figure 3. The detection situation of CBZ in apple by Raman qualitative method and reference method. Note: (**A**,**B**) are the POD curve and the dPOD curve of the Raman qualitative detection method and the reference method, respectively.

(2) Consistency evaluation of qualitative method among different labs

Table 4 shows the qualitative POD of SERS detection method between two different labs with different additive concentrations. The environments of lab I and lab II are different, and the same Raman instrument was always used during the experiment and the acquisition parameters were the same. The results showed that the POD of the Raman qualitative method in the two labs was 0 when the additive concentration was not more than 0.05 mg/kg. When the added concentration was in the range of 0.05 mg/kg to 2.5 mg/kg, the POD between the two labs was different, and the POD of lab II was higher than that of lab I. When the additive concentration was no less than 2.5 mg/kg, the POD between the two labs was 100%, and the dPOD was 0, which shows that the Raman qualitative method has the same detection results among different labs and has good repeatability.

Table 4. The POD of CBZ in apple by Raman qualitative method among different labs.

Lab	I					II					**Difference in POD (dPOD)**
Concentration (mg/kg)	x	N	POD	LCL	UCL	x	N	POD	LCL	UCL	
0.00	0	10	0.000	0.000	0.280	0	10	0.000	0.000	0.280	0.000
0.05	0	50	0.000	0.000	0.070	0	50	0.000	0.000	0.070	0.000
0.10	40	50	0.800	0.670	0.890	48	50	0.960	0.865	0.989	0.160
0.50	39	40	0.975	0.870	1.000	40	40	1.000	0.910	1.000	0.025
2.50	40	40	1.000	0.910	1.000	40	40	1.000	0.910	1.000	0.000
5.00	40	40	1.000	0.910	1.000	40	40	1.000	0.910	1.000	0.000

Note: x is the number of positive samples with positive results detected; N is the total number of samples; POD is the probability of detection; LCL is the lower limit of the 95% confidence interval; UCL is the upper limit of the 95% confidence interval.

The POD curve and the dPOD curve between different labs were obtained based on the POD of the qualitative methods between the different laboratories (Figure 4). Compared with the analysis results in Figure 3A, it can be seen from Figure 4A that the false positive rate of the method between different labs is 0, and the specificity is 100%. When the concentration is from 0.05 mg/kg to 2.5 mg/kg, the method had higher false negatives in lab I and higher sensitivity in lab II. When the concentration is no less than 2.5 mg/kg, the dPOD is 0 (Figure 4B), indicates that the detection results of this method between different labs are consistent. In other words, at the MRL (5.0 mg/kg) of CBZ, the Raman qualitative detection method can meet the requirements of limited detection and the results are not affected by the environment.

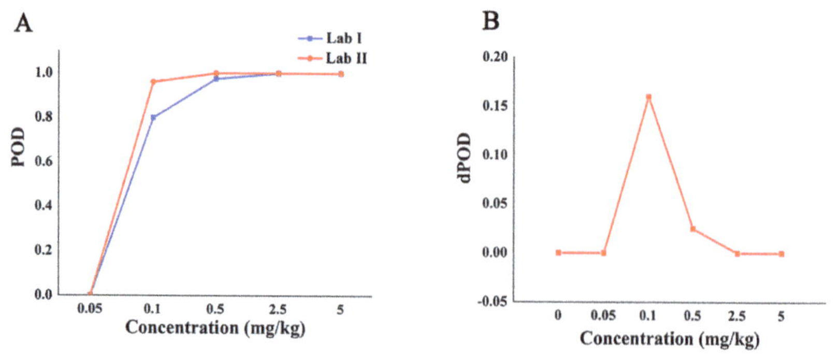

Figure 4. The detection situation of CBZ in apple by Raman qualitative method in different labs. Note: (**A**,**B**) are the POD curve and the dPOD curve of the Raman qualitative detection method between different labs, respectively.

2.2. Establishment of a Semi-Quantitative Analysis Method for Carbendazim in Apple

The SERS rapid detection method allows accurate and qualitative analysis based on the characteristic peaks of CBZ. However, as CBZ is a finite contaminant in food, screening for finite contaminant concentrations is necessary for practical applications. In order to further judge whether the concentration of SERS qualitative detection reaches the MRL of CBZ, a semi-quantitative analysis method of CBZ based on Raman intensity threshold was developed. The method can determine whether the concentration of CBZ in the sample exceeds the MRL according to the characteristic peak intensity, which can avoid complex quantitative analysis and meet the needs of actual finite contaminant detection.

2.2.1. Establishment and Screening of Semi-Quantitative Models

The intensity distribution of the characteristic peaks of CBZ was obtained at the MRL level (5 mg/kg) based on the Raman spectral information of the semi-quantitative model training set (Figure 5). The actual intensity distribution was fitted with the theoretical Gaussian distribution, and the theoretical threshold for semi-quantitative analysis was obtained. The results showed that the actual distribution curve of Raman intensity at 630 cm^{-1} fit well with the theoretical Gaussian curve, and the fitting degree of the remaining five characteristic peaks was poor. When the CI was higher than 95%, the semi-quantitative model had a Raman intensity threshold of 1.4×10^4 at 630 cm^{-1}, indicating that the concentration of CBZ in apple was no less than 5 mg/kg when the intensity of characteristic peak at 630 cm^{-1} of CBZ in apple was greater than 1.4×10^4.

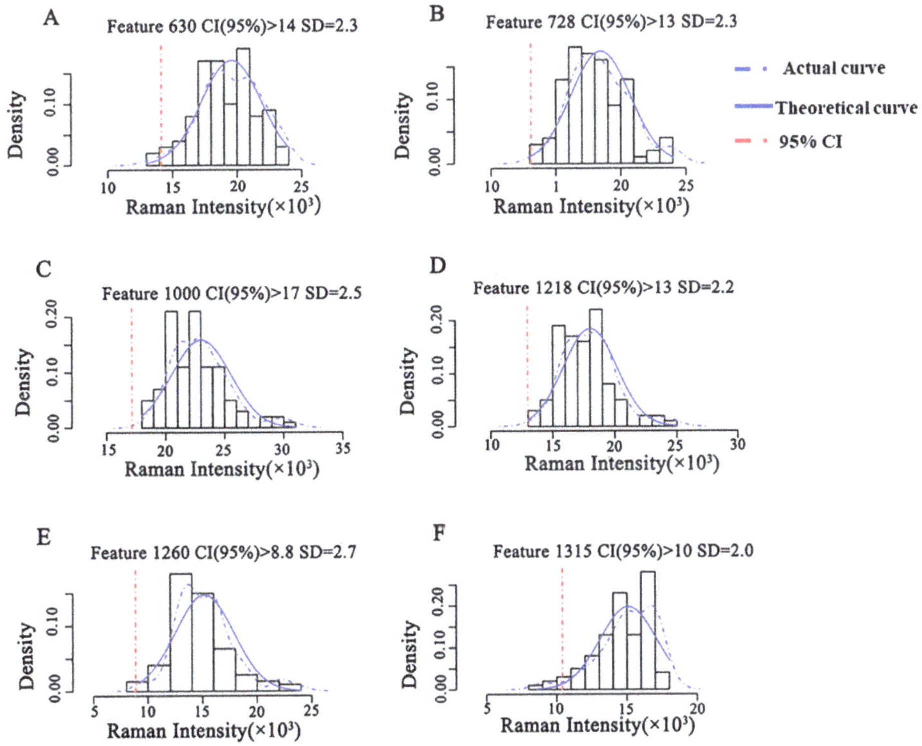

Figure 5. Semi-quantitative model of CBZ in apple at the MRL (5 mg/kg). Note: (**A**–**F**) represent the semi-quantitative models established by different Raman shifts, respectively. (**A**) 630 cm^{-1}; (**B**) 728 cm^{-1}; (**C**) 1000 cm^{-1}; (**D**) 1218 cm^{-1}; (**E**) 1260 cm^{-1}; (**F**) 1315 cm^{-1}. SD: Standard deviation.

The concentration discrimination results of the three test sets of the semi-quantitative model are shown in Figure 6 based on these semi-quantitative models of CBZ in apple at the MRL (5 mg/kg). As shown in Figure 6, the discriminant results of the semi-quantitative models established by different characteristic peaks are different. The semi-quantitative model established at 630 cm^{-1} can well distinguish 0.5 mg/kg, 2.5 mg/kg and 5 mg/kg, and the concentration distribution results discriminated by the model are consistent with the reality. The test sets of the semi-quantitative models established at the remaining characteristic peaks have different degrees of overlap, and the results of model discrimination do not match the actual ones.

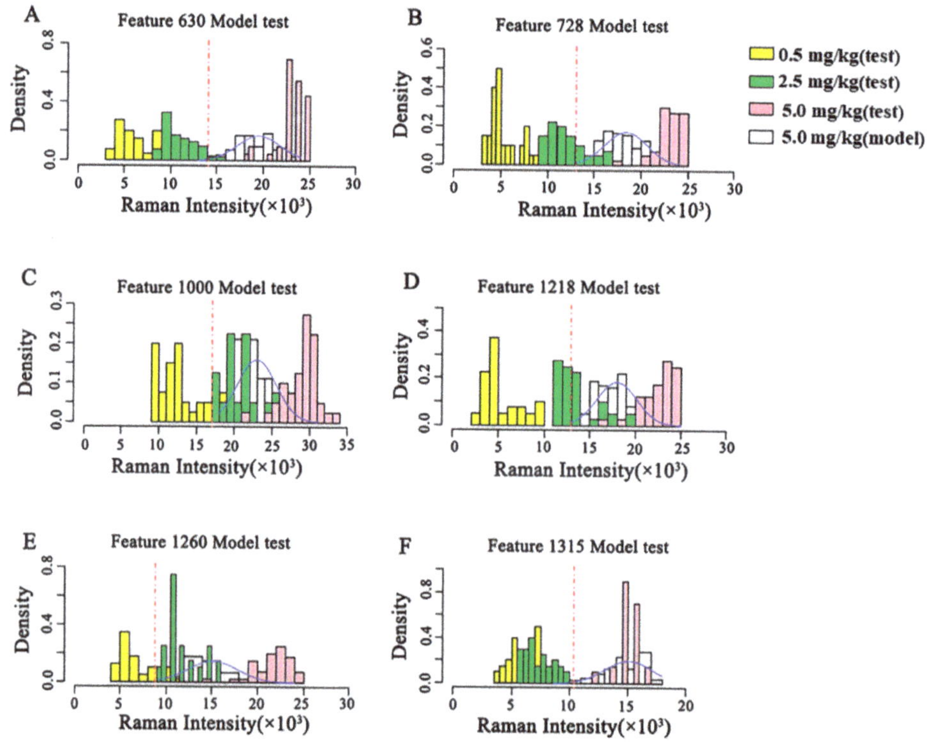

Figure 6. Validation of the semi-quantitative model on the MRL (5 mg/kg) of CBZ in apple. (**A**–**F**) represent the validation of semi-quantitative models established by different Raman shifts, respectively. (**A**) 630 cm^{-1}; (**B**) 728 cm^{-1}; (**C**) 1000 cm^{-1}; (**D**) 1218 cm^{-1}; (**E**) 1260 cm^{-1}; (**F**) 1315 cm^{-1}. SD: Standard deviation.

According to the Raman intensity thresholds of different semi-quantitative models (Figure 5) and the validation results of different semi-quantitative models (Figure 6), the POD that the concentration of CBZ in the three test sets exceeds the MRL and the scores of different semi-quantitative models are shown in Table 5. According to the calculation Formula (1) of the score, POD3 and score raised along with decrease of POD1 and POD2. Therefore, the higher the score of the model, the lower the false positive rate of the model, the higher the sensitivity, the better the model. The semi-quantitative model scores at 630 cm^{-1} and 1315 cm^{-1} are 97 and 100, respectively. However, the semi-quantitative model at 1315 cm^{-1} cannot distinguish the spiked samples at 0.5 mg/kg and 2.5 mg/kg (Figure 6). The semi-quantitative model at 630 cm^{-1} was well fitted (Figure 5) and the concentration distribution is consistent with reality, so the semi-quantitative model at 630 cm^{-1} is chosen as the optimal model for the semi-quantitative analysis of CBZ in apple.

The model has a false positive rate of 0 at 0.5 mg/kg, a false positive rate of 5% at 2.5 mg/kg, and a POD of 100% at 5 mg/kg.

Table 5. Scores for semi-quantitative models.

Concentration (mg/kg)	0.5		2.5		5		Score (S)
Peaks (cm^{-1})	x	POD1	x	POD2	x	POD3	
630	0	0	2	0.050	40	1	97
728	0	0	11	0.275	40	1	83
1000	8	0.200	39	0.975	40	1	44
1218	0	0	29	0.725	40	1	60
1260	9	0.225	40	1	40	1	43
1315	0	0	0	0	40	1	100

Note: x is the number of positive samples with positive results detected; POD is the probability of detection.

2.2.2. Consistency Evaluation of Semi-Quantitative Methods among Different Labs

Table 6 shows the detection probability (whether the concentration of CBZ exceeds the MRL) of the semi-quantitative method in different labs under different additive concentrations. As shown in Table 6, when the added concentration is less than 5 mg/kg, the POD of the semi-quantitative method is different among different labs, and the POD of lab II is higher than that of lab I. When the additive concentration is 5 mg/kg, the POD of the semi-quantitative methods among different labs is 100%, and the dPOD was 0. Although the semi-quantitative model has higher false positives in lab II, it can accurately determine whether the concentration of CBZ exceeds the MRL between the two labs, and the repeatability is good.

Table 6. Detection situation between different labs by semi-quantitative method for CBZ in apple.

Lab	I					II					Difference in POD (dPOD)
Concentration (mg/kg)	x	N	POD	LCL	UCL	x	N	POD	LCL	UCL	
0.50	0	40	0.000	0	0.088	7	40	0.175	0.087	0.320	0.175
2.50	25	40	0.625	0.470	0.758	34	40	0.850	0.710	0.930	0.225
5.00	40	40	1.000	0.912	1.000	40	40	1.000	0.912	1.000	0.000

Note: x is the number of positive samples with positive results detected; N is the total number of samples; POD is the probability of detection; LCL is the lower limit of the 95% confidence interval; UCL is the upper limit of the 95% confidence interval.

According to the POD of the semi-quantitative method between labs, the POD curve and the dPOD curve between different labs were obtained (Figure 7). It can be seen from Figure 7A that when the added concentration is less than 5 mg/kg, the POD of the semi-quantitative method in different labs is different which suggested that the sensitivity of lab II is higher than that of lab I if sensitivity is different. The POD of both labs is 100% at 5 mg/kg. As shown in Figure 7B, when the concentration of CBZ in the sample is no less than 5 mg/kg, dPOD = 0, which indicates that the semi-quantitative method has the same detection situation among different labs. Therefore, the semi-quantitative method can effectively distinguish whether the additive concentration of the blind sample exceeds the MRL according to the intensity of the characteristic peak at 630 cm^{-1}, and the results are not affected by the environment.

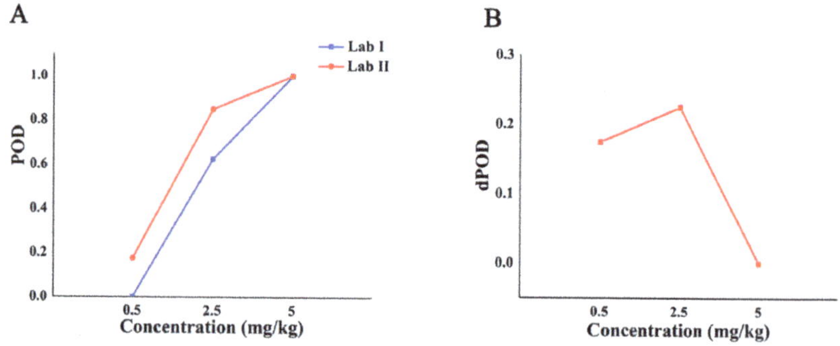

Figure 7. The detection situation of CBZ in apple by semi-quantitative method in different labs. Note: (**A**,**B**) are the POD curve and dPOD curve of semi-quantitative detection method between different labs, respectively.

3. Discussion

CBZ is widely used in agriculture and is a common pesticide residue that threatens human and animal health. At present, some classical detection methods such as HPLC [7] can provide accurate qualitative and quantitative analysis of CBZ, but these traditional methods all involve complex pretreatment process, long detection cycles and complicated instrumentation. In order to meet the needs of rapid screening of finite contaminants in practical applications, it is necessary to develop a rapid detection method with a short detection cycle and simple operation.

SERS has been widely performed in the qualitative, quantitative and semi-quantitative analysis of food finite contaminants due to its specificity, sensitivity, non-destructive sample, and no interference from aqueous solutions [39]. The basis of SERS qualitative analysis is based on the Raman characteristic peaks of the target, selecting the characteristic shifts of the spiked sample and the target but not in the blank matrix as the characteristic peak of the target. In this study, the peak intensities at 630 cm^{-1}, 728 cm^{-1}, 1000 cm^{-1}, 1218 cm^{-1}, 1260 cm^{-1} and 1315 cm^{-1} in the CBZ standard and spiked samples were obvious. Therefore, these six Raman shifts were selected as the Raman characteristic peaks of CBZ in apple. The characteristic peaks of CBZ obtained here are the same as those of existing research [34]. For example, the peak at 630 cm^{-1} is related to the C–C–C in-plane bending and the peak at 728 cm^{-1} is attributed to the out-of-plane bending of the C-H bond in the benzene ring. The LOD of CBZ was 0.5 mg/kg, which was lower than the MRL of CBZ in apple (5 mg/kg). At the same time, it is four times lower than the LOD (2 mg/kg) of previous research methods [40].

Quantitative analysis of SERS is of great significance in the detection of finite contaminants in food. However, the Raman intensity stability of SERS is easily affected by enhanced matrix activity and environment, which makes quantitative analysis difficult. To overcome these issues, existing researches mainly focused on developing curing techniques to improve the stability and reproducibility of reinforced matrices. For example, Sivashanmugan et al. [41]. developed novel Au nanodot arrays on graphene substrates for highly active enhanced Raman scattering. By using Rhodamine 6G (R6G) as a molecular probe, the LOD was as low as 10^{-12} M and the Raman enhancement factor was as high as 10^{8}. Wu et al. developed a simple and effective SERS tape based on biconical gold nanoparticles (BP-AuNPs) for monitoring methyl parathion residues on the surfaces of vegetables and fruits [21]. In real world applications, the screening of finite contaminants in food mainly depends on the MRL. If the added concentration of the sample is higher than the MRL, it is judged as a non-conforming product, otherwise it is judged as a qualified product. In order to fulfill the detection requirements of finite contaminants, we developed a threshold-based semi-quantitative analysis method for finite contaminants, which reduces the difficulty and cost of developing new materials and improves the detection efficiency. The model established at 630 cm^{-1} was selected as the basis for semi-quantitative analysis of CBZ in apple after screening and verification. When the Raman intensity at 630 cm^{-1} was greater than 1.4×10^4, the concentration of CBZ in the sample was higher than the MRL (5 mg/kg). When the additive concentration was 5 mg/kg, the POD of this semi-quantitative method was 100%. The semi-quantitative method developed in this study only requires modelling based on a large number of samples from MRL, and the semi-quantitative results are determined by the Raman characteristic peak intensity. Compared to existing CBZ semi-quantitative analysis methods of CBZ, such as PLS-DA [42], this semi-quantitative does not require complex classification models and classification parameters such as the variable importance of variables in projection fraction, so it is simpler and more tractable. In order to ensure the SERS rapid detection method and semi-quantitative analysis method satisfy the evaluation standards of rapid detection methods (released by the State Food and Drug Administration in 2017). POD curve and dPOD curve were constructed for different methods or different labs within a certain concentration range, and the consistency of the methods was determined based on whether the POD is the same. The results showed that at the MRL level of CBZ (5 mg/kg), the SERS qualitative detection method was consistent among different labs, and the results were the same as the reference method. Compared with 'Technology specification for the evaluation of food rapid detection products' [37] , this evaluation method can show the change of sensitivity with concentration, and the LOD is well defined. It can compare the consistency of each concentration interval within the detection concentration range and can be applied to the consistency analysis between methods, environments, and instruments. The obtained POD curve can display the results visually, which is more statistically significant.

The semi-quantitative analysis method based on SERS technology can perform rapid primary screening of samples according to the characteristic peak intensity of CBZ at 630 cm^{-1} in samples, which can improve the detection efficiency, and can be extended to other finite contaminants such as melamine in liquid milk and other pesticide residues in food. However, the training set of the semi-quantitative model requires a large amount of sample size (the number of samples >50), which leads to consume most of the time and energy before modeling. How to use less time to obtain the more sample information and reduce the preparation time is the bottleneck and future development direction of this research. Raman hyperspectral imaging technology is an advanced non-destructive testing technology that combines conventional imaging and spectroscopy to collect Raman spectral information of each pixel in space, so as to conduct qualitative, quantitative and localized analysis of samples [43]. Compared with SERS, the advantage of Raman hyperspectral imaging technology is that it can continuously collect a large number of spectral information through an automated sample platform, so as to obtain more sample information in less time. For example, Yang et al. applied Raman hyperspectral imaging technology to continuously collect spectral information of 100 pixels in 10 min [44]. However, it would have taken at least 50 min to collect the spectral information with SERS. Therefore, the semi-quantitative method developed in this study can be combined with the Raman hyperspectral imaging to shorten the sample information acquisition time and further improve the efficiency of sample screening.

4. Materials and Methods

4.1. Samples, Reagents and Instruments

Sample: Apple (commercially available); Reagents: Carbendazim solid standard, ethanol (AR), dichloromethane (AR), NaCl solution (1 mol/L), nano-gold solution (Shanghai Oceanhood opto-electronics tech Co., Ltd., Shanghai, China).

Instruments: Portable Raman spectrometer (SEED 3000, Shanghai Oceanhood opto-electronics tech Co., Ltd., China), precision electronic balance (Sartorius, Germany), vortex mixer (VORTEX-GENIE2, Scientific Industries, New York, NY, USA), Eppendorf 5810R centrifuge (20050647GZ, Eppendorf, Germany).

4.2. *Methods*

4.2.1. Sample Preparation

Apple pulp was put into a 50 mL tube, and it was broken up with a homogenizer. 2 g of pulp was weighed and added with 20 μL of standard solutions of CBZ with different concentrations, so that the concentrations of CBZ in samples were 5 mg/kg, 2.5 mg/kg, 1 mg/kg, 0.5 mg/kg, 0.1 mg/kg, 0.05 mg/kg. The samples with 3 mL of ethanol (5%, v/v) were vortexed for 3 min and centrifuged at 4000 rpm for 2 min. 2 mL of sample supernatant was mixed with 2 mL of dichloromethane in a 5.0 mL tube. After layering, 1 mL of liquid in the lower layer was taken into a gas-phase injection vial, dried with a nitrogen blower, and reconstituted with 500 μL of ethanol (20%, v/v). The reconstituted liquid was vortexed for 1 min and used for SERS detection.

4.2.2. SERS Detection

50 μL reconstituted solution with 200 μL of nano-gold solution and 50 μL of NaCl solution (1 mol/L) were added to the detection vial. The samples were quickly mixed with a pipette and then tested on the SEED 3000. To avoid the aggregation of gold nanoparticles, the SERS detection was completed within 1 min. The excitation wavelength of the Raman spectrometer was 785 nm, the wavelength range of data acquisition was 200–4000 cm^{-1}, the integration time was 1 s, and the laser power was 200 mw. Spectral data were collected by Uspecral-PRO software. (Shanghai Oceanhood opto-electronics tech Co., Ltd., Shanghai, China).

4.2.3. Data Processing

(1) Screening of Raman characteristic peaks of CBZ in apple

Based on the positions of the Raman characteristic peak of CBZ standard, the Raman shift peaks (the number varies from 2 to 6) specific to the blank samples were selected as the characteristic peaks of CBZ, and a library of Raman characteristic peak was established. After obtaining the Raman spectral of the sample, the search for the first order derivative peak and matching of the characteristic peak were performed. When the Raman characteristic peaks of the sample had all the characteristic peaks of CBZ, it was determined that the sample contains CBZ. The tolerance range of Raman shift is 3–10 cm^{-1} can be considered as the same characteristic peak.

(2) Establishment and screening of semi-quantitative models

The semi-quantitative model of the target at a specific concentration (such as MRL, etc.) is used to determine whether the concentration of target in the sample exceeds a specific concentration. To establish a semi-quantitative model at a specific concentration, it is first necessary to determine the applicable concentration of the Raman detection method and to confirm that the discriminant concentration is within the detection range of the current method. In practice, it is necessary to obtain Raman spectral data of spiked samples at specific concentrations to establish semi-quantitative models at different characteristic peaks. Then the optimal semi-quantitative model was determined according to the fitting of Gaussian distribution and the discrimination of different additive concentrations. These steps for establishing a semi-quantitative model at a specific concentration are as follows:

1. The establishment of semi-quantitative models. Raman spectral data of spiked samples (number of samples > 50) at specific concentrations were collected as a training set for a semi-quantitative model. A histogram of the Raman intensity of each Raman characteristic peak at a specific additive concentration was obtained and the distribution of the intensity was viewed. If the intensity of the characteristic peak does not obey the Gaussian distribution, Raman intensity of the characteristic peak is not only affected by the random error of detection, so it is not suitable for the semi-quantitative model and should be eliminated. For the characteristic peaks whose Raman intensity follows a Gaussian distribution, calculate the intensity mean and standard deviation, determine the confidence level, and draw the confidence interval (CI). The Raman

intensity corresponding to the lower limit of the confidence interval (CI > 95%) is used as the semi-quantitative threshold.

2. Screening of semi-quantitative models. Raman data of spiked samples (number of samples > 20) at low concentration, half of specific concentration and specific concentration were collected as the test set for the semi-quantitative models of different characteristic peaks. The threshold value of the semi-quantitative model was used to determine whether the concentration in the sample exceeds a specific concentration, the POD was calculated, and the semi-quantitative model score under different characteristic peaks was computed. The higher the score of the model, the more accurate the semi-quantitative model will be. The model with the highest score and the Raman intensity conforming to the Gaussian distribution was selected as the optimal semi-quantitative model.

The formula for calculating the score is as follows: S is the score of the semi-quantitative model, POD1 is the probability of detection at low concentration, POD2 is the probability of detection at half of specific concentration, POD3 is the probability of detection at specific concentration.

$$S = 2^{(POD^3 - POD^2 - POD^1)} \times 50 \tag{1}$$

3. Result determination of semi-quantitative models. When the qualitative determination result is that the sample contains target substance, if the characteristic peak intensity of the target substance exceeds the semi-quantitative threshold, it is determined that the concentration of the target substance is not lower than a specific concentration. On the contrary, the concentration is lower than a specific concentration.

(3) Evaluation of method based on POD model

The evaluation of Raman rapid detection method based on POD model specifically includes three steps: establishment of POD model, determination of blind sample results and calculation of evaluation result.

1. Establishment of POD model. According to the qualitative results of samples at blank, low concentration, half of a specific concentration, a specific concentration, etc. (the number of samples > 20), the number of test samples, positive samples and negative samples at each concentration were counted. The POD was calculated, and POD curve of the POD changing with the added concentration was drawn.

2. Determination of blind sample results [45]. According to the qualitative discrimination results of spiked blind samples (the number of samples >20) in different labs and reference methods, the number of test samples, positive samples and negative samples under each concentration was counted. At the same time, the number of samples exceeding a specific concentration is calculated from the semi-quantitative identification of the low concentration, half of the specific concentration and the spiked sample at the specific concentration.

3. Calculation of evaluation result. The POD of qualitative and semi-quantitative methods was drawn, and it was established for the method evaluation system of Raman detection. Sensitivity, specificity, and LOD were evaluated by the POD model, and the consistency with the reference method and the inter-laboratory consistency were evaluated by the difference of POD (dPOD).

The POD and its confidence interval are calculated as follows: x is the number of positive samples with positive results detected, N is the total number of samples, POD is the probability of detection, LCL is the lower limit of the 95% confidence interval, and UCL is the upper limit of the 95% confidence interval:

(1) when x = 0,

$$POD = 0,$$

$$LCL = 0,$$

$$UCL = 3.8415/(N + 3.8415) \tag{2}$$

(2) When x = N,

$$\begin{aligned} POD &= 1, \\ LCL &= N/(N+3.8415), \\ UCL &= 1; \end{aligned} \tag{3}$$

(3) When 0 < x < N,

$$POD = x/N, \tag{4}$$

$$LCL = \frac{x + 1.9207 - 1.9600\sqrt{x - \frac{x^2}{N} + 0.9604}}{N + 3.8415} \tag{5}$$

$$UCL = \frac{x + 1.9207 + 1.9600\sqrt{x - \frac{x^2}{N} + 0.9604}}{N + 3.8415} \tag{6}$$

The dPOD and its confidence interval are calculated as follows: dPOD is the difference of POD, I is the method to be evaluated/laboratory, II is the reference method/laboratory:

$$dPOD = POD_I - POD_{II}, \tag{7}$$

$$LCL = dPOD - \sqrt{(POD_I - LCL_I)^2 + (POD_{II} - UCL_{II})^2} \tag{8}$$

$$UCL = dPOD + \sqrt{(POD_I - UCL_I)^2 + (POD_{II} - LCL_{II})^2}; \tag{9}$$

5. Conclusions

This study carried out qualitative and semi-quantitative analysis of CBZ in apple and the method was evaluated by a binary output-based validation system. The LOD of the qualitative method was 0.5 mg/kg. When the additive concentration was 5 mg/kg, the POD of the semi-quantitative method was 100%. When the added concentration was greater than 5 mg/kg, the SERS method and the reference method are consistent and the detection results are not affected by the lab. The threshold-based semi-quantitative method proposed can quickly determine whether the finite contaminants in blind samples exceed the MRL. Meanwhile, the evaluation method based on binary output provided a reference for the evaluation system of Raman spectroscopy rapid detection technology, which is of great significance for food rapid detection technology and has some significance in other rapid detection fields.

6. Patents

The work reported in this manuscript resulted in a patent, which has been granted under the patent number ZL201911005222.1

Author Contributions: Conceptualization, data curation, formal analysis, investigation, methodology, software, validation, visualization, writing—original draft preparation, writing-review and editing, Q.Y.; Conceptualization, methodology, software, visualization, H.L.; formal analysis, software, validation, writing-review and editing, J.M.; validation, N.C. and C.Z.; resources, project administration, funding acquisition, D.G.; formal analysis, supervision, writing-review and editing, B.N.; validation, Z.Z.; resources, supervision, project administration, funding acquisition, writing-review and editing, X.D.; formal analysis, supervision, writing-review and editing, Q.C. All authors have read and agreed to the published version of the manuscript.

Funding: This research study was supported by the National Key Research and Development Plan Project, grant number 2018YFC1603503; Shanghai Agricultural Research Project, grant number 19391901500; Science and Technology Joint Project of the Yangtze River Delta, grant number 19395810100; and Shanghai Technical Standard Project, grant number 18DZ2201200.

Institutional Review Board Statement: Not applicable.

Informed Consent Statement: Not applicable.

Data Availability Statement: The dataset used and/or analyzed during the current study are available from the corresponding author on reasonable request.

Acknowledgments: The authors wish to thank anonymous reviewers for their constructive comments on the presentation of this article.

Conflicts of Interest: The authors declare no conflict of interest.

Sample Availability: Samples of the compounds are not available from the authors.

References

1. Liu, Z.Y.; Chen, Y.; Han, J.H.; Chen, D.; Yang, G.Q.; Lan, T.T.; Li, J.M.; Zhang, K.K. Determination, dissipation dynamics, terminal residues and dietary risk assessment of thiophanate-methyl and its metabolite carbendazim in cowpeas collected from different locations in China under field conditions. *J. Sci. Food Agric.* **2021**, *101*, 5498–5507. [CrossRef] [PubMed]
2. Singh, S.; Singh, N.; Kumar, V.; Datta, S.; Wani, A.B.; Singh, D.; Singh, K.; Singh, J. Toxicity, monitoring and biodegradation of the fungicide carbendazim. *Environ. Chem. Lett.* **2016**, *14*, 317–329. [CrossRef]
3. Prashantkumar, W.; Sethi, R.S.; Pathak, D.; Rampal, S.; Saini, S.P.S. Testicular damage after chronic exposure to carbendazim in male goats. *Toxicol. Environ. Chem.* **2012**, *94*, 1433–1442. [CrossRef]
4. Daundkar, P.S.; Rampal, S. Evaluation of ameliorative potential of selenium on carbendazim induced oxidative stress in male goats. *Environ. Toxicol. Pharmacol.* **2014**, *38*, 711–719. [CrossRef] [PubMed]
5. *GB 2763-2021*; National Food Safety Standard—Maximum Residue Limits for Pesticides in Food. National Health Commission: Beijing, China, 2021.
6. EU. *On Maximum Residue Levels of Pesticides in or on Food and Feed of Plant and Animal Origin and Amending Council Directive 91/414/EEC*; (EC) NO 396/2005; EU: Brussels, Belgium, 2005.
7. Zhao, K.; Che, J.; Huang, A.; Shi, F. Determination of Residual Carbendazim in the Sugar Orange by HPLC. *Hubei Agric. Sci.* **2010**, *49*, 1193–1195.
8. Economou, A.; Botitsi, H.; Antoniou, S.; Tsipi, D. Determination of multi-class pesticides in wines by solid-phase extraction and liquid chromatography-tandem mass spectrometry. *J. Chromatogr. A* **2009**, *1216*, 5856–5867. [CrossRef]
9. de Macedo, J.F.; Alves, A.A.C.; Sant'Anna, M.V.S.; Cunha, F.G.C.; Oliveira, G.d.A.R.; Liao, L.M.; Sussuchi, E.M. Electrochemical determination of carbendazim in grapes and their derivatives by an ionic liquid-modified carbon paste electrode. *J. Appl. Electrochem.* **2022**, *52*, 729–742. [CrossRef]
10. Jiang, X.; Li, D.; Xu, X.; Ying, Y.; Li, Y.; Ye, Z.; Wang, J. Immunosensors for detection of pesticide residues. *Biosens. Bioelectron.* **2008**, *23*, 1577–1587. [CrossRef]
11. Lee, H.S.; Rahman, M.M.; Chung, H.S.; Kabir, H.; Yoon, K.S.; Cho, S.K.; Abd El-Aty, A.M.; Shim, J.H. An effective methodology for simultaneous quantification of thiophanate-methyl, and its metabolite carbendazim in pear, using LC-MS/MS. *J. Chromatogr. b-Anal. Technol. Biomed. Life Sci.* **2018**, *1095*, 1–7. [CrossRef]
12. Liu, Z.; Liu, W.; Wu, Q.; Zang, X.; Zhou, X.; Zeng, X.; Wang, Z. Determination of carbendazim and thiabendazole in apple juice by hollow fibre-based liquid phase microextraction-high performance liquid chromatography with fluorescence detection. *Int. J. Environ. Anal. Chem.* **2012**, *92*, 582–591. [CrossRef]
13. Langer, J.; Jimenez de Aberasturi, D.; Aizpurua, J.; Alvarez-Puebla, R.A.; Auguié, B.; Baumberg, J.J.; Bazan, G.C.; Bell, S.E.J.; Boisen, A.; Brolo, A.G.; et al. Present and Future of Surface-Enhanced Raman Scattering. *ACS Nano* **2020**, *14*, 28–117. [CrossRef] [PubMed]
14. Zhu, J.; Agyekum, A.A.; Kutsanedzie, F.Y.H.; Li, H.; Chen, Q.; Ouyang, Q.; Jiang, H. Qualitative and quantitative analysis of chlorpyrifos residues in tea by surface-enhanced Raman spectroscopy (SERS) combined with chemometric models. *LWT* **2018**, *97*, 760–769. [CrossRef]
15. Craig, A.P.; Franca, A.S.; Irudayaraj, J. Surface-enhanced Raman spectroscopy applied to food safety. *Annu. Rev. Food Sci. Technol.* **2013**, *4*, 369–380. [CrossRef] [PubMed]
16. Jiang, L.; Hassan, M.M.; Ali, S.; Li, H.; Sheng, R.; Chen, Q. Evolving trends in SERS-based techniques for food quality and safety: A review. *Trends Food Sci. Technol.* **2021**, *112*, 225–240. [CrossRef]
17. Pilot, R. SERS detection of food contaminants by means of portable Raman instruments. *J. Raman Spectrosc.* **2018**, *49*, 954–981. [CrossRef]
18. Liu, B.; Zheng, S.; Li, H.; Xu, J.; Tang, H.; Wang, Y.; Wang, Y.; Sun, F.; Zhao, X. Ultrasensitive and facile detection of multiple trace antibiotics with magnetic nanoparticles and core-shell nanostar SERS nanotags. *Talanta* **2022**, *237*, 122955. [CrossRef]

19. Dugandžić, V.; Kupfer, S.; Jahn, M.; Henkel, T.; Weber, K.; Cialla-May, D.; Popp, J. A SERS-based molecular sensor for selective detection and quantification of copper(II) ions. *Sens. Actuators B Chem.* **2019**, *279*, 230–237. [CrossRef]
20. Wei, C.; Li, M.; Zhao, X. Surface-Enhanced Raman Scattering (SERS) With Silver Nano Substrates Synthesized by Microwave for Rapid Detection of Foodborne Pathogens. *Front. Microbiol.* **2018**, *9*, 2857. [CrossRef]
21. Wu, H.; Luo, Y.; Hou, C.; Huo, D.; Zhou, Y.; Zou, S.; Zhao, J.; Lei, Y. Flexible bipyramid-AuNPs based SERS tape sensing strategy for detecting methyl parathion on vegetable and fruit surface. *Sens. Actuators B Chem.* **2019**, *285*, 123–128. [CrossRef]
22. Chen, X.; Lin, M.; Sun, L.; Xu, T.; Lai, K.; Huang, M.; Lin, H. Detection and quantification of carbendazim in Oolong tea by surface-enhanced Raman spectroscopy and gold nanoparticle substrates. *Food Chem.* **2019**, *293*, 271–277. [CrossRef]
23. Zhou, H.L.; Li, X.D.; Wang, L.H.; Liang, Y.F.; Jialading, A.; Wang, Z.S.; Zhang, J.G. Application of SERS quantitative analysis method in food safety detection. *Rev. Anal. Chem.* **2021**, *40*, 173–186. [CrossRef]
24. Li, D.; Yao, D.; Li, C.; Luo, Y.; Liang, A.; Wen, G.; Jiang, Z. Nanosol SERS quantitative analytical method: A review. *TrAC Trends Anal. Chem.* **2020**, *127*, 115885. [CrossRef]
25. Long, Y.; Li, H.; Du, Z.; Geng, M.; Liu, Z. Confined Gaussian-distributed electromagnetic field of tin(II) chloride-sensitized surface-enhanced Raman scattering (SERS) optical fiber probe: From localized surface plasmon resonance (LSPR) to waveguide propagation. *J. Colloid Interface Sci.* **2021**, *581*, 698–708. [CrossRef] [PubMed]
26. Wilrich, P.-T. The determination of precision of qualitative measurement methods by interlaboratory experiments. *Accredit. Qual. Assur.* **2010**, *15*, 439–444. [CrossRef]
27. Fleiss, J.L.; Levin, B.; Paik, M.C. *Statistical Methods for Rates and Proportions*, 2nd ed.; John Wiley & Sons: Hoboken, NJ, USA, 2013.
28. Wehling, P.; LaBudde, R.A.; Brunelle, S.L.; Nelson, M.T. Probability of detection (POD) as a statistical model for the validation of qualitative methods. *J. AOAC Int.* **2011**, *94*, 335–347. [CrossRef]
29. van Wieringen, W.N.; de Mast, J. Measurement System Analysis for Binary Data. *Technometrics* **2008**, *50*, 468–478. [CrossRef]
30. Gondim, C.d.S.; Junqueira, R.G.; de Souza, S.V.C.; Callao, M.P.; Ruisánchez, I. Determining performance parameters in qualitative multivariate methods using probability of detection (POD) curves. Case Study Two Common Milk Adulterants. *Talanta* **2017**, *168*, 23–30. [CrossRef]
31. Jarvis, B.; Wilrich, C.; Wilrich, P.T. Estimation of the POD Function and the LOD of a Binary Microbiological Measurement Method from an Interlaboratory Experiment. *J. AOAC Int.* **2019**, *102*, 1617–1623. [CrossRef]
32. Uhlig, S.; Frost, K.; Colson, B.; Simon, K.; Mäde, D.; Reiting, R.; Gowik, P.; Grohmann, L. Validation of qualitative PCR methods on the basis of mathematical–statistical modelling of the probability of detection. *Accredit. Qual. Assur.* **2015**, *20*, 75–83. [CrossRef]
33. Syed Akbar Ali, M.S.; Kumar, A.; Rao, P.B.; Tammana, J.; Balasubramaniam, K.; Rajagopal, P. Bayesian synthesis for simulation-based generation of probability of detection (PoD) curves. *Ultrasonics* **2018**, *84*, 210–222. [CrossRef]
34. Furini, L.N.; Sanchez-Cortes, S.; López-Tocón, I.; Otero, J.C.; Aroca, R.F.; Constantino, C.J.L. Detection and quantitative analysis of carbendazim herbicide on Ag nanoparticles via surface-enhanced Raman scattering. *J. Raman Spectrosc.* **2015**, *46*, 1095–1101. [CrossRef]
35. Ren, X.; Feng, X.; Li, X.; Li, X. Preparation of silver with an ultrathin molecular imprinted layer for detection of carbendazim by SERS. *Chem. Pap.* **2021**, *75*, 6477–6485. [CrossRef]
36. Sharma, V.; Krishnan, V. Fabrication of highly sensitive biomimetic SERS substrates for detection of herbicides in trace concentration. *Sens. Actuators B Chem.* **2018**, *262*, 710–719. [CrossRef]
37. *DB36/T 1334-2020*; Technology Specification for the Evaluation of Food Rapid Detection Products. Jiangxi Provincial Administration for Market Regulation: Nanchang, China, 2020.
38. *GB/T 23380-2009*; Determination of Carbendazim Residues in Fruits and Vegetables—HPLC Method. Standardization Administration of China: Beijing, China, 2009.
39. Neng, J.; Zhang, Q.; Sun, P.L. Application of surface-enhanced Raman spectroscopy in fast detection of toxic and harmful substances in food. *Biosens. Bioelectron.* **2020**, *167*, 112480. [CrossRef]
40. Shen, Z.D.; Fan, Q.Z.; Yu, Q.; Wang, R.; Wang, H.; Kong, X.M. Facile detection of carbendazim in food using TLC-SERS on diatomite thin layer chromatography. *Spectrochim. Acta Part A-Mol. Biomol. Spectrosc.* **2021**, *247*, 119037. [CrossRef]
41. Sivashanmugan, K.; Nguyen, V.-H.; Nguyen, B.-S. Tailoring a novel Au nanodot arrays on graphene substrate for a highly active Surface-Enhanced Raman Scattering (SERS). *Mater. Lett.* **2020**, *271*, 127807. [CrossRef]
42. Valderrama, L.; Valderrama, P.; Carasek, E. A semi-quantitative model through PLS-DA in the evaluation of carbendazim in grape juices. *Food Chem.* **2022**, *368*, 130742. [CrossRef]
43. Lohumi, S.; Lee, H.; Kim, M.S.; Qin, J.; Cho, B.-K. Raman hyperspectral imaging and spectral similarity analysis for quantitative detection of multiple adulterants in wheat flour. *Biosyst. Eng.* **2019**, *181*, 103–113. [CrossRef]
44. Yang, Q.; Niu, B.; Gu, S.; Ma, J.; Zhao, C.; Chen, Q.; Guo, D.; Deng, X.; Yu, Y.; Zhang, F. Rapid Detection of Nonprotein Nitrogen Adulterants in Milk Powder Using Point-Scan Raman Hyperspectral Imaging Technology. *ACS Omega* **2022**, *7*, 2064–2073. [CrossRef]
45. Brain, C.W.; Mi, J. On some properties of the quantiles of the chi-square distribution and their applications to interval estimation. *Commun. Stat. Theory Methods* **2001**, *30*, 1851–1867. [CrossRef]

molecules

Review

A Review of The Application of Spectroscopy to Flavonoids from Medicine and Food Homology Materials

Lin Zou [1], Huijun Li [1], Xuejie Ding [1], Zifan Liu [1], Dongqiong He [2], Jamal A. H. Kowah [2], Lisheng Wang [2,*], Mingqing Yuan [1,*] and Xu Liu [1,*]

1 College of Medicine, Guangxi University, Nanning 530004, China
2 College of Chemistry and Chemical Engineering, Guangxi University, Nanning 530004, China
* Correspondence: lswang@gxu.edu.cn (L.W.); yuanmingqing1985@163.com (M.Y.); xuliu1978@gxu.edu.cn (X.L.); Tel.: +86-13737071312 (L.W.); +86-18589921890 (M.Y.); +86-15807710318 (X.L.)

Abstract: Medicinal and food homology materials are a group of drugs in herbal medicine that have nutritional value and can be used as functional food, with great potential for development and application. Flavonoids are one of the major groups of components in pharmaceutical and food materials that have been found to possess a variety of biological activities and pharmacological effects. More and more analytical techniques are being used in the study of flavonoid components of medicinal and food homology materials. Compared to traditional analytical methods, spectroscopic analysis has the advantages of being rapid, economical and free of chemical waste. It is therefore widely used for the identification and analysis of herbal components. This paper reviews the application of spectroscopic techniques in the study of flavonoid components in medicinal and food homology materials, including structure determination, content determination, quality identification, interaction studies, and the corresponding chemometrics. This review may provide some reference and assistance for future studies on the flavonoid composition of other medicinal and food homology materials.

Keywords: flavonoid; medicine and food homology; spectroscopy

Citation: Zou, L.; Li, H.; Ding, X.; Liu, Z.; He, D.; Kowah, J.A.H.; Wang, L.; Yuan, M.; Liu, X. A Review of The Application of Spectroscopy to Flavonoids from Medicine and Food Homology Materials. *Molecules* **2022**, *27*, 7766. https://doi.org/10.3390/molecules27227766

Academic Editors: Weiying Lu and Yanping Chen

Received: 31 August 2022
Accepted: 27 October 2022
Published: 11 November 2022

1. Introduction

In recent years, as people have become more conscious of food safety and health, more and more people are looking at foods with health benefits and therapeutic effects. In Chinese traditional medicine concepts, this is referred to as "medicine and food come from the same source" or "medicinal and food homology". In the earliest records of the Yellow Emperor's Internal Classic, food can be used as a complementary function to medicine to treat disease. In the Tang dynasty, the "Qianjin recipe", and then in the Yuan dynasty, the "principles of correct diet"(Yin Shan Zheng Yao), the efficacy and contraindications of food were documented, and the theoretical system of the medicinal and food homology was gradually improved [1]. In November 2021, China's National Health and Wellness Commission issued the "Regulations on the Management of the Catalogue of Substances that are Traditionally Both Food and Chinese Medicinal Herbs", listing 110 medicinal and food homology materials [2].

These herbs, which have been popular in China for millennia, possess high medical value. Unfortunately, however, their clinical development remains highly limited. An important influencing factor is that the composition of herbal medicines is often not a single but a complex mixture, thus making quality control of herbal medicines more difficult. Therefore, qualitative identification and quantitative analysis of herbs is essential to ensure quality standards and safe efficacy. For medicinal and food homology materials, it is even more important to carry out quality assessments to ensure safety due to the particular characteristics of the materials that differ from those of ordinary food products. The Pharmacopoeia of the People's Republic of China and the Law of the People's Republic of

China on the Administration of Medicinal Products have strict regulations on medicinal and food homology materials.

Secondary metabolites in medicinal plants include flavonoids, polysaccharides, terpenoids, quinones, steroids and others, of which flavonoids are one of the most important active ingredients [3]. Flavonoids are a large class of polyphenols, which are widely present in various plants in free form or as glycosides. The basic skeletal structure of flavonoids consists of two benzene rings (A and B) which are linked by an oxygen-containing heterocyclic pyran ring (C) [4]. Flavonoids can be divided into different six major groups depending on the unsaturation of the linking chain, the connection between the rings and the chemical structure, including flavanones, isoflavonoids, flavanols, flavonols, flavones and anthocyanidins [5] (Figure 1).

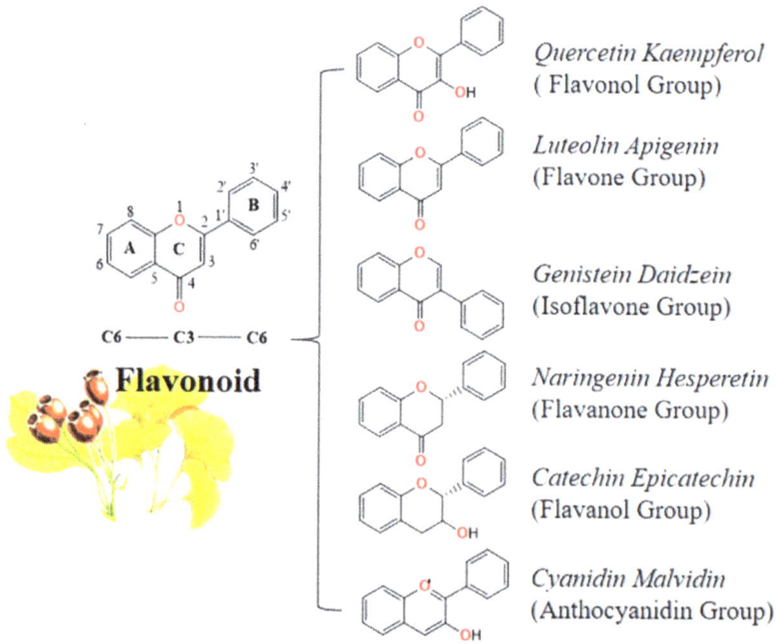

Figure 1. Classification of flavonoids.

Flavonoids derived from food and herbs have been found to contain a variety of pharmacological activities, such as antioxidant [6], anti-inflammatory [4], anti-cancer [5,7], modulation of intestinal immune function [8] and cardiovascular protection [9] (Figure 2).

The flavonoids extracted from various herbal medicines are numerous and have complex and diverse structures. Therefore, a range of analytical technical tools and methods, such as high-performance liquid chromatography (HPLC), gas chromatography (GC), mass spectrometry (MS) and spectroscopy, are used to determine the basic structure and characterization of flavonoid components. Traditional analytical techniques, such as HPLC [10], GC [11], liquid chromatography-mass spectrometry (LC-MS) [12,13], gas chromatography-mass spectrometry (GC-MS) [14], have been used to determine the concentration as well as identify the structure of the substance. However, these methods are expensive, time-consuming, require complex operation skills and consume large amounts of solvents and reagents. In addition, these methods usually measure a small number of samples to represent a larger sample, so it is not suitable for large-scale measurement applications. At present, various spectroscopic techniques are frequently used for the detection and identification of biological molecules, such as UV, NIR, fluorescence spectroscopy, terahertz time-domain spectroscopy, etc.

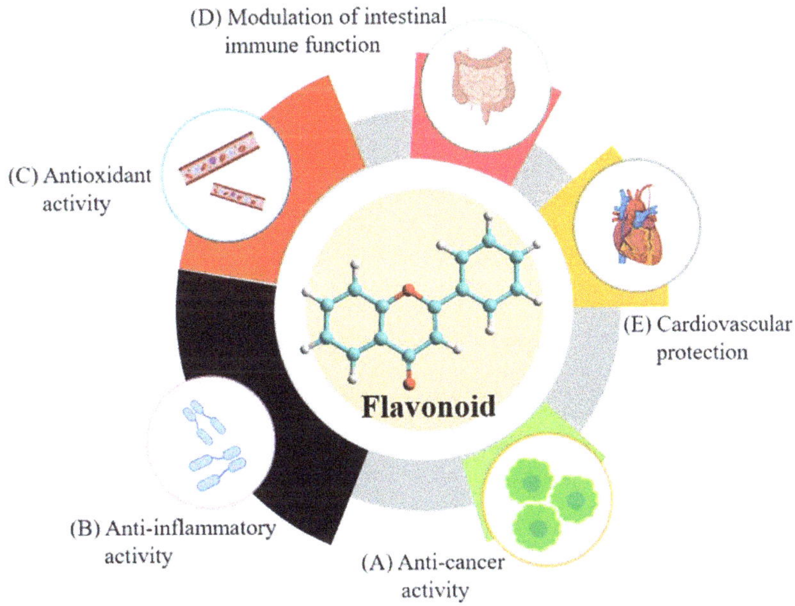

Figure 2. Pharmacological activities of flavonoids.

Compared to these techniques, spectroscopic analysis has the advantage of being relatively simple, inexpensive and does not require extended preparation methods of samples or other chemicals [15]. On the other hand, because each compound has its own specific characteristic spectral information, it is easy to differentiate classes of functional groups, the structure of bonds within compounds and their conformation by corresponding spectral profiles, which has led to the widespread use of spectroscopic analysis for the development and study of flavonoid components [16].

This paper provides a review of spectroscopic techniques used in recent years in the quality control and development applications of flavonoid in medicinal and food homology materials and details the advantages and research progress of each technique. These findings provide an important scientific basis for future research and development of flavonoid components in medicinal and food homology materials.

2. Qualitative and Quantitative Analysis

The analytical identification of active compounds extracted from herbal medicines also depends on various analytical techniques. The detection of flavonoids in herbal medicines has received increasing attention. Many studies related to the qualitative and quantitative analysis of flavonoid components in medicinal and food homologous materials have been published. Compared with the traditional time-consuming, labor-intensive and expensive chromatography methods, the spectroscopic method is simple, fast and environmentally friendly [17]. Spectral analysis is based on their characteristic spectra and is used to study the structure of compounds or to determine their chemical composition [18]. Various organic compounds with different structures have their own unique characteristic spectra. Table 1 has summarised the spectroscopic techniques used for quantitative and qualitative analysis of medicinal and food homology materials.

Table 1. Sources, flavonoid composition, spectroscopy techniques and chemometrics of medicinal and food homology materials.

Sources	Flavonoid Types	Compound Name	Analysis Technique	Purpose	Chemometrics	Year	References
Platycladi Cacumen	21	Myricitrin; quercitrin; afzelin; amentoflavone	NMR	Chemical structures	OPLSR	2022	[19]
Glycyrrhiza uralensis	22	Apigenin 6-C-α-L-rhamnoside-8-C-(6′-(3-hydroxy-3-methylglutaroyl)-β-D-glucoside); apigenin 6-C-α-L-glucoside-8-C-(6′-O-(3-hydroxyl-3-methylglurtaroyl)-β-D-glucoside); apigenin 6-C-α-L-arabinoside-8-C-(6′-O-(3-hydroxyl-3-methylglurtaroyl)-β-D-glucoside)	NMR	Chemical structures	OPLS-DA	2022	[20]
Astragalus membranaceus	14	7-β-D-glucopyranosyloxy-astrapterocarpan; (3R,4R)-7-(2-O-β-erythro-D-apiofuranosyl-β-D-glucopyranosyloxy)-3′,4′-dimethoxyl-pterocarpan	NMR	Chemical structures	\	2022	[21]
Capsella bursa-pastoris(L.)	4	Isoorientin; isoorientin-2″-O-α-L-arabinopyranosyl; isoorientin-2″-O-α-L-xylose; kaempferol-3-O-β-D-glucoside,	NMR; UV-Vis	Chemical structures; Quantitatively analyse	\	2021	[22]
Glycyrrhiza uralansis	10	Diosmetin; echinatin; Licofuranol A; licofuranol B; calycosin; luteolin; scopoletin; glycypytilbene B,	NMR; UV-Vis	Chemical structures	\	2019	[23]
Radix puerariae	2	Daidzein; puerarin	NMR	NMR chemical shifts; NMR shielding parameters	\	2019	[24]
Psoralea corylifolia L.	62	Bavaisoflavone; bavachinone; bavaflavone	NMR; Circular dichroism spectra	Chemical structures	\	2022	[25]
Flavonoids	10	Baicalein; baicalin; apigenin; quercetin; naringenin; hesperetin; daidzein; genistein; puerarin; gastrodin	Terahertz time-domain spectroscopy(THz-TDS)	Qualitative identification; quantitative analysis	PCA; SVM; PLSR; ANN	2020	[26]
Flavonols	3	Myricetin; quercetin; kaempferol	Terahertz spectroscopy	Qualitative identification; quantitative analysis	PLSR; least squares SVM	2018	[27]
Citri reticulatae pericarpium	6	Nobiletin; sinensetin; 3,5,6,7,8,3′,4′-heptamethoxyflavone; tangeretin; hesperidin	Fluorescence spectroscopy	Quality evaluation; identify storage year	PLS-DA	2020	[28]
Pueraria lobata	3	Puerarin, daidzin, daidzein	UV	Quantitative analysis	\	2016	[29]
Pueraria lobata	6	Lobatflavate, 3S,4R-tuberosin, daidzein, puerarin, daidzin, ononin	IR; UV; NMR	Chemical structures	\	2017	[30]

Table 1. *Cont.*

Sources	Flavonoid Types	Compound Name	Analysis Technique	Purpose	Chemometrics	Year	References
Trollius europaeus	10	Orientin; isoorientin; vitexin; isovitexin; orientin 2′-O-β-arabinopyranoside; orientin 2′-O-β-glucopyranoside; vitexin 2′-O-β-arabinopyranoside; vitexin 2′-O-β-galactopyranoside	NMR; UV; MS	Chemical structures	\	2018	[31]
Marine red alga acanthophora spicifera	1	Apigenin	IR; UV; MS	Chemical structures	\	2016	[32]
Flavonoid	1	Luteolin	UV-vis	Chemical structures	\	2021	[33]
Black wolfberry	Total flavonoids	Rutin	NIR	Quantitative analysis	PLS	2018	[34]
Pueraria lobata	Total isoflavonoid	Puerarin; daidzin	NIR	Quantitative analysis	PLS	2014	[35]
Propolis	Total flavones; flavonols; flavanones	Quercetin; rutin; pinocembrin	NIR	Quantitative analysis	PLS	2017	[36]
Poplar propolis	2	Chrysin; galangin	Vis-NIR	Quantitative analysis	PLS; ANN; MLR; LS-SVM	2013	[37]
Pueraria lobat	Total flavonoids	Puerarin	NIR; UV-Vis	Quantitative analysis	SEPA-PLS	2022	[38]
Black goji berries	Total flavonoids	Quercetin	Near-infrared hyperspectral imaging (NIR-HSI)	Quantitative analysis	PCA; PLS; LS-SVM	2020	[39]
Chrysanthemum morifolium	Total flavonoids	Rutin	NIR-HSI	Quantitative analysis	PLS; LS-SVM	2018	[40]

Notes: Orthogonal partial least squares regression (OPLSR); Partial least squares (PLS); Orthogonal partial least squares discriminate analysis (OPLS-DA); Principal component analysis (PCA); Support vector machine (SVM), Partial least-squares regression (PLSR); Artificial neural network (ANN); Multiple linear regression (MLR); Sampling error profile analysis (SEPA).

2.1. Nuclear Magnetic Resonance

Nuclear magnetic resonance (NMR) spectroscopy can detect and quantify molecular interactions to accurately characterize the structure of flavonoid molecules in complex samples, although it has disadvantages such as high cost and may not be suitable for all applications [41]. Duan et al. isolated and purified 21 compounds from the Platycladi Cacumen extract and identified the chemical structures of several flavonoids by NMR [19]. Shang et al. studied the chemical composition and biological activities in four different parts of the roots, stems, leaves and seeds of *Glycyrrhiza uralensis* [20]. Three flavonoid C-glycosides were identified by NMR spectroscopic analysis. Bao et al. isolated 14 isoflavone components from the root of *Astragalus membranaceus*, the planar structures of which were determined by detailed analysis of 1D and 2D NMR data [21]. Xie et al. first analyzed four major flavonoids in *Capsella bursa-pastoris* (L.) by UPLC, 1H-NMR and 13C-NMR spectroscopic techniques to investigate the effect of *Capsella bursa-pastoris* (L.) on the clinical treatment of cataracts [22]. Wang et al. isolated and identified four new compounds and eight known flavonoids from the leaves of *Glycyrrhiza uralansis* and determined their structures by high-resolution electrospray mass spectrometry (HR-ESI-MS) and UV spectrum and 1H-nuclear magnetic resonance (NMR) and 13C-NMR spectroscopy [23]. Yi et al. determined the 1H and 13C NMR chemical shifts and NMR shielding parameters of daidzein and puerarin, the major flavonoid active ingredients in Radix puerariae, by NMR analysis [24]. Xu et al. reported 109 chemical constituents, including 21 flavonoids, 6 flavones and 35 isoflavonoids, from the mature fruit of *Psoralea corylifolia* L. The structures of the compounds were elucidated by 1H NMR, 13C NMR, 2D NMR and Rh2(OCOCF3)4 and Mo2(OAc)4-induced circular dichroism spectroscopic methods [25].

2.2. Terahertz Time-Domain Spectroscopy

Terahertz (THz) spectroscopy is an emerging and powerful investigation technology that contains abundant information about the physics, chemistry and structure of materials. In contrast to conventional far-infrared spectroscopy, terahertz spectroscopy exploits a part of the electromagnetic spectrum that lies between the microwave and infrared regions. In the terahertz band, biological molecules conduct complex molecular vibrations such as rotations, low frequency bond vibrations, hydrogen bonds and van der Waals forces. Based on terahertz characteristic spectra, biomolecules can be identified effectively, especially for those with similar chemical structures. In recent decades, terahertz spectroscopy has been widely used in the fields of physics, chemistry, materials science and biomedicine on account of its rapid, safe and non-destructive advantages [42,43]. However, terahertz spectroscopy also has limitations in terms of limited penetration, scattering effects and limited sensitivity [44]. Furthermore, the cost of terahertz spectroscopy is still high compared to other applications and the current terahertz database on natural active compounds is incomplete and needs further improvement and development.

Yin et al. used terahertz time-domain spectroscopy (THz-TDS) for the qualitative identification and quantitative analysis of ten common flavonoids, including baicalein, baicalin, apigenin, quercetin, naringenin, hesperetin, daidzein, genistein, puerarin, and gastrodin [26]. These flavonoids with similar molecular structures had significantly different characteristic absorption peaks in the terahertz band and were distinguished by the terahertz absorption spectrum. Yan et al. identified three flavonols with similar structures, including myricetin, quercetin, and kaempferol, and determined their concentrations by terahertz spectroscopy [27].

2.3. Fluorescence Spectroscopy

Fluorescence spectroscopy is now increasingly used for quality control and monitoring analysis of food and pharmaceuticals as a sensitive, simple and fast detection technique [45]. Flavonoid components will produce fluorescence when exposed to constant excitation light due to their native fluorescence properties. Shan et al. used a fluorescence spectrophotometer to obtain synchronous fluorescence spectra of tea infusion to quantify the content of flavonoids in green tea. The excitation/emission spectra of flavonols were in the range of 365–390 nm/450–470 nm and those of flavanols were in the range of 480–500 nm/510–520 nm [46]. In recent years, as nanotechnology becomes more sophisticated, various new nano-fluorescent sensor-based fluorescence spectroscopy has found more applications in quality assessment [47]. Lan et al. used a nanomaterial-based fluorescent sensor combined with spectral splicing to successfully evaluate the quality of *Citri Reticulatae Pericarpium* and identify their storage year [28]. Specifically, nanogold particles and cadmium telluride quantum dots were chosen as nanosensors and mixed with aqueous extracts of *Citri Reticulatae Pericarpium* to collect fluorescence quenching spectra. Then, the self-fluorescence and fluorescence quenching spectra of the same sample were combined to integrate the spectra of different fluorescence sensing systems at the same coordinate axes to obtain spliced spectra. This new strategy achieved accurate recognition of different *Citri Reticulatae Pericarpium* samples by identifying the interaction between the nanoparticles and the fluorescent components in the *Citri Reticulatae Pericarpium* sample.

2.4. UV Spectrophotometry

Ultraviolet-visible (UV-Vis) spectrophotometry is used to qualitatively and quantitatively analyze compounds by utilizing the absorption spectra of compounds that will absorb energy in the UV or visible region and undergo an electron energy leap. In addition to simple operation and fast analysis speed, UV-Vis has the advantage of greater sensitivity and selectivity [48].

In a study to determine whether *Pueraria lobata* could protect human umbilical vein endothelial cells (HUVECs) from apoptosis, Gao et al. used UV spectrophotometry and HPLC to determine the content of isoflavones in the ethanolic extracts of *Pueraria lobata*. Puerarin, daidzin and daidzein were found to be the major isoflavonoid components of *Pueraria lobata*, accounting for 84.94% of the extract [29]. Wang et al. isolated four known isoflavone analogs, a new isoflavone and a new flavone hydrate from *Pueraria lobata* (Willd.) ohwi, and their structures were characterized by IR, UV, HR-ESI-MS, 1D and 2D NMR spectroscopic methods [34]. Xie et al. measured the total flavonoids content of *Capsella bursa-pastoris* (L.) extract to 65.18$\pm$2.16% based on UV-VIS spectrophotometry [22]. Witkowska-Banaszczak et al. used NMR, UV spectroscopy and electrospray ionization tandem mass spectrometry (ESI-MS/MS) for the structural identification of 10 flavonoids in the extracts from the flowers of *Trollius europaeus* [31]. El Shoubaky et al. isolated a flavonoid from the acetone extract of marine red alga *Acanthophora spicifera* and identify the structure of the flavone compound by infrared, mass and UV spectroscopy. This flavonoid was confirmed to be apigenin and showed promising analgesic, anti-inflammatory and anti-proliferative activities [32]. Luteolin, a flavonoid widely occurring in natural plants, has a variety of activities including anti-inflammatory, cardioprotective and can interact with certain metals and biomolecules. Jomova et al. used UV-vis spectroscopy to characterize the interaction between Cu(II) and luteolin in their study of the effect of luteolin on DNA damage in the copper-catalyzed Fenton reaction [33].

2.5. Near Infrared Spectroscopy

Near-infrared (NIR) spectroscopy is a simple, fast, accurate and non-destructive technique that has been used in a variety of fields for process analysis and quality control in recent years [49–51]. Compared to the mid-infrared (MIR) range, the shorter NIR wavelengths increase the depth of penetration. NIR spectroscopy covers the wavelength range from 800 to 2500 nm and mainly records the spectral bands corresponding to the molecular vibrations of hydrogen bonds (e.g., C-H, N-H, O-H) to obtain the characteristic information of the hydrogen-containing groups in compounds [52]. Compared to IR spectroscopy, which requires the sample powder and KBr powder to be mixed and ground and pressed into tablets first, IR spectroscopy requires no sample preparation and will not produce any waste products [17]. However, NIR spectra usually rely on reference methods and need to be combined with chemometrics to build models [53].

Arslan et al. performed quantitative analyses of flavonoid components in black wolfberry by Fourier-transform near-infrared (NIR) spectroscopy combined with chemometric algorithms [34]. Wang et al. used near-infrared (NIR) spectroscopy to quantitatively monitor the content of flavonoid active ingredients in the water-ethanol extraction process of *Pueraria lobata* [35]. Betances-Salcedo et al. analysed the total contents of flavones and flavonols, flavanones and dihydroflavonols by using the methods of NIR methodology in 99 samples of propolis from Spain and Chile [36].

In general, herbal medicines contain a large number of active ingredients and other components that make spectroscopic measurements very complex, and it is difficult to achieve accurate measurements from NIR spectroscopy results alone. Therefore, multiple techniques are often used to obtain more complete and comprehensive information. UV-Vis detects only information about specific groups in the molecule, so UV-Vis cannot obtain information about all active compounds, but it is possible to get rid of the interference of solutions during spectroscopic analysis. In contrast, NIR is greatly affected by the solution, but more signals of the compound can be obtained. Combining NIR and UV-Vis, therefore, gives more comprehensive and effective information in a dual spectrum. Nie et al. developed a rapid method for the determination of flavonoid compounds (chrysin and galangin) in poplar propolis by means of visible and near-infrared spectroscopy (Vis-NIR) and a total of 114 propolis samples from China were analysed [37].

The effective extraction of Chinese medicinal ingredients is the primary premise behind identifying the structure and studying the biological activity, which directly determines the quality of the preparations in production and the actual clinical efficacy. Therefore, a reasonable extraction time is crucial, too long or too short an extraction time will affect the extraction yield. Therefore, a simple, rapid and efficient means of monitoring and analysis is required in the extraction process of herbal medicines.

In this context, Xu et al. developed a dual-spectrum portable spectrometer technology based on near-infrared (NIR) and ultraviolet-visible (UV-Vis) spectroscopy. With the spectral acquisition analysis, a quantitative analysis model of total flavonoids in *Pueraria lobata* was developed to enable online monitoring of the extraction process of flavonoid active compounds from *Pueraria lobata* [38]. In particular, they have chosen to use cylindrical cuvettes instead of traditional quartz cuvettes or fibre optic probes when measuring spectroscopy, making it easier to clean the solid residues produced during the extraction process, while reducing the cost of detection and facilitating its widespread application. The results showed that the dual-spectrum online monitoring system is simple to operate, has fast sampling speed, low cost and provides more comprehensive information than individual NIR or UV–Vis spectrum. It is a promising tool for the quantitative analysis of some traditional Chinese medicines with complex compositions. It should be noted, however, that if the compound component to be measured does not absorb in the UV–Vis region, then the dual-spectrum system will not be able to identify it.

2.6. Hyperspectral Imaging

Hyperspectral imaging (HSI) combines imaging and spectroscopic techniques so that spectral and spatial information about the sample can be obtained simultaneously. Compared to conventional analytical methods such as liquid chromatography, hyperspectral imaging is faster and more non-destructive and has been used to detect various compositions of samples, such as food [54], agriculture [55] and herbal medicine [56,57].

Near-infrared (NIR) spectroscopy generally collects spectral information from small sampling points and is unable to obtain the spectra of the entire sampled region; the final spectra of the sample are represented by the average spectra measured from multiple small sampling points. In hyperspectral imaging, a 3D data cube ($x \times y \times \lambda$) is created by combining two spatial (x; y) and one wavelength (λ) dimension, where images are collected as a function of wavelength [57–59]. Hyperspectral imaging has the advantage of acquiring spectral information from the entire sampling area within the hyperspectral images. As a result, hyperspectral imaging enables more representative spectral information to be obtained than near-infrared spectra acquired from a single point. In addition, based on the character that each pixel within the hyperspectral images has a spectrum, a distribution map can be formed to explore the composition distribution differences within and among samples [60–62].

Zhang et al. determined the total flavonoid content of dried black goji berries by near-infrared hyperspectral imaging (NIR-HSI) [39]. In the work of He et al. near-infrared hyperspectral imaging was used to determine the total polysaccharide and total flavonoid content of *Chrysanthemum morifolium* [40]. Compared to reference methods using UV–Vis spectroscopy for total polysaccharides and total flavonoids, hyperspectral imaging was more environmentally friendly, and more efficient in handling large numbers of samples and online predictions could be made for different *Chrysanthemum morifolium*.

3. Identification of Quality

Generally, the quality of herbs can be strictly influenced by the culture conditions, growth year, and geographic origin, thus resulting in significant differences in their medicinal quality and clinical efficacy [63]. Differences in culture conditions, such as light, temperature, and air humidity, can affect the number of secondary metabolites accumulated in herb plants, while the content of these active ingredients increases with the number of growth years. The composition and content of active compounds in the same herb may vary depending on the geographical origins [64]. Hence, the cultivation methods, growth years and geographical origin of herbs have attracted more and more attention from consumers. It is necessary to establish a reliable and effective analytical method for identification.

3.1. Cultivation Methods and Growth Years

Hai et al. collected a total of 320 samples of *Dendrobium huoshanense* (DHS), which were mainly divided into greenhouse cultivation and wild-like cultivation, each cultivation containing four growth years [65]. It is difficult to identify and distinguish between samples of *Dendrobium huoshanense* from different cultivation methods and growth years due to the high similarity of these samples and the limited information obtained from a single spectrum. Therefore, Hai et al. synthesized metallized tetraphenylporphyrin (ZnTPP) to obtain nano-effect near and mid-infrared spectroscopy by axial coordination, hydrogen bonds or electrostatic interactions with flavonoids, amplifying the differences in spectral signals of *Dendrobium huoshanense* samples. The nano-effect near and mid-infrared spectral data were then fused to obtain nano-effect feature fusion spectra, which successfully identified samples of *Dendrobium huoshanense* (DHS) with different growth years and cultivation methods, achieving 100% accuracy. Tetraphenyl zinc porphyrin (ZnTPP) is a metalloporphyrin formed by the introduction of Zn^{2+} into the cavity in the porphyrin ring. ZnTPP achieved complementary properties of both substances, with highly large π-conjugated properties and excellent optical properties [66,67]. The results showed that the addition of ZnTPP increased the peak difference, and the overlap of the nano-effect mid-infrared spectra was reduced compared to the original spectra. For the nano-effect NIR spectra, the peaks were more dispersed overall and the difference in spectral properties increased. It indicated that the nano-effect of ZnTPP amplifies the variability of the near and mid-infrared spectra.

3.2. Geographic Origin

NIR spectroscopy, with the advantages of simple, rapid, high efficiency and no sample preparation, has been a powerful analytical tool in use for identifying and analysing the geographical origin of food and herbal medicines [68]. Chen et al. collected a total of 250 raw material samples of notoginseng from four main geographical origins (Yunnan, Xizang, Guangxi and Guizhou provinces of China). The geographical origin of notoginseng was successfully identified by near-infrared spectroscopy [69].

However, for those herbs with complex and similar compositions, their NIR spectral peaks often overlap severely and some significant differences in the spectral peak cannot be obtained, and the classification accuracy is low. Therefore, the original NIR spectra are difficult to be used directly for the quality evaluation of herbal medicines.

At present, a large number of sensors based on nanomaterials are widely used to monitor food safety and authenticity due to their low cost, high sensitivity, and convenience [70–72]. As a very popular nanomaterial, carbon dots(CDs) have many merits, such as remarkable optical properties, excellent biocompatibility, good stability, and environmental friendliness, and they have been widely used in chemical sensors. Porphyrins possess a rigid macrocyclic structure, a large π-conjugation system, and various functional groups and substituents, enabling them to recognize multifarious molecules.

In order to solve this dilemma, Long et al. completed the identification of the geographical origin of lily using carbon dot-tetramethoxyporphyrin nanocomposite(CDs-TMPP)-based nano-effect near-infrared spectroscopy sensor [73]. Long et al. first collected a total of 120 lily samples from 12 different geographic origins in China and then collected nano-effect NIR spectra in the presence of CDs-TMPP. The results show that the nano-effect spectroscopy sensor method had better classification performance compared with the original NIR spectra, with 100% accuracy in identifying the geographical origin of the lily samples. The active components in the lily interact with CDs-TMPP through hydrogen bonding, electrostatic interaction, and steric hindrance, enhancing the differences in near-infrared spectroscopy peaks of lily samples from different geographical sources. In previous studies, Lv et al. developed a near-infrared spectroscopic sensor combined with porphyrin to realize the identification of dendrobes from 12 different geographical sources [74]. Compared with traditional NIR spectra, the NIR spectra with TMPP can obtain more characteristic information, which greatly improves the accuracy of model identification. After adding TMPP, the accuracy rate of NIR spectra reached 100%. The possible mechanism was that the π-π conjugated system and the methoxy groups of TMPP interact with the chemical components of dendrobes, which increase the specificity of NIR spectra.

4. Interaction Studies

Spectroscopy techniques can be used not only for quantitative and qualitative analysis, but also in combination with bioactivity analysis to study the interaction of flavonoids bound to biological macromolecules, including lipid membranes, human serum albumin, and hyaluronidase.

4.1. Interaction of Flavonoids with Membrane Lipids

The amphiphilic character of flavonoids allows them to intercalate into or bind with lipid bilayers. Many of the biological effects of flavonoids have been assumed to result from interactions with the membranes [75–77]. In previous studies, IR spectroscopy has been used extensively to elucidate the interactions between flavonoids and membrane lipids, but the conclusions drawn from IR spectroscopy about the way flavonoids are incorporated onto or into lipid bilayers are rather vague and are generally not far reaching [78,79]. The main reason lies in the fact that the actual state in which the flavonoid molecules are bound to lipids in a buffer solution is not known.

When there are interactions between flavonols and their surroundings, changes in molecular structure may occur, resulting in changes in the spectrum. Such structural changes such as structural rearrangements, conformational changes and deprotonation may give rise to hydroxyl and carbonyl vibrations. The vibrations of anyone hydroxyl group do not occur in isolation but are mixed with ring vibrations and other neighboring hydroxyl groups to form very complex vibrational patterns. The structural changes caused by the interactions do not remain in a narrow part of the spectrum, but actually affect the whole spectrum. This leads to very different spectra of structurally similar compound molecules. Therefore, it can hardly be satisfactorily to discuss the spectra of flavonoids in terms of characteristic vibrations of either the hydroxyl or carbonyl group. Baranović et al. performed the infrared spectroscopic analysis of seven flavones (flavone, 3-and 5-hydroxyflavone, chrysin, apigenin, fisetin and luteolin) and five flavonols (galangin, kaempferol, quercetin, morin and myricetin) and reported that hydroxyl and carbonyl vibrations in the interaction of these flavonoids with membrane lipids [80]. By monitoring spectral changes brought about by the interaction of flavonoids with membrane lipids it revealed structural details of how the flavonoids incorporate onto or into the lipid bilayer.

4.2. Interaction of Flavonoids with HSA

Human serum albumin (HSA), the most abundant protein in plasma, has the capability of binding reversibly to a large variety of drugs via its binding sites. After binding to human serum albumin, the drug compound is transported to various locations in the body for release. Therefore, the strength of the drug compound's ability to bind with human serum albumin determines the therapeutic effect [81]. A large part of the general population is exposed to some flavonoids in their daily diet. Consequently, the risk of interaction between the binding levels of these flavonoids and human serum albumin is much higher. This interaction is likely to alter the binding of another drug to human serum albumin, thus affecting the pharmacological effect of another drug.

Quercetin (QUE) is one of the most abundant flavonoids in the human diet and is also the main active ingredient in many herbal medicines, with a variety of pharmacological effects on the human organism [82]. Diosmin (DIO), a flavonoid commonly found in citrus fruits, is the active component of many drugs, especially ones used in the treatment of various blood vessel disorders [83]. Catechins (CAT) are also common plant-derived flavonoids with anti-inflammatory and antioxidant potential and have a wide range of applications in the pharmaceutical and food industries. Tigecycline (TGC), is an antibiotic drug commonly used in clinical practice. Sovrlić et al. investigated the effect of flavonoids (catechins, quercetin and diosmin) on the binding of antibiotics (tigecycline) to HSA using multiple spectroscopic measurements, as well as their effects on the structure of the active site and the nature of interactions [84]. The formation of triple complexes of HSA–TGC–FLAVs with high binding affinity was demonstrated by UV–Vis absorption spectroscopy and fluorescence analysis. The conformational changes of HSA were analyzed by simultaneous fluorescence spectroscopy, Fourier transform infrared spectroscopy and circular dichroism, and it was found that the triple complex of HSA–TGC–FLAVs did not affect the microenvironment around the tryptophan (Trp) and tyrosine (Tyr) residues of HSA.

4.2.1. UV–Vis Absorption Spectra

UV–Vis absorption spectroscopy, a convenient, rapid and effective technology, is commonly used to study protein-drug interactions and complex formation [85]. Red-shift and blue-shift changes observed in UV–Vis spectra can explain the interaction mode between drugs and proteins. UV–Vis spectral results showed that the absorption intensity of HSA- flavonoids (QUE, CAT and DIO) decreased as the concentration of TGC increased, demonstrating the formation of triple protein–drug–drug complexes between HSA, TGC and flavonoids (QUE, CAT and DIO) [84].

4.2.2. Fluorescence Quenching Measurements

The presence of Trp and Tyr amino acids in the HSA structure provides HSA with fluorescent properties. This fluorescence property is sensitive to the microenvironment of the HSA molecule and when the local environment of the HSA molecule changes, such as protein denaturation or biomolecular binding, its fluorescence is quenched [86,87]. In previous studies, Matei et al. found that the fluorescence of HSA was quenched in a concentration-dependent manner upon interaction with kaempferol by fluorescence measurements [88]. Kaempferol, a natural flavonoid compound with a wide range of biological activities, is widely found in plants and foods. Therefore, the change in the fluorescence intensity of HSA can determine whether small molecules are bound to it, affecting the environment around the Trp and Tyr amino acid residues. The results showed the fluorescence quenching of HSA-flavonoids (QUE, CAT and DIO) in the presence of varying concentrations of TGC, indicating the formation of the HSA-TGC-flavonoids (QUE, CAT and DIO) complex [84].

4.2.3. Synchronous Fluorescence Spectra

The synchronous fluorescence method allows information on the molecular environment in the vicinity of the Trp and Tyr fluorophores of HSA to be obtained and to determine whether polarity change around the microenvironment has occurred by measuring the position of its maximum emission wavelength. Synchronous fluorescence of HSA can provide characteristic information around tyrosine (Tyr) and tryptophan (Trp) residues when the scanning wavelength intervals ($\Delta\lambda$) are fixed at 15 nm and 60 nm, respectively [89]. The red or blue shift in the maximum fluorescence emission (wavelength) of HSA indicates enhanced hydrophilicity or hydrophobicity of the microenvironment around Tyr or Trp residues, respectively [84]. In the findings of Matei et al., no significant shifts in the position of the maximum emission wavelength were registered, but the fluorescence burst of Trp was stronger than that of Tyr, suggesting that the binding site of kaempferol to HSA is nearer to the Trp residue [88]. The results of the synchronous fluorescence spectra of the ternary HSA–TGC–flavonoids (QUE, CAT and DIO) systems showed no significant changes in the maximum emission wavelength of Tyr and Trp residues, indicating that the interaction of HSA-TGC with flavonoids does not affect the conformation of the micro-region of Tyr and Trp [84].

4.2.4. Circular Dichroism Measurements

It is known that protein–ligand interactions can alter the secondary structure, resulting in changes in the protein conformation, which are reflected by the circular dichroism (CD) spectrum. The circular dichroism spectrum of HSA observed two negative minima at 208 and 222 nm, and represents α- helix structure transition of $\pi \rightarrow \pi^*$ and $n \rightarrow \pi^*$. In the findings of Matei et al., the percentage of α-helices decreased progressively upon binding of kaempferol to HSA, compared to free HSA molecules, indicating a degree of protein folding [88]. Sovrlić et al. found that the CD spectra for the ternary HSA-TGC-flavonoids (QUE, CAT and DIO) systems showed no significant change in signal and shape from the original spectra, indicating that the binding of HSA–TGC–flavonoids had a negligible effect on the secondary structure of the protein [84].

4.2.5. Fourier Transform Infrared Spectroscopy (FT-IR)

The investigation of the secondary structure of HSA was performed using the FT-IR spectroscopic technique [88]. The protein amide I band at 1650–1654 cm^{-1} and amide II bands at 1548–1560 cm^{-1} are attributed to the secondary structure of all proteins. FT-IR spectra of HSA showed no significant difference between the ternary HSA-TGC-FLAV system and free HSA, indicating no conformational change in the HSA protein [84].

4.3. *Interaction of Flavonoids with Hyaluronidase*

It has been reported that there were several enzymes known to be involved in promoting inflammatory pathways, of which hyaluronidase (HAase) is one of the most important enzymes, cleaving hyaluronic acid in the extracellular matrix and improving the permeability of cell membranes and blood vessels [90]. During the development of inflammation, the level of HAase in the body increases dramatically. It has been found that some flavonoids exhibit strong anti-inflammatory effects while being able to inhibit the activation of HAase [91].

4.3.1. Fluorescence Spectra

In the study by Zeng et al., the interaction between eight flavonoids(apigenin, luteolin, keampferol, quercetin, morin, naringenin, daidzein, genistein) and HAase was investigated by fluorescence spectroscopic and molecular modeling methods [90]. The results showed that the eight flavonoids formed flavonoid–HAase complexes mainly by binding to HAase interactions through electrostatic forces, hydrophobic interactions and hydrogen bonding. According to synchronous and three-dimensional fluorescence spectra, the presence of flavonoids significantly altered the microenvironment and conformation of HAase, leading to reduced enzyme activity.

Li et al. studied the binding of three flavonoids(baicalin, liquiritin and isoliquiritigenin), extracted from *Scutellaria baicalensis Georgi* and *Glycyrrhiza uralensis* to hyaluronidase by steady state fluorescence, time-resolved fluorescence and circular dichroism (CD) spectroscopy [92]. The results of the fluorescence spectra showed that when baicalin and liquiritin were bound to HAase, the strongest fluorescence emission peak of HAase was red-shifted and fluorescence quenching was stronger. When isoliquiritigenin interacted with HAase, it acted as quencher to decrease the fluorescence intensity of HAase with no significant change in the position of the emission peak. The fluorescence quenching mechanism of HAase by the three flavonoids is a static quenching procedure.

4.3.2. Synchronous Fluorescence Spectra

Synchronous fluorescence spectra present information about the molecular microenvironment in the vicinity of the fluorophore by measuring the emission wavelength shift. The wavelength interval ($\Delta\lambda$) between the excitation wavelength and the emission wavelength is fixed individually at 15 and 60 nm, which gives the characteristic information of tyrosine (Tyr) or tryptophan (Trp), respectively.

Synchronous fluorescence spectra showed that with the addition of baicalin, the maximum emission wavelengths of Trp- and Tyr- are both observed to have a redshift, indicating that the polarity around Trp-and Tyr-residues increases and the hydrophobicity decreases. For the liquiritin–HAase system, the maximum emission wavelength of Trp-residues has an obvious red shift, but that of Tyr- has no obvious change. The results suggested that the interaction of liquiritin with HAase increases the polarity and decreases the hydrophobicity around Trp-residue, but has no effect on the microenvironment around Tyr-residue. For the isoliquiritigenin–HAase system, the synchronous fluorescence peaks of Trp- and Tyr- do not change significantly, indicating that the microenvironment of Trp-and Tyr-is not disturbed by isoliquiritigenin [92].

4.3.3. Circular Dichroism

Circular dichroism (CD) is usually executed to investigate the secondary structural changes of protein because of its accuracy and sensitivity [93]. The results of CD spectroscopy showed that the binding of flavonoids(baicalin, liquiritin and isoliquiritigenin) with HAase leaded to changes in the secondary structure of HAase with an increase in the α-helix content in HAase.

5. Chemometrics

Due to the complex composition of herbal raw materials, the obtained spectra are difficult to identify, have severe overlaps of spectrum bands and contain a lot of useless information, which brings a great challenge for the identification of spectral information. Therefore, it is necessary to analyse the spectral information of the samples by chemometric methods in order to obtain valid spectral information, improve the selectivity of fluorescence spectroscopy and achieve rapid multi-component analysis. Table 1 has summarised the chemometric methods used for the spectroscopic analysis of medicinal and food homology materials.

5.1. PLS-DA Analysis

Partial least squares discriminant analysis (PLS-DA) is a very common classification method that is used in various fields of analysis [94–96]. PLS-DA is based on the classical PLS regression algorithm, which combines PLS regression with discriminant analysis to look for latent variables (LV) with a maximum covariance with the dependent Y variable.

The number of LVs is usually determined based on the optimal correct classification rate of the cross-validation procedure [97]. The dependent variable Y is a dummy matrix composed of binary values, with the one-hot encoding used to represent the class belonging of the samples. In dummy matrix Y, a value of 1 means that the sample belongs to a specific class and 0 means the opposite. However, the estimated response values of PLS regression models are usually not exactly equal to 0 or 1. When the estimated response value for a sample is closer to 1, the sample is considered to belong to the corresponding class, and the opposite is true when it is closer to 0.

The model based on the traditional raw fusion spectrum is poor, resulting in low accuracy in the training and prediction sets. It means that traditional spectra struggle to identify DHS from different cultivation years. To enable feature fusion spectra of DHS samples of different growth years to be identified by classification, PLS-DA was used to build a PLS-DA model based on nano-effects feature fusion spectra. Of these, 70% of the DHS samples were classified as the training set and 30% were classified as the prediction set, with the training and prediction sets proceeding by random classification. The results showed that the PLS-DA model based on nano-effects feature fusion spectroscopy can achieve accurate discrimination of DHS samples of different growth years with 100% accuracy [65].

Long et al. used partial least squares discriminant analysis (PLS-DA) to identify the geographical origin of the lily based on the collected nano-effect NIR spectroscopy, achieving 100% classification accuracy [73]. Lan et al. collected nanomaterial-based fluorescence splicing spectra of *Citri Reticulatae Pericarpium* samples and achieved 100% sample species identification and 98.04% storage year identification by partial least squares discriminant analysis [28]. In order to classify dendrobe of different geographical origins, Lv et al. used partial least squares to process the NIR spectral data of dendrobe samples and achieved 100% accuracy in identifying the origin of the samples [74]. Arslan et al. used partial least squares(PLS) to establish the quantitative analysis models of the total flavonoid content and other components in black goji berries based on near-infrared fourier transform spectroscopy [34]. Yin et al. developed a quantitative prediction terahertz time-domain spectroscopy model of the ternary flavonoid mixtures (genistein, naringenin, daidzein) by means of partial least squares regression, achieving simultaneous prediction of the concentrations of these three analytes [26]. Wang et al. used partial least squares regression (PLSR) to develop a calibration model for near-infrared spectroscopy that could be used to rapidly monitor the concentration of isoflavone compounds during *Pueraria lobata* extraction [35]. Betances-Salcedo et al. used a modified partial least squares (MPLS) regression method to develop a data calibration model to evaluate the NIR spectral data of 99 propolis samples, which well quantified the composition of flavonoids and flavonols, flavanones and dihydroflavonols in propolis [36]. The MPLS model calculates and standardized the NIR residuals for each factor and wavelength and is typically more stable and accurate than the standard PLS algorithm.

5.2. OPLS-DA Analysis

Orthogonal partial least squares discriminant analysis (OPLS-DA) is a very effective supervised analysis method that is commonly used to deal with classification and discrimination problems [98]. The orthogonal partial least squares discriminant analysis (OPLS-DA) adds a positive exchange algorithm to the partial least squares discriminant analysis (PLS-DA), which can filter out signals that are not relevant in the model's classification matrix [99]. Therefore, the OPLS-DA model is able to maximize the classification differences between groups and can better identify differences in chemical composition between groups compared to the PLS-DA model. The classification indicators of the OPLS-DA model include accumulated explanatory power parameters (R2X, R2Y) and predictive ability parameters (Q2) [100]. Among them, R2X and R2Y respectively represent the percentage of X and Y matrix information that the OPLS-DA classification model can explain, and Q2 is calculated through cross-validation to evaluate the predictive ability of the OPLS-DA model.

The closer these indicators are to 1, the better the OPLS-DA model fits the data and predictive power. When these indicators are greater than 0.5 the model is considered to have good results. In the results of Hai et al., an OPLS-DA model for nano-effects feature fusion spectroscopy was developed with model parameters including R2X, R2Y and Q2 are both greater than 0.9, indicating that the model has good goodness-of-fit and prediction ability to successfully distinguish between different cultivation mode of DHS samples [65].

5.3. VIP Value Analysis

The variable importance for the projection (VIP) value describes the importance of each variable in the models, and variables with a VIP score above 1 will be considered to contribute significantly to the PLS-DA model, thereby classifying and screening important spectral information [101]. Screening valid information through VIP can improve the model classification accuracy for subsequent chemometric analysis.

In order to improve the accuracy of identifying the cultivation methods and growth years of *Dendrobium huoshanense* (DHS), Hai et al. processed the spectral data by chemometrics after obtaining nano-effect near-infrared spectroscopy and nano-effect mid-infrared spectroscopy of 320 DHS samples [65]. After screening by variable importance for the projection (VIP greater than 1), multivariate data extraction and integration are performed to obtain the feature vectors of the fusion spectra. The feature vectors were then combined with partial least squares discriminant analysis (PLS-DA) and orthogonal partial least squares discriminant analysis (OPLS-DA) to identify DHS in different growth years and cultivation methods.

5.4. Data Fusion Strategy

A data fusion strategy is the integration of data from all sources to provide complementary data on the overall chemical signature [102]. It can be divided into three levels: low, medium and high. Compared with single data analysis, the data fusion strategy provides more efficient and accurate chemometric characterization, which is advantageous for tracing geographical origin [103] and quality identification [104]. Low-level data fusion is the concatenation of signals from different analysis instruments to form a new matrix where the rows represent the number of samples to be analysed and the columns represent the signal variables. Medium-level data fusion extracts the desired features from the signals of different analysis instruments and then concatenated these features into a new matrix for multivariate classification analysis. High-level data fusion calculates the classification results for each dataset separately and then finishes the combined analysis by assigning ratios based on discriminant accuracy.

The original data of the DHS samples were screened with VIP to retain nearly half of the important variables [65]. The extracted feature vectors were then subjected to data fusion to obtain the NIR-MIR fusion spectra and nano-effect NIR-MIR fusion spectra, which contain more comprehensive and valid information compared to the original spectra.

5.5. Sampling Error Profile Analysis (SEPA) Method

The sampling error profile analysis (SEPA) method is based on the Monte Carlo sampling(MCS) strategy and error profile analysis and can be used in outlier detection, CV, pretreatment method and wavelength selection, and model evaluation [105]. Multiple sub-models and their resulting sub-errors were obtained from the MCS. Error distribution analysis was performed on these sub-errors, from which the median, variance and bias of the errors were estimated. In spectral analysis, the choice of the number of latent variables (LV) and the evaluation of the model are essential for the optimization of the spectral analysis model. Cross validation (CV) is a commonly used method for selecting the number of LVs and can be used for model optimization. In both model optimization and evaluation, error analysis of statistical models is required. The median is more robust and can evaluate the model more accurately instead of the mean. The variance and bias of the model determine the predictive power of the model. As the complexity of the model increases, the bias of the model becomes smaller and the variance higher, but the generalization ability decreases and the prediction error becomes more significant, resulting in over-fitting of the data. Therefore, an optimal model should have a good complexity and a small prediction error, which means that the model is accurate and robust. Meanwhile, after the data have been fitted, the parameters selected in the model need to be tested for adequacy. The SEPA method can make the model more predictive and stable.

Xu et al. established a SEPA-PLS model based on the near-infrared (NIR) and ultraviolet–visible (UV–Vis) dual spectra of *Pueraria lobata* and achieved online quantitative monitoring of total flavonoids during the extraction of *Pueraria lobata* with good accuracy and precision [38]. SEPA-PLS models were developed using individual near-infrared (NIR) and ultraviolet-visible (UV–Vis) spectra from the extraction process of *Pueraria lobata*. Possible outliers were screened by error analysis. Finally, NIR and UV–Vis spectral data were fused to construct the dual-spectrum model.

5.6. Principal Component Analysis

Principal Component Analysis (PCA) is a classical multivariate statistical and data processing method that is widely used in a variety of analytical fields [106,107]. PCA can be used to extract feature variables, reduce the dimensionality and removes overlapping information from the data set without reducing the variance, and highlights the characteristics of the data through principal components. PCA is an orthogonal transformation method that changes the original correlated variables to uncorrelated components, which are named principal components (PCs), namely, a linear combination of the original variables. PCs are arranged in descending order of variance. When the cumulative variance contribution rate is more than 85%, these PCs can be considered to be able to replace the original data set [108].

Yin et al. collected terahertz time-domain spectroscopy (THz-TDS) of 10 common flavonoids, including baicalein, baicalin, apigenin, quercetin, naringenin, hesperetin, daidzein, genistein, puerarin, and gastrodin. Then the THz-TDS of these flavonoids were qualitatively identified and quantitatively analysed by chemometric methods, including principal component analysis (PCA), support vector machine (SVM), partial least squares regression (PLSR) and artificial neural networks (ANN) [26]. They performed a PCA method of the THz-TDS data of all samples and extracted the top five principal components representing the important information of the original data based on the three-dimensional score graph.

5.7. Support Vector Machine

Support vector machine (SVM) is a promising classification and regression technique for solving linear and non-linear multivariate calibration problems with excellent generalization capabilities [109]. Compared with other statistical methods, SVM does not require a large number of training samples for modeling [108]. Generally, PCA is combined with SVM techniques in order to improve the prediction accuracy of classification models.

Yin et al. used the THz-TDS spectral feature variables extracted by principal component analysis (PCA) as input variables of SVM to classify and identify 10 flavonoids. The model had 100% classification accuracy compared to the original spectral results [26]. Yan et al. performed quantitative analyses of three structurally similar flavonols by terahertz spectroscopy combined with partial least squares regression (PLSR) and least squares support vector machine (LS-SVM) [31]. The LS-SVM model demonstrated better results compared to the PLSR model for myricetin, quercetin, and kaempferol, respectively. He et al. used partial least squares (PLS) and least squares support vector machine (LS-SVM) to build prediction models and combined with near-infrared hyperspectral imaging to determine the total polysaccharides and total flavonoids content in *Chrysanthemum morifolium*, and obtained good prediction results [40]. Zhang et al. performed feature extraction by principal component analysis (PCA), developed partial least squares (PLS) and least squares support vector machine (LS-SVM) models, and used near-infrared hyperspectral imaging (NIR-HSI) techniques to determine the total flavonoid content in black goji berries [39].

5.8. Artificial Neural Network

Artificial neural network (ANN) is a commonly used non-linear econometric method, often used to solve machine learning problems such as regression and classification [110]. ANN is based on the operating principles of biological nerve cells and uses mathematical expressions to simulate the signal transmission between neurons, thus constructing interconnected hierarchical artificial neural networks. In spectral analysis, spectral data are introduced into the ANN model as an input layer and the output layer is the predicted result. The performance of the model is evaluated by the coefficient of determination and root mean square in the calibration (RMSEC) and the prediction set (RMSEP) [111]. Yin et al. used ANN regression models combined with THz-TDS for the quantitative detection of ternary mixtures of flavonoids (genistein, naringenin, daidzein) and showed good predictions [26]. Nie et al. used visible and near-infrared spectroscopy (Vis–NIR) combined with several chemometric models, including partial least squares (PLS), artificial neural networks (ANN), multiple linear regression(MLR) and least square-support vector machine (LS-SVM), to perform a rapid determination of the content of chrysin and galangin in poplar propolis. Among them, the ANN model achieved the best results [37].

5.9. Soft Independent Method of Class Analogy

The soft independent method of class analogy (SIMCA) is a classical class modeling technique that incorporates principal component analysis (PCA) to reduce the dimensions of the spectral data and provides a high dimensional variations classification [112]. The SIMCA model contains a collection of mutually independent PCA datasets. The training set is modeled by PCA and new samples can be fitted to the model and classified according to their similarity or dissimilarity to the training set. Chen et al. used partial least squares discriminant analysis (PLSDA) and soft independent modeling of class analogy (SIMCA) to construct the discriminant models that combined NIR spectral data to classify 250 notoginseng samples from different geographical origins, and the models achieved 100% sensitivity and 100% specificity on both the training and test sets [69].

6. Conclusions

The development and application of natural products have always been an area of significant interest to researchers. The abundant pharmacological activities of natural flavonoids demonstrate their adequate potential for future development and application in therapeutic drugs, functional foods and cosmetic additives. Phytochemical and pharmacological studies of medicinal and food homology materials are becoming increasingly attractive because of their combination of medicinal and food health effects. Nevertheless, to date, there are few reviews on the application of spectroscopic analysis of medicine and food homology flavonoids. This paper reviewed common spectroscopic methods applied to the analytical study of flavonoid components in medicinal and food homology materials, including qualitative and quantitative analysis of the compound structure and content, identification of herbal cultivation methods, growth years and geographical origin, interaction studies when combined with biomolecules, and chemometric methods used in combination with spectroscopic techniques.

Nuclear magnetic resonance (NMR) spectroscopy is a common and effective method for accurately identifying the chemical structure of multi-component complex samples. However, NMR has limited applicability in other analytical applications and is expensive. Terahertz spectroscopy is a very powerful technique that has emerged in recent years and has the advantages of being fast, safe and non-destructive, enabling the effective identification of structurally similar biomolecules. However, terahertz spectroscopy also has limitations in terms of limited penetration, scattering effects and limited sensitivity, and is very costly. Fluorescence spectroscopy is a sensitive, simple and rapid detection technique that is commonly used for quality control of food products and monitoring and analysis of the environment. In recent years, as nanotechnology becomes more sophisticated, various new nano-fluorescent sensor-based fluorescence spectroscopy has found more applications in quality assessment. Near-infrared (NIR) spectroscopy is a simple, fast, accurate, non-destructive and will not produce waste products technique that has been used in a variety of fields for process analysis and quality control in recent years. However, NIR spectra usually rely on reference methods and need to be combined with chemometrics to build models. In addition to simple operation and fast analysis speed, UV–Vis has the advantage of greater sensitivity and selectivity. UV–Vis detects only information about specific groups in the molecule, so UV–Vis cannot obtain information about all active compounds, but it is possible to get rid of the interference of solutions during spectroscopic analysis. In contrast, NIR is greatly affected by the solution, but more signals of the compound can be obtained. Combining NIR and UV–Vis, therefore, gives more comprehensive and effective information in a dual spectrum. Hyperspectral imaging (HSI) combines imaging and spectroscopic techniques so that spectral and spatial information about the sample can be obtained simultaneously. In addition, hyperspectral imaging enables more representative spectral information to be obtained than near-infrared spectra acquired from a single point. Compared to conventional analytical methods such as liquid chromatography, hyperspectral imaging is faster and more non-destructive.

A growing number of flavonoid components of medicinal and food homology materials have been isolated, identified and studied, based on the development of various techniques of spectroscopic analysis and other analytical methods. Nevertheless, continuous efforts are still needed to develop more analytical techniques with development potential to ensure quality control of medicinal and food homology materials and further clarify the potential molecular mechanisms. This will provide favorable conditions for better product development and market application of medicinal and food homology materials. This review may contribute to a rapid understanding of the application of spectroscopic techniques to flavonoid components in medicinal and food homology materials and provides an important reference for the research and development of flavonoids in medicinal and food homology materials.

Author Contributions: Conceptualization, X.L.; formal analysis, X.L.; investigation, M.Y.; resources, L.W.; data curation, L.Z.; writing—original draft preparation, L.Z.; writing—review and editing, X.L., H.L., X.D., Z.L. and L.Z.; visualization, D.H.; supervision, J.A.H.K. and X.L.; project administration, X.L.; funding acquisition, X.L. and All authors have read and agreed to the published version of the manuscript.

Funding: This work was financially supported by Guangxi Innovation-Driven Development Special Fund Project (Funding Number: GUANGXI AA18242040, Funder: Guangxi Science and Technology Department), Guangxi Natural Science Foundation (Funding Number: 2020GXNSFAA297178, Funder: Guangxi Science and Technology Department) and National Natural Science Foundation of China (Funding Number: 22078073, Funder: National Natural Science Foundation of China). This work was also supported by Fund of High-level talents of Youjiang Medical College for Nationalities (Funding Number: YY2021SK02, Funder: Youjiang Medical College) and Fund of Science and Technology Planning Project of Nanning Science and Technology Bureau (Funding Number: 2020023, Funder: Nanning Science and Technology Bureau).

Institutional Review Board Statement: Not applicable.

Informed Consent Statement: Not applicable.

Data Availability Statement: Not applicable.

Acknowledgments: The authors gratefully thank the financial support provided by Guangxi Innovation-Driven Development Special Fund Project, Guangxi Natural Science Foundation, National Natural Science Foundation of China, Fund of High-level talents of Youjiang Medical College and Fund of Science and Technology Planning Project of Nanning Science and Technology Bureau.

Conflicts of Interest: The authors declare no conflict of interest.

References

1. Hou, Y.; Jiang, J.-G. Origin and concept of medicine food homology and its application in modern functional foods. *Food Funct.* **2013**, *4*, 1727–1741. [CrossRef]
2. Yang, G.; Su, F.; Min, C. Origin and Prospect of Homology Medicine and Food. *Mod. Chin. Med.* **2021**, *23*, 1851–1856.
3. Shan, F.; Huang, L.-Q.; Guo, J.; Chen, M. History and development of "One Root of Medicine and Food". *Chin. Bull. Life Sci.* **2015**, *27*, 1061–1069.
4. Maleki, S.J.; Crespo, J.F.; Cabanillas, B. Anti-inflammatory effects of flavonoids. *Food Chem.* **2019**, *299*, 125124. [CrossRef]
5. Kopustinskiene, D.M.; Jakstas, V.; Savickas, A.; Bernatoniene, J. Flavonoids as Anticancer Agents. *Nutrients* **2020**, *12*, 457. [CrossRef] [PubMed]
6. Eren-Guzelgun, B.; Ince, E.; Gurer-Orhan, H. In vitro antioxidant/prooxidant effects of combined use of flavonoids. *Nat. Prod. Res.* **2017**, *32*, 1446–1450. [CrossRef] [PubMed]
7. Liskova, A.; Koklesova, L.; Samec, M.; Smejkal, K.; Samuel, S.M.; Varghese, E.; Abotaleb, M.; Biringer, K.; Kudela, E.; Danko, J.; et al. Flavonoids in Cancer Metastasis. *Cancers* **2020**, *12*, 1498. [CrossRef] [PubMed]
8. Pei, R.; Liu, X.; Bolling, B. Flavonoids and gut health. *Curr. Opin. Biotechnol.* **2020**, *61*, 153–159. [CrossRef] [PubMed]
9. Sánchez, M.; Romero, M.; Gómez-Guzmán, M.; Tamargo, J.; Pérez-Vizcaino, F.; Duarte, J. Cardiovascular Effects of Flavonoids. *Curr. Med. Chem.* **2019**, *26*, 6991–7034. [CrossRef]
10. Razack, S.; Kumar, K.H.; Nallamuthu, I.; Naika, M.; Khanum, F. Antioxidant, Biomolecule Oxidation Protective Activities of Nardostachys jatamansi DC and Its Phytochemical Analysis by RP-HPLC and GC-MS. *Antioxidants* **2015**, *4*, 185–203. [CrossRef]
11. Liu, C.; Wen, W.; Shao, J.; Zhao, W.; Qi, K.; Yang, J.; Pan, Y. Fast and comprehensive characterization of chemical ingredients in traditional Chinese herbal medicines by extractive atmospheric pressure photoionization (EAPPI) mass spectrometry. *Rapid Commun. Mass Spectrom.* **2017**, *31*, 1491–1498. [CrossRef] [PubMed]
12. Qian, Z.; Huang, Q.; Feng, X.; Li, D. Comparison of the chemical-antioxidant profiles of different parts of *Citrus reticulata* Blanco (Rutaceae) based on OLE-HPLC-DAD-MS/MS-ABTS assay. *J. Food Meas. Charact.* **2021**, *15*, 3873–3883. [CrossRef]
13. Xie, B.; Wu, X.F.; Luo, H.T.; Huang, X.L.; Huang, F.; Zhang, Q.Y.; Zhou, X.; Wu, H.Q. Chemical profiling and quality evaluation of Pogostemon cablin Benth by liquid chromatography tandem mass spectrometry combined with multivariate statistical analysis. *J. Pharm. Biomed. Anal.* **2022**, *209*, 114526. [CrossRef] [PubMed]
14. Xie, Y.; Ge, S.; Jiang, S.; Liu, Z.; Chen, L.; Wang, L.; Chen, J.; Qin, L.; Peng, W. Study on biomolecules in extractives of *Camellia oleifera* fruit shell by GC-MS. *Saudi J. Biol. Sci.* **2017**, *25*, 234–236. [CrossRef]
15. Geraldes, C.F.G.C. Introduction to Infrared and Raman-Based Biomedical Molecular Imaging and Comparison with Other Modalities. *Molecules* **2020**, *25*, 5547. [CrossRef] [PubMed]
16. Krysa, M.; Szymańska-Chargot, M.; Zdunek, A. FT-IR and FT-Raman fingerprints of flavonoids—A review. *Food Chem.* **2022**, *393*, 133430. [CrossRef]
17. Lu, J.; Xiang, B.; Liu, H.; Xiang, S.; Xie, S.; Deng, H. Application of two-dimensional near-infrared correlation spectroscopy to the discrimination of Chinese herbal medicine of different geographic regions. *Spectrochim. Acta Part A Mol. Biomol. Spectrosc.* **2007**, *69*, 580–586. [CrossRef]
18. Chang, X.; Chen, X.; Guo, Y.; Gong, P.; Pei, S.; Wang, D.; Wang, P.; Wang, M.; Chen, F. Advances in Chemical Composition, Extraction Techniques, Analytical Methods, and Biological Activity of Astragali Radix. *Molecules* **2022**, *27*, 1058. [CrossRef]
19. Duan, P.-B.; Xiao, P.-T.; Yang, X.; Hao, J.-H.; Li, K.; Liu, E.H. Screening of hypoglycemic components in Platycladi Cacumen by phytochemical investigation, spectrum-effect relationship, and chemometric methods. *J. Sep. Sci.* **2022**, *45*, 2591–2602. [CrossRef]
20. Shang, Z.; Tian, Y.; Yi, Y.; Li, K.; Qiao, X.; Ye, M. Comparative bioactivity evaluation and chemical profiling of different parts of the medicinal plant *Glycyrrhiza uralensis*. *J. Pharm. Biomed. Anal.* **2022**, *215*, 114793. [CrossRef]
21. Bao, X.-F.; Cao, P.-H.; Zeng, J.; Xiao, L.-M.; Luo, Z.-H.; Zou, J.; Wang, C.-X.; Zhao, Z.-X.; Zhou, Z.-Q.; Zhi, H.; et al. Bioactive pterocarpans from the root of *Astragalus membranaceus var. mongholicus*. *Phytochemistry* **2022**, *200*, 113249. [CrossRef] [PubMed]
22. Xie, L.-K.; Xu, X.-J.; Wu, X.; Wang, M.-J.; Gao, C.-F.; Wang, D.-M.; Ren, S.-M.; Pan, Y.-N.; Liu, X.-Q. *Capsella bursa-pastoris* (L.) Medic. extract alleviate cataract development by regulating the mitochondrial apoptotic pathway of the lens epithelial cells. *J. Ethnopharmacol.* **2021**, *284*, 114783. [CrossRef] [PubMed]
23. Wang, L.; Zhang, K.; Han, S.; Zhang, L.; Bai, H.; Bao, F.; Zeng, Y.; Wang, J.; Du, H.; Liu, Y.; et al. Constituents Isolated from the Leaves of *Glycyrrhiza uralansis* and Their Anti-Inflammatory Activities on LPS-Induced RAW264.7 Cells. *Molecules* **2019**, *24*, 1923. [CrossRef] [PubMed]
24. Yi, Y.; Adrjan, B.; Li, J.; Hu, B.; Roszak, S. NMR studies of daidzein and puerarin: Active anti-oxidants in traditional Chinese medicine. *J. Mol. Model.* **2019**, *25*, 202. [CrossRef] [PubMed]
25. Xu, Q.; Zhang, Y.; He, Z.; Liu, Z.; Zhang, Y.; Xu, W.; Yang, X. Constituents promoting osteogenesis from the fruits of *Psoralea corylifolia* and their structure-activity relationship study. *Phytochemistry* **2022**, *198*, 113143. [CrossRef]
26. Yin, M.; Wang, J.; Huang, H.; Huang, Q.; Fu, Z.; Lu, Y. Analysis of Flavonoid Compounds by Terahertz Spectroscopy Combined with Chemometrics. *ACS Omega* **2020**, *5*, 18134–18141. [CrossRef]
27. Yan, L.; Liu, C.; Qu, H.; Liu, W.; Zhang, Y.; Yang, J.; Zheng, L. Discrimination and Measurements of Three Flavonols with Similar Structure Using Terahertz Spectroscopy and Chemometrics. *J. Infrared Millim. Terahertz Waves* **2018**, *39*, 492–504. [CrossRef]

28. Lan, W.; Wang, S.; Wu, Y.; Chen, H.; Yang, J.; Wei, L.; Xie, H.; Li, S.; Guo, L.; Fu, H. A novel fluorescence sensing strategy based on nanoparticles combined with spectral splicing and chemometrics for the recognition of *Citrus reticulata* 'Chachi' and its storage year. *J. Sci. Food Agric.* **2020**, *100*, 4199–4207. [CrossRef]
29. Gao, Y.; Wang, X.; He, C. An isoflavonoid-enriched extract from *Pueraria lobata* (kudzu) root protects human umbilical vein endothelial cells against oxidative stress induced apoptosis. *J. Ethnopharmacol.* **2016**, *193*, 524–530. [CrossRef]
30. Wang, X.-L.; Jiao, F.-R.; Yu, M.; Lin, L.-B.; Xiao, J.; Zhang, Q.; Wang, L.; Duan, D.-Z.; Xie, G. Constituents with potent α-glucosidase inhibitory activity from *Pueraria lobata* (Willd.) ohwi. *Bioorganic Med. Chem. Lett.* **2017**, *27*, 1993–1998. [CrossRef]
31. Witkowska-Banaszczak, E, Flavonoids from *Trollius europaeus* flowers and evaluation of their biological activity. *J. Pharm. Pharmacol.* **2018**, *70*, 550–558. [CrossRef] [PubMed]
32. El Shoubaky, G.A.; Abdel-Daim, M.M.; Mansour, M.H.; Salem, E.A. Isolation and Identification of a Flavone Apigenin from Marine Red Alga *Acanthophora spicifera* with Antinociceptive and Anti-Inflammatory Activities. *J. Exp. Neurosci.* **2016**, *10*, JEN-S25096. [CrossRef] [PubMed]
33. Jomova, K.; Hudecova, L.; Lauro, P.; Simunková, M.; Barbierikova, Z.; Malcek, M.; Alwasel, S.H.; Alhazza, I.M.; Rhodes, C.J.; Valko, M. The effect of Luteolin on DNA damage mediated by a copper catalyzed Fenton reaction. *J. Inorg. Biochem.* **2021**, *226*, 111635. [CrossRef] [PubMed]
34. Arslan, M.; Xiaobo, Z.; Xuetao, H.; Elrasheid Tahir, H.; Shi, J.; Khan, M.R.; Zareef, M. Near infrared spectroscopy coupled with chemometric algorithms for predicting chemical components in black goji berries (*Lycium ruthenicum Murr.*). *J. Near Infrared Spectrosc.* **2018**, *26*, 275–286. [CrossRef]
35. Wang, P.; Zhang, H.; Yang, H.; Nie, L.; Zang, H. Rapid determination of major bioactive isoflavonoid compounds during the extraction process of kudzu (*Pueraria lobata*) by near-infrared transmission spectroscopy. *Spectrochim. Acta Part A Mol. Biomol. Spectrosc.* **2014**, *137*, 1403–1408. [CrossRef]
36. Betances-Salcedo, E.; Revilla, I.; Vivar-Quintana, A.M.; González-Martín, M.I. Flavonoid and Antioxidant Capacity of Propolis Prediction Using Near Infrared Spectroscopy. *Sensors* **2017**, *17*, 1647. [CrossRef]
37. Nie, P.; Xia, Z.; Sun, D.W.; He, Y. Application of visible and near infrared spectroscopy for rapid analysis of chrysin and galangin in Chinese propolis. *Sensors* **2013**, *13*, 10539–10549. [CrossRef]
38. Xu, Z.; Talpur, Z.H.; Yang, W.; Xiong, Y.; Wu, T.; Zhang, Y.; Shen, X.; Du, Y. Dual-spectrum online monitoring of puerarin and total flavonoids contents during the extraction process of *Pueraria lobata*. *Talanta* **2022**, *248*, 123608. [CrossRef]
39. Zhang, C.; Wu, W.; Zhou, L.; Cheng, H.; Ye, X.; He, Y. Developing deep learning based regression approaches for determination of chemical compositions in dry black goji berries (*Lycium ruthenicum Murr.*) using near-infrared hyperspectral imaging. *Food Chem.* **2020**, *319*, 126536. [CrossRef]
40. He, J.; Chen, L.; Chu, B.; Zhang, C. Determination of Total Polysaccharides and Total Flavonoids in *Chrysanthemum morifolium* Using Near-Infrared Hyperspectral Imaging and Multivariate Analysis. *Molecules* **2018**, *23*, 2395. [CrossRef]
41. Pellecchia, M.; Bertini, I.; Cowburn, D.; Dalvit, C.; Giralt, E.; Jahnke, W.; James, T.L.; Homans, S.W.; Kessler, H.; Luchinat, C.; et al. Perspectives on NMR in drug discovery: A technique comes of age. *Nat. Rev. Drug Discov.* **2008**, *7*, 738–745. [CrossRef] [PubMed]
42. Gong, A.; Qiu, Y.; Chen, X.; Zhao, Z.; Xia, L.; Shao, Y. Biomedical applications of terahertz technology. *Appl. Spectrosc. Rev.* **2019**, *55*, 418–438. [CrossRef]
43. Sterczewski, L.A.; Nowak, K.; Szlachetko, B.; Grzelczak, M.P.; Szczesniak-Siega, B.; Plinska, S.; Malinka, W.; Plinski, E.F. Chemometric Evaluation of THz Spectral Similarity for the Selection of Early Drug Candidates. *Sci. Rep.* **2017**, *7*, 14583. [CrossRef]
44. Afsah-Hejri, L.; Hajeb, P.; Ara, P.; Ehsani, R.J. A Comprehensive Review on Food Applications of Terahertz Spectroscopy and Imaging. *Compr. Rev. Food Sci. Food Saf.* **2019**, *18*, 1563–1621. [CrossRef] [PubMed]
45. Hu, L.; Ma, S.; Yin, C.; Liu, Z. Quality evaluation and traceability of *Bletilla striata* by fluorescence fingerprint coupled with multi-way chemometrics analysis. *J. Sci. Food Agric.* **2018**, *99*, 1413–1424. [CrossRef] [PubMed]
46. Shan, J.; Wang, X.; Russel, M.; Zhao, J. Rapid determination of flavonoids in green tea by synchronous fluorescence spectra coupled with chemometrics. *Spectrosc. Lett.* **2017**, *50*, 501–506. [CrossRef]
47. Lv, M.; Liu, Y.; Geng, J.; Kou, X.; Xin, Z.; Yang, D. Engineering nanomaterials-based biosensors for food safety detection. *Biosens. Bioelectron.* **2018**, *106*, 122–128. [CrossRef]
48. Liu, L.C.; Wang, X.Y.; Li, L.N.; Yang, L.; Wang, Z.T. Research advances of chemical constituents and analytical methods of *Citri Reticulatae Pericarpium Viride* and *Citri Reticulatae Pericarpium*. *Zhongguo Zhong Yao Za Zhi = Zhongguo Zhongyao Zazhi = China J. Chin. Mater. Med.* **2022**, *47*, 2866–2879.
49. Rohman, A.; Nugroho, A.; Lukitaningsih, E. Sudjadi Application of Vibrational Spectroscopy in Combination with Chemometrics Techniques for Authentication of Herbal Medicine. *Appl. Spectrosc. Rev.* **2014**, *49*, 603–613. [CrossRef]
50. Feng, Y.; Lei, D.; Hu, C. Rapid identification of illegal synthetic adulterants in herbal anti-diabetic medicines using near infrared spectroscopy. *Spectrochim. Acta. Part A Mol. Biomol. Spectrosc.* **2014**, *125*, 363–374. [CrossRef]
51. Li, W.; Cheng, Z.; Wang, Y.; Qu, H. A study on the use of near-infrared spectroscopy for the rapid quantification of major compounds in Tanreqing injection. *Spectrochim. Acta. Part A Mol. Biomol. Spectrosc.* **2013**, *101*, 1–7. [CrossRef] [PubMed]
52. Luypaert, J.; Massart, D.L.; Vander Heyden, Y. Near-infrared spectroscopy applications in pharmaceutical analysis. *Talanta* **2007**, *72*, 865–883. [CrossRef] [PubMed]
53. Manley, M. Near-infrared spectroscopy and hyperspectral imaging: Non-destructive analysis of biological materials. *Chem. Soc. Rev.* **2014**, *43*, 8200–8214. [CrossRef] [PubMed]

54. Erkinbaev, C.; Henderson, K.; Paliwal, J. Discrimination of gluten-free oats from contaminants using near infrared hyperspectral imaging technique. *Food Control.* **2017**, *80*, 197–203. [CrossRef]
55. Lowe, A.; Harrison, N.; French, A.P. Hyperspectral image analysis techniques for the detection and classification of the early onset of plant disease and stress. *Plant Methods* **2017**, *13*, 80. [CrossRef]
56. Karim, S.; Qadir, A.; Farooq, U.; Shakir, M.; Laghari, A.A. Hyperspectral Imaging: A Review and Trends towards Medical Imaging. *Curr. Med. Imaging* **2022**. [CrossRef]
57. Sandasi, M.; Vermaak, I.; Chen, W.; Viljoen, A.M. Hyperspectral imaging and chemometric modeling of echinacea—A novel approach in the quality control of herbal medicines. *Molecules* **2014**, *19*, 13104–13121. [CrossRef]
58. He, J.; He, Y.; Zhang, C. Determination and Visualization of Peimine and Peiminine Content in *Fritillaria thunbergii* Bulbi Treated by Sulfur Fumigation Using Hyperspectral Imaging with Chemometrics. *Molecules* **2017**, *22*, 1402. [CrossRef]
59. Zhang, C.; Wang, Q.; Liu, F.; He, Y.; Xiao, Y. Rapid and non-destructive measurement of spinach pigments content during storage using hyperspectral imaging with chemometrics. *Measurement* **2016**, *97*, 149–155. [CrossRef]
60. Zhang, N.; Liu, X.; Jin, X.; Li, C.; Wu, X.; Yang, S.; Ning, J.; Yanne, P. Determination of total iron-reactive phenolics, anthocyanins and tannins in wine grapes of skins and seeds based on near-infrared hyperspectral imaging. *Food Chem.* **2017**, *237*, 811–817. [CrossRef]
61. Mo, C.; Kim, M.S.; Kim, G.; Lim, J.; Delwiche, S.R.; Chao, K.; Lee, H.; Cho, B.-K. Spatial assessment of soluble solid contents on apple slices using hyperspectral imaging. *Biosyst. Eng.* **2017**, *159*, 10–21. [CrossRef]
62. Munera, S.; Besada, C.; Aleixos, N.; Talens, P.; Salvador, A.; Sun, D.-W.; Cubero, S.; Blasco, J. Non-destructive assessment of the internal quality of intact persimmon using colour and VIS/NIR hyperspectral imaging. *LWT Food Sci. Technol.* **2016**, *77*, 241–248. [CrossRef]
63. Zhou, Y.; Zuo, Z.; Xu, F.; Wang, Y. Origin identification of Panax notoginseng by multi-sensor information fusion strategy of infrared spectra combined with random forest. *Spectrochim. Acta. Part A Mol. Biomol. Spectrosc.* **2020**, *226*, 117619. [CrossRef] [PubMed]
64. Yuan, Y.; Tang, X.; Jia, Z.; Li, C.; Ma, J.; Zhang, J. The Effects of Ecological Factors on the Main Medicinal Components of Dendrobium officinale under Different Cultivation Modes. *Forests* **2020**, *11*, 94. [CrossRef]
65. Hai, C.; Long, W.; Suo, Y.; Lu, H.; Chen, H.; Yang, X.; Yang, J.; Fu, H. Nano-effect multivariate fusion spectroscopy combined with chemometrics for accurate identification the cultivation methods and growth years of *Dendrobium huoshanense*. *Microchem. J.* **2022**, *179*, 107556. [CrossRef]
66. Ding, Y.; Zhu, W.-H.; Xie, Y. Development of Ion Chemosensors Based on Porphyrin Analogues. *Chem. Rev.* **2016**, *117*, 2203–2256. [CrossRef]
67. Kumar, G.A.; Thomas, V.; Jose, G.; Unnikrishnan, N.V.; Nampoori, V.P.N. Optical properties of porphyrins in borate glassy matrix. *Mater. Chem. Phys.* **2002**, *73*, 206–211. [CrossRef]
68. Liu, Y.; Xia, Z.; Yao, L.; Wu, Y.; Li, Y.; Zeng, S.; Li, H. Discriminating geographic origin of sesame oils and determining lignans by near-infrared spectroscopy combined with chemometric methods. *J. Food Compos. Anal.* **2019**, *24*, 250–256. [CrossRef]
69. Chen, H.; Lin, Z.; Tan, C. Fast discrimination of the geographical origins of notoginseng by near-infrared spectroscopy and chemometrics. *J. Pharm. Biomed. Anal.* **2018**, *161*, 239–245. [CrossRef]
70. Wang, S.; Chen, H.; Xie, H.; Wei, L.; Xu, L.; Zhang, L.; Lan, W.; Zhou, C.; She, Y.; Fu, H. A novel thioctic acid-carbon dots fluorescence sensor for the detection of Hg2+ and thiophanate methyl via S-Hg affinity. *Food Chem.* **2020**, *346*, 128923. [CrossRef]
71. Zhai, X.; Li, Z.; Shi, J.; Huang, X.; Sun, Z.; Zhang, D.; Zou, X.; Sun, Y.; Zhang, J.; Holmes, M.; et al. A colorimetric hydrogen sulfide sensor based on gellan gum-silver nanoparticles bionanocomposite for monitoring of meat spoilage in intelligent packaging. *Food Chem.* **2019**, *290*, 135–143. [CrossRef] [PubMed]
72. Wei, L.; Hu, O.; Chen, H.; Yang, T.; Fan, Y.; Xu, L.; Zhang, L.; Lan, W.; She, Y.; Fu, H. Variety identification and age prediction of Pu-erh tea using graphene oxide and porphyrin complex based mid-infrared spectroscopy coupled with chemometrics. *Microchem. J.* **2020**, *158*, 105255. [CrossRef]
73. Long, W.; Hu, Z.; Wei, L.; Chen, H.; Liu, T.; Wang, S.; Guan, Y.; Yang, X.; Yang, J.; Fu, H. Accurate identification of the geographical origins of lily using near-infrared spectroscopy combined with carbon dot-tetramethoxyporphyrin nanocomposite and chemometrics. *Spectrochim. Acta. Part A Mol. Biomol. Spectrosc.* **2022**, *271*, 120932. [CrossRef]
74. Lv, C.; He, Y.; Kang, C.; Zhou, L.; Wang, T.; Yang, J.; Guo, L. Tracing the Geographical Origins of Dendrobe (*Dendrobium* spp.) by Near-Infrared Spectroscopy Sensor Combined with Porphyrin and Chemometrics. *J. Anal. Methods Chem.* **2020**, *2020*, 8879957. [CrossRef] [PubMed]
75. Pawlikowska-Pawlega, B.; Gruszecki, W.I.; Misiak, L.; Paduch, R.; Piersiak, T.; Zarzyka, B.; Pawelec, J.; Gawron, A. Modification of membranes by quercetin, a naturally occurring flavonoid, via its incorporation in the polar head group. *Biochim. Biophys. Acta* **2007**, *1768*, 2195–2204. [CrossRef] [PubMed]
76. Wesołowska, O.; Hendrich, A.B.; Łaniapietrzak, B.; Wiśniewski, J.; Molnar, J.; Ocsovszki, I.; Michalak, K. Perturbation of the lipid phase of a membrane is not involved in the modulation of MRP1 transport activity by flavonoids. *Cell. Mol. Biol. Lett.* **2009**, *14*, 199–221. [CrossRef] [PubMed]
77. Selvaraj, S.; Krishnaswamy, S.; Devashya, V.; Sethuraman, S.; Krishnan, U.M. Influence of membrane lipid composition on flavonoid-membrane interactions: Implications on their biological activity. *Prog. Lipid Res.* **2015**, *58*, 1–13. [CrossRef]

78. Cieślik-Boczula, K.; Maniewska, J.; Grynkiewicz, G.; Szeja, W.; Koll, A.; Hendrich, A.B. Interaction of quercetin, genistein and its derivatives with lipid bilayers—An ATR IR-spectroscopic study. *Vib. Spectrosc.* **2012**, *62*, 64–69. [CrossRef]
79. Pawlikowska-Pawlęga, B.; Misiak, L.E.; Zarzyka, B.; Paduch, R.; Gawron, A.; Gruszecki, W.I. FTIR, (1)H NMR and EPR spectroscopy studies on the interaction of flavone apigenin with dipalmitoylphosphatidylcholine liposomes. *Biochim. Biophys. Acta* **2013**, *1828*, 518–527. [CrossRef]
80. Baranović, G.; Šegota, S. Infrared spectroscopy of flavones and flavonols. Reexamination of the hydroxyl and carbonyl vibrations in relation to the interactions of flavonoids with membrane lipids. *Spectrochim. Acta. Part A Mol. Biomol. Spectrosc.* **2018**, *192*, 473–486. [CrossRef]
81. Ali, M.S.; Al-Lohedan, H.A. Interaction of human serum albumin with sulfadiazine. *J. Mol. Liq.* **2014**, *197*, 124–130. [CrossRef]
82. Singh, P.; Arif, Y.; Bajguz, A.; Hayat, S. The role of quercetin in plants. *Plant Physiol. Biochem. PPB* **2021**, *166*, 10–19. [CrossRef] [PubMed]
83. Huwait, E.; Mobashir, M. Potential and Therapeutic Roles of Diosmin in Human Diseases. *Biomedicines* **2022**, *10*, 1076. [CrossRef] [PubMed]
84. Sovrlić, M.; Mrkalić, E.; Jelić, R.; Ćendić Serafinović, M.; Stojanović, S.; Prodanović, N.; Tomović, J. Effect of Caffeine and Flavonoids on the Binding of Tigecycline to Human Serum Albumin: A Spectroscopic Study and Molecular Docking. *Pharmaceuticals* **2022**, *15*, 266. [CrossRef] [PubMed]
85. Ali, M.S.; Al-Lohedan, H.A. Spectroscopic and computational evaluation on the binding of safranal with human serum albumin: Role of inner filter effect in fluorescence spectral correction. *Spectrochim. Acta. Part A Mol. Biomol. Spectrosc.* **2018**, *203*, 434–442. [CrossRef]
86. Li, X.; Wang, S. Study on the interaction of (+)-catechin with human serum albumin using isothermal titration calorimetry and spectroscopic techniques. *New J. Chem.* **2014**, *39*, 386–395. [CrossRef]
87. Liu, T.; Liu, M.; Guo, Q.; Liu, Y.; Zhao, Y.; Wu, Y.; Sun, B.; Wang, Q.; Liu, J.; Han, J. Investigation of binary and ternary systems of human serum albumin with oxyresveratrol/piceatannol and/or mitoxantrone by multipectroscopy, molecular docking and cytotoxicity evaluation. *J. Mol. Liq.* **2020**, *311*, 113364. [CrossRef]
88. Matei, I.; Hillebrand, M. Interaction of kaempferol with human serum albumin: A fluorescence and circular dichroism study. *J. Pharm. Biomed. Anal.* **2010**, *51*, 768–773. [CrossRef]
89. Maji, A.; Beg, M.; Mandal, A.K.; Das, S.; Jha, P.K.; Kumar, A.; Sarwar, S.; Hossain, M.; Chakrabarti, P. Spectroscopic interaction study of human serum albumin and human hemoglobin with Mersilea quadrifolia leaves extract mediated silver nanoparticles having antibacterial and anticancer activity. *J. Mol. Struct.* **2017**, *1141*, 584–592. [CrossRef]
90. Zeng, H.J.; Ma, J.; Yang, R.; Jing, Y.; Qu, L.B. Molecular Interactions of Flavonoids to Hyaluronidase: Insights from Spectroscopic and Molecular Modeling Studies. *J. Fluoresc.* **2015**, *25*, 941–959. [CrossRef]
91. Lee, J.H.; Kim, G.H. Evaluation of antioxidant and inhibitory activities for different subclasses flavonoids on enzymes for rheumatoid arthritis. *J. Food Sci.* **2010**, *75*, H212–H217. [CrossRef]
92. Li, X.; Liu, H.; Yang, Z.; Duan, H.; Wang, Z.; Cheng, Z.; Song, Z.; Wu, X. Study on the interaction of hyaluronidase with certain flavonoids. *J. Mol. Struct.* **2021**, *1241*, 130686. [CrossRef]
93. Greenfield, N.J. Using circular dichroism spectra to estimate protein secondary structure. *Nat. Protoc.* **2006**, *1*, 2876–2890. [CrossRef]
94. Yan, S.-M.; Hu, Z.-F.; Wu, C.-X.; Jin, L.; Chen, G.; Zeng, X.-Y.; Zhu, J.-Q. Electronic Tongue Combined with Chemometrics to Provenance Discrimination for a Green Tea (Anji-White Tea). *J. Food Qual.* **2017**, *2017*, 3573197. [CrossRef]
95. Ballabio, D.; Consonni, V. Classification tools in chemistry. Part 1: Linear models. PLS-DA. *Anal. Methods* **2013**, *5*, 3790–3798. [CrossRef]
96. Yun, Y.-H.; Wei, Y.-C.; Zhao, X.-B.; Wu, W.-J.; Liang, Y.-Z.; Lu, H.-M. A green method for the quantification of polysaccharides in *Dendrobium officinale*. *RSC Adv.* **2015**, *5*, 105057–105065. [CrossRef]
97. Long, W.J.; Wu, H.L.; Wang, T.; Dong, M.Y.; Chen, L.Z.; Yu, R.Q. Fast identification of the geographical origin of *Gastrodia elata* using excitation-emission matrix fluorescence and chemometric methods. *Spectrochim. Acta. Part A Mol. Biomol. Spectrosc.* **2021**, *258*, 119798. [CrossRef]
98. Yang, Q.; Lin, S.S.; Yang, J.T.; Tang, L.J.; Yu, R.Q. Detection of inborn errors of metabolism utilizing GC-MS urinary metabolomics coupled with a modified orthogonal partial least squares discriminant analysis. *Talanta* **2017**, *165*, 545–552. [CrossRef]
99. Xia, Z.; Liu, X.; Tong, L.; Wang, H.; Feng, M.; Xi, X.; He, P.; Qin, X. Comparison of chemical constituents of *Bupleurum marginatum var. stenophyllum* and *Bupleurum chinense* DC. using UHPLC–Q-TOF–MS based on a metabonomics approach. *Biomed. Chromatogr.* **2021**, *35*, e5133. [CrossRef]
100. Fu, H.; Wei, L.; Chen, H.; Yang, X.; Kang, L.; Hao, Q.; Zhou, L.; Zhan, Z.; Liu, Z.; Yang, J.; et al. Combining stable C, N, O, H, Sr isotope and multi-element with chemometrics for identifying the geographical origins and farming patterns of Huangjing herb. *J. Food Compos. Anal.* **2021**, *102*, 103972. [CrossRef]
101. Jin, Q.; Jiao, C.; Sun, S.; Song, C.; Cai, Y.; Lin, Y.; Fan, H.; Zhu, Y. Metabolic Analysis of Medicinal *Dendrobium officinale* and *Dendrobium huoshanense* during Different Growth Years. *PLoS ONE* **2016**, *11*, e0146607. [CrossRef]
102. Wu, X.M.; Zhang, Q.Z.; Wang, Y.Z. Traceability of wild Paris polyphylla Smith var. yunnanensis based on data fusion strategy of FT-MIR and UV-Vis combined with SVM and random forest. *Spectrochim. Acta. Part A Mol. Biomol. Spectrosc.* **2018**, *205*, 479–488. [CrossRef]

103. Li, Y.; Zhang, J.Y.; Wang, Y.Z. FT-MIR and NIR spectral data fusion: A synergetic strategy for the geographical traceability of *Panax notoginseng*. *Anal. Bioanal. Chem.* **2018**, *410*, 91–103. [CrossRef]
104. Sun, W.; Zhang, X.; Zhang, Z.; Zhu, R. Data fusion of near-infrared and mid-infrared spectra for identification of rhubarb. *Spectrochim. Acta. Part A Mol. Biomol. Spectrosc.* **2017**, *171*, 72–79. [CrossRef]
105. Chen, W.; Du, Y.; Zhang, F.; Zhang, R.; Ding, B.; Chen, Z.; Xiong, Q. Sampling error profile analysis (SEPA) for model optimization and model evaluation in multivariate calibration. *J. Chemom.* **2017**, *32*, e2933. [CrossRef]
106. Yang, J.; Zhang, D.; Frangi, A.F.; Yang, J.Y. Two-dimensional PCA: A new approach to appearance-based face representation and recognition. *IEEE Trans. Pattern Anal. Mach. Intell.* **2004**, *26*, 131–137. [CrossRef]
107. Anderson, S.L.; Rovnyak, D.; Strein, T.G. Identification of Edible Oils by Principal Component Analysis of 1H NMR Spectra. *J. Chem. Educ.* **2017**, *94*, 1377–1382. [CrossRef]
108. Liang, J.; Guo, Q.; Chang, T.; Li, K.; Cui, H.-L. Reliable origin identification of Scutellaria baicalensis based on terahertz time-domain spectroscopy and pattern recognition. *Optik* **2018**, *174*, 7–14. [CrossRef]
109. Kim, H.-C.; Pang, S.; Je, H.-M.; Kim, D.; Yang Bang, S. Constructing support vector machine ensemble. *Pattern Recognit.* **2003**, *36*, 2757–2767. [CrossRef]
110. Shi, X.; Gan, X.; Wang, X.; Peng, J.; Li, Z.; Wu, X.; Shao, Q.; Zhang, A. Rapid detection of *Ganoderma lucidum* spore powder adulterated with dyed starch by NIR spectroscopy and chemometrics. *LWT Food Sci. Technol.* **2022**, *167*, 113829. [CrossRef]
111. Jintao, X.; Quanwei, Y.; Chunyan, L.; Yun, J.; Shuangxi, W.; Mingxiang, Z.; Peng, L. Rapid and Simultaneous Determination of Three Active Components in Raw and Processed Root Samples of Scutellaria baicalensis by Near-infrared Spectroscopy. *Planta Med.* **2019**, *85*, 72–80. [CrossRef]
112. Yang, I.C.; Tsai, C.-Y.; Hsieh, K.-W.; Yang, C.-W.; Ouyang, F.; Martin Lo, Y.; Chen, S. Integration of SIMCA and near-infrared spectroscopy for rapid and precise identification of herbal medicines. *J. Food Drug Anal.* **2013**, *21*, 268–278. [CrossRef]

molecules

Article

Fast and Non-Destructive Profiling of Commercial Coffee Aroma under Three Conditions (Beans, Powder, and Brews) Using GC-IMS

Yanping Chen, He Chen, Dandan Cui, Xiaolei Fang, Jie Gao and Yuan Liu *

Department of Food Science & Technology, School of Agriculture & Biology, Shanghai Jiao Tong University, Shanghai 200240, China
* Correspondence: y_liu@sjtu.edu.cn; Tel.: +86-021-34208536

Abstract: The flavor of coffee can be affected by the preparation parameters. In this investigation, the flavor profiles of three coffee brands under three conditions (bean, powder, and brew) were analyzed by gas chromatography—ion mobility spectrometry (GC-IMS) and the electronic nose (E-nose). The flavor results were further studied using multiple factor analysis (MFA). A total of 117 peaks were identified in all coffee samples by GC-IMS, and the principal component analysis (PCA) showed these coffee samples could be grouped and separated. A total of 37 volatile organic compounds (VOCs) were selected as biomarkers to distinguish coffee samples, including 5 aldehydes, 10 ketones, 8 alcohols, 2 acids, 4 esters, 5 furans, and 3 other compounds. The comparison between E-nose and GC-IMS data using partial least squares regression (PLSR) and MFA showed GC-IMS could present very close sample spaces. Compared with E-nose, GC–IMS could not only be used to classify coffee samples in a very short time but also provide VOC bio-markers to discriminate coffee samples.

Keywords: coffee; E-nose; GC-IMS; volatile organic components

Citation: Chen, Y.; Chen, H.; Cui, D.; Fang, X.; Gao, J.; Liu, Y. Fast and Non-Destructive Profiling of Commercial Coffee Aroma under Three Conditions (Beans, Powder, and Brews) Using GC-IMS. *Molecules* **2022**, *27*, 6262. https://doi.org/10.3390/molecules27196262

Academic Editor: Michael C. Qian

Received: 24 August 2022
Accepted: 17 September 2022
Published: 23 September 2022

1. Introduction

Coffee, originating from Ethiopia, has become one of the most popular drinks in the world. The typical coffee flavor is mainly caused by volatile organic compounds (VOCs) generated during the roasting process at a high temperature (above 200 °C), when flavor precursors inside raw coffee beans undergo the Maillard reaction, Strecker degradation, caramelization, and pyrolysis [1]. The attractive coffee flavors include nutty, caramel, floral, and chocolate, which might owe to aldehydes, ketones, alcohols, sulfur-containing compounds, and volatile heterocyclic compounds. The sulfur-containing compounds originate from the thermal degradation of sulfur-containing amino acids in the presence of sugars, and heterocyclic compounds, such as pyrazine, pyrrole, and pyridine, are yielded by the Strecker reaction between amino acids and aldehydes [2,3].

Coffee flavor can be identified by many analytical methods. The most commonly used techniques, such as solid-phase micro-extraction–gas chromatography–mass spectrometry (SPME-GC-MS), and gas chromatography–olfactometry (GC-O) can be used to predict volatile compounds (e.g., aldehydes and pyrazines) in single-roasted coffee beans [4]. In one study, ultra-performance liquid chromatography–time-of-flight mass spectrometry (UPLC-TOF-MS) revealed a marker substance for a coffee beverage spiked with acetaldehyde [5]. Flash gas chromatography electronic nose (FGC E-nose) can obtain a more convincing flavor assessment than a single assessment of aroma or volatile profiles [6].

The electronic nose (E-nose) is a novel instrument that imitates the human olfactory process to analyze and recognize complex odors efficiently. Previous studies have shown good discrimination of roasted coffee, instant coffee, and coffee products with different roasting degrees using E-nose [7]. However, compared to gas chromatography–mass spectrometry (GC-MS), E-nose can only respond to profiles from sensor arrays, and the volatiles contributing to the odors cannot be identified. Originating in the 1970s [8], ion mobility spectrometry (IMS) coupled with GC can be used to screen and distinguish food samples in a very short time. It contains a low detection limit at the trace level (ng/L) without pretreatment. A comparison between the principal component analysis (PCA) of GC-IMS and two-dimensional gas chromatography–mass spectrometry (GC×GC-MS) volatiles and multiple factor analysis (MFA) results showed that GC-IMS could be used as a fast way to discriminate dry-cured ham products in different regions in China [9]. The comparison between GC-IMS and ultra-fast GC E-nose showed they both could be used to discriminate between raw and cooked fish [10]. GC-IMS was used to identify the authenticity of Iberian ham samples and reported VOCs such as octan-2-one, trans-2-octenal, and nonan-2-one as biomarkers [11]. In addition, a total of 26 volatile metabolites (aldehydes, ketones, alcohols, and esters) were identified to facilitate the characterization of specific attributes of olive oils [12].

The flavor of coffee is affected by many parameters, such as natural environmental factors (light intensity, plant density, and rainfall) and the processing conditions (roasting temperature, and roasting time) [13]. The varied preparation methods of coffee brews also result in varied coffee flavors. The extract levels of flavor components depend on the ground particle size of roasted coffee beans and the temperature of hot water poured over coffee powder [3,14]. The cations contained inside coffee beans and dissolved inside water for coffee brew extraction may have an influence on the coffee aroma [15]. Therefore, the objectives of this study were (1) to distinguish commercial coffee under three conditions (bean, powder, and brew) using GC-IMS, (2) to identify potential biomarkers to separate those samples using GC-IMS, and (3) to compare the discrimination ability between GC-IMS and E-nose. This study could develop a fast method for the discrimination of coffee with varied flavors.

2. Results and Discussion

2.1. GC-IMS Results

The VOCs isolated from the three commercial coffee samples (S, C, and P) in the bean, powder, and brew forms are shown in Figure 1. The Figure 1 showed the background is blue. The red vertical line at the abscissa, 1.0, is the reactive ion peak (RIP). The X-axis and Y-axis represent the drift time and retention time, respectively [16]. The signals were mostly located within the retention time range, from 100 to 400 s, and a drift time from 1.0 to 1.5 riprel. Each detected VOC produced a signal that presented as a colored spot. The concentration variances of each VOC depended on the signal intensity shown in the varied colors, with red representing higher intensity and blue meaning not present.

A total of 117 signals were detected in all samples (data not shown). To visualize the vast amount of data by reducing dimensions, PCA was performed based on the identified signals using GC-IMS. The PCA was a multivariate statistical analysis technique. Through dimensional reduction analysis, the selected PCs could represent the contribution rates in different samples [17]. Figure 2 shows the score plots obtained via PCA, and each dot represents a sample. The dots of the same color represent the replicates of the sample. The first and second principal components (PC1 and PC2) explained 45.8% (Figure 2A) variance and the second and third principal components (PC2 and PC3) explained 40.3% (Figure 2B) variance.

From the results (Figure 2A), it can be seen that three brands of coffee beans scattered along PC1 and PC2. The powder was located in a similar position along the positive side of PC1 and grouped into three categories along PC2. The coffee brew was mostly located on the positive side of PC1, and the coffee bean was located on the negative side of PC1. In Figure 2B, the coffee brew is located on the upper positive side of PC3, the coffee bean is located on the negative side of PC3 (−0.5 to 0), and the coffee powder is located on the negative downside of PC3 (−0.5 to −1.0). The distribution diagram shows that the different statuses of the coffee samples could be distinguished by PC1 and PC2 (Figure 2A) and PC2 and PC3 (Figure 2B), i.e., bean, powder, and brew.

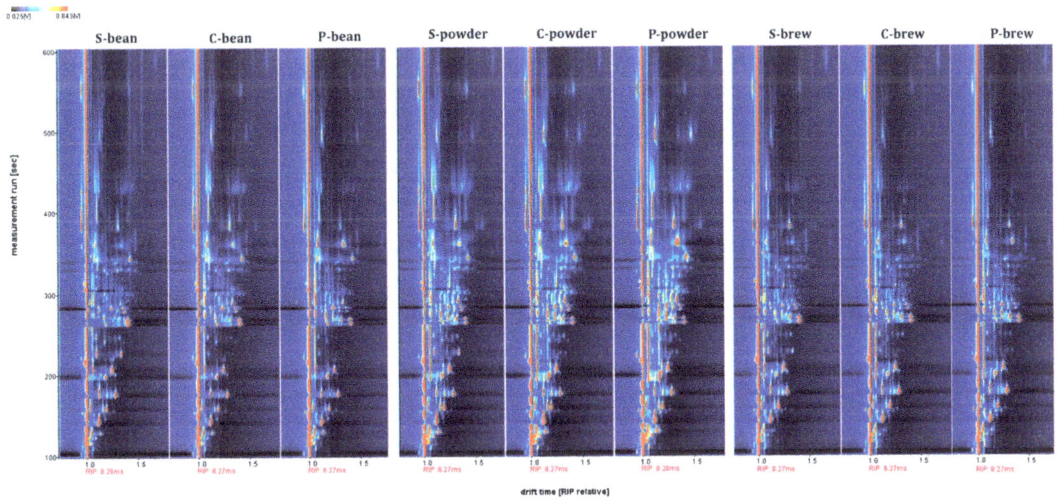

Figure 1. Fingerprints of VOCs isolated from three commercial coffee samples (S, C, and P) presented in bean, powder, and brew, respectively. The ordinate shows the retention time, and the abscissa shows the drift time. The background of the plot is blue, and the red vertical line on the left side represents the reactive ion peak (RIP, normalized drift time of 8.27 ms).

Figure 2B shows the coffee bean, powder, and brew of three brands (C, P, and S) to be separated in PC2. In Figure 2B, brand C is located more separately from brand P and brand S, including its bean, powder, and brew forms, indicating that the flavors of brand P and brand S are more similar. The results showed coffee samples could be distinguished by brands and status. This suggested that their flavor profiles were different from each other and could be distinguished by their brands. Figure 2 indicates that GC-IMS can differentiate coffee samples of different brands in bean, powder, and brew using PCA.

2.2. Biomarkers Identified Using GC-IMS

A total of 37 compounds were tentatively identified as important markers for discrimination, including 5 aldehydes, 10 ketones, 8 alcohols, 2 acids, 4 esters, 5 furans, and 3 other compounds (Table 1). The dimer eluted after the monomer, included butanal, butanone, 2-pentanone, ethanol, 2-ethylfuran, and furfural (Table 1), because the molecular mass of the dimer was higher than that of the monomer counterparts [18]. We found that butanal dimer, pentanal, 2-propanone, 2,3-butanedione, butanone monomer, butanone dimer, methyl isobutyl, 2-propanol, furfural dimer, and ethyl pyrazine had higher signal intensities (>60) in each sample. These compounds commonly existed in all three brands of coffee in the bean, powder, and brew forms and developed the typical coffee flavor. The 2,3-butanedione contributed a buttery, fruity, and caramel-like aroma [19,20]. Furfural was most likely generated by Amadori rearrangement through the 3-deoxyosone pathway or, alternatively, via the oxidation of furfuryl alcohol with the almond aroma [21]. Ethyl pyrazine contributed a roasted and nutty odor [22].

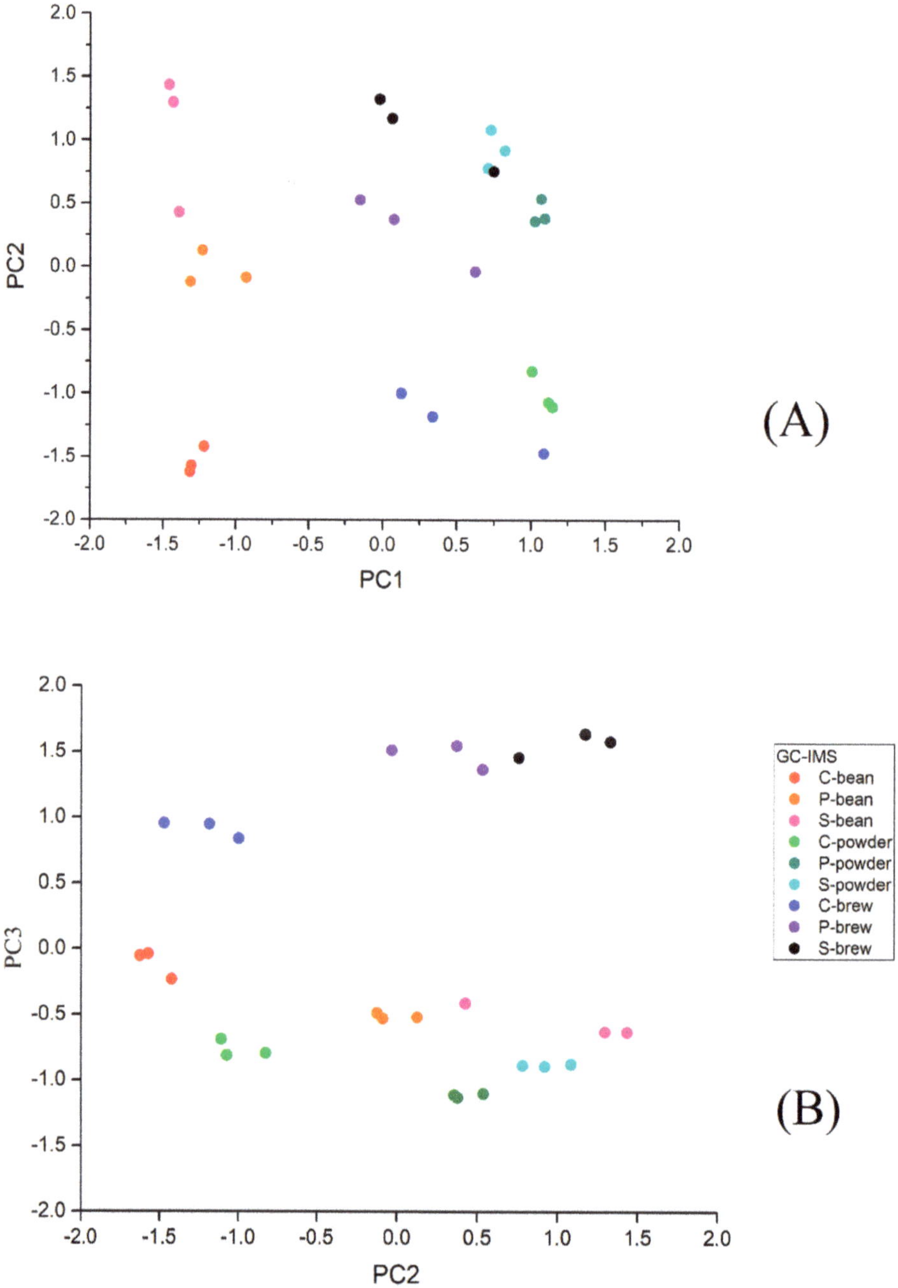

Figure 2. Principal component analysis of the flavor profiles of three commercial coffee samples (S, C, and P) in three conditions (bean, powder, and brew) obtained by GC−IMS on PC1−PC2 (**A**) and PC2−PC3 (**B**).

Table 1. Compounds in three brands (P, S, C) in coffee bean, powder, and brew identified by GC-IMS.

No.	Compounds	CAS	RI	Dt/ Riprel	C-Bean	P-Bean	S-Bean	C-Powder	P-Powder	S-Powder	C-Brew	P-Brew	S-Brew
	Aldehydes (5)												
1	butanal monomer	123-72-8	546	1.3	21 ± 2^{a}	23 ± 3^{a}	96 ± 6^{d}	27 ± 3^{ab}	23 ± 2^{ab}	94 ± 6^{d}	46 ± 4^{c}	30 ± 3^{b}	98 ± 3^{d}
2	butanal dimer	123-72-8	615	1.1	86 ± 5^{b}	80 ± 7^{ab}	93 ± 9^{c}	98 ± 2^{c}	83 ± 0^{b}	75 ± 1^{a}	99 ± 1^{c}	96 ± 1^{c}	94 ± 1^{c}
3	pentanal	110-62-3	660	1.2	85 ± 7^{bc}	95 ± 7^{c}	84 ± 21^{bc}	43 ± 5^{a}	63 ± 3^{ab}	90 ± 8^{c}	81 ± 19^{bc}	87 ± 16^{c}	85 ± 12^{bc}
4	propanal	123-38-6	805	1.1	11 ± 2^{a}	56 ± 5^{c}	91 ± 15^{e}	35 ± 3^{b}	79 ± 2^{d}	96 ± 6^{e}	15 ± 3^{a}	62 ± 4^{c}	97 ± 5^{e}
5	methional	3268-49-3	916	1.1	52 ± 6^{b}	91 ± 11^{de}	77 ± 1^{cd}	71 ± 1^{c}	94 ± 3^{de}	99 ± 2^{e}	33 ± 10^{a}	44 ± 21^{ab}	50 ± 13^{ab}
	Ketones (10)												
6	2-propanone	67-64-1	500	1.1	85 ± 1^{a}	92 ± 7^{bc}	88 ± 3^{ab}	99 ± 0^{d}	99 ± 1^{d}	98 ± 0^{cd}	95 ± 1^{cd}	98 ± 0^{cd}	99 ± 1^{d}
7	2,3-butanedione	431-03-8	571	1.2	91 ± 4^{bc}	94 ± 6^{bc}	80 ± 15^{b}	97 ± 3^{c}	86 ± 2^{bc}	60 ± 5^{a}	94 ± 6^{bc}	85 ± 9^{bc}	80 ± 11^{b}
8	butanone monomer	78-93-3	590	1.1	62 ± 5^{a}	83 ± 15^{bc}	83 ± 14^{bc}	95 ± 7^{c}	85 ± 9^{c}	65 ± 8^{ab}	78 ± 12^{abc}	89 ± 9^{c}	93 ± 8^{c}
9	butanone dimer	78-93-3	587	1.1	80 ± 5^{a}	80 ± 9^{a}	92 ± 7^{bcd}	94 ± 2^{cd}	97 ± 3^{d}	89 ± 1^{bc}	86 ± 1^{ab}	95 ± 0^{cd}	99 ± 1^{d}
10	2,3-pentanedione	600-14-6	671	1.3	90 ± 10^{de}	66 ± 5^{c}	25 ± 5^{a}	99 ± 1^{e}	89 ± 2^{de}	77 ± 4^{cd}	41 ± 11^{b}	33 ± 11^{ab}	28 ± 9^{a}
11	2-pentanone monomer	107-87-9	686	1.1	23 ± 6^{a}	66 ± 7^{cd}	86 ± 19^{e}	40 ± 5^{b}	72 ± 3^{d}	92 ± 7^{e}	55 ± 1^{c}	87 ± 1^{e}	98 ± 2^{e}
12	2-pentanone dimer	107-87-9	685	1.4	40 ± 1^{a}	74 ± 2^{bc}	90 ± 14^{cd}	75 ± 3^{bc}	97 ± 4^{d}	78 ± 2^{bcd}	60 ± 16^{b}	83 ± 16^{cd}	86 ± 15^{cd}
13	3-pentanone	96-22-0	693	1.4	36 ± 9^{a}	78 ± 6^{b}	86 ± 20^{bc}	88 ± 3^{bc}	97 ± 4^{c}	74 ± 2^{b}	74 ± 1^{b}	96 ± 1^{c}	98 ± 2^{c}
14	3-hydroxy-2-butanone	513-86-0	701	1.3	93 ± 6^{cd}	91 ± 8^{cd}	60 ± 1^{b}	98 ± 2^{d}	91 ± 1^{cd}	84 ± 2^{c}	15 ± 7^{a}	14 ± 7^{a}	15 ± 7^{a}
15	methyl isobutyl	108-10-1	723	1.2	63 ± 2^{a}	97 ± 2^{d}	95 ± 4^{d}	71 ± 9^{abc}	93 ± 3^{d}	94 ± 6^{d}	69 ± 13^{ab}	86 ± 14^{cd}	81 ± 13^{bcd}
	Alcohols (8)												
16	ethanol monomer	64-17-5	482	1.0	99 ± 1^{ef}	95 ± 3^{def}	92 ± 6^{d}	99 ± 1^{f}	93 ± 2^{de}	84 ± 1^{c}	94 ± 6^{def}	48 ± 3^{a}	54 ± 3^{b}
17	ethanol dimer	64-17-5	482	1.1	81 ± 4^{c}	51 ± 4^{b}	44 ± 3^{a}	90 ± 2^{d}	92 ± 3^{d}	97 ± 3^{e}	99 ± 1^{e}	80 ± 3^{c}	87 ± 2^{d}
18	isopropyl alcohol	67-63-0	506	1.2	52 ± 6^{bc}	77 ± 26^{cd}	68 ± 16^{c}	97 ± 4^{e}	40 ± 16^{ab}	28 ± 11^{a}	95 ± 5^{de}	60 ± 5^{bc}	65 ± 1^{c}
19	2-propanol	123-38-6	546	1.3	58 ± 2^{a}	80 ± 18^{bc}	74 ± 9^{b}	92 ± 1^{cd}	98 ± 2^{d}	93 ± 1^{d}	87 ± 0^{cd}	97 ± 1^{d}	99 ± 1^{d}
20	butanol	71-36-3	660	1.2	25 ± 1^{a}	28 ± 5^{a}	49 ± 1^{b}	82 ± 15^{cd}	71 ± 4^{c}	53 ± 2^{b}	92 ± 9^{d}	82 ± 5^{cd}	94 ± 3^{d}
21	2-methyl-1-propanol	78-83-1	673	1.2	77 ± 6^{b}	83 ± 10^{b}	91 ± 9^{b}	96 ± 3^{b}	73 ± 5^{b}	40 ± 9^{a}	86 ± 9^{b}	77 ± 28^{b}	80 ± 21^{b}
22	3-methyl-2-butanol	598-75-4	692	1.2	98 ± 3^{e}	82 ± 3^{bc}	68 ± 2^{a}	98 ± 2^{e}	93 ± 2^{de}	89 ± 1^{cd}	94 ± 8^{de}	76 ± 7^{b}	64 ± 6^{a}
23	2-methyl-1-butanol	137-32-6	776	1.5	21 ± 7^{a}	67 ± 12^{b}	97 ± 5^{c}	84 ± 14^{c}	95 ± 2^{c}	64 ± 10^{b}	50 ± 6^{b}	92 ± 9^{c}	86 ± 10^{c}

Table 1. *Cont.*

No.	Compounds	CAS	RI	Dt/ Riprel	C-Bean	P-Bean	S-Bean	C-Powder	P-Powder	S-Powder	C-Brew	P-Brew	S-Brew
	Acids (2)												
24	acetic acid	64-19-7	635	1.2	40 ± 5^{a}	74 ± 4^{b}	90 ± 9^{c}	72 ± 8^{b}	92 ± 6^{c}	93 ± 8^{c}	65 ± 8^{b}	87 ± 8^{c}	95 ± 8^{c}
25	propanoic acid	79-09-4	694	1.1	48 ± 3^{a}	77 ± 2^{c}	91 ± 10^{d}	45 ± 3^{a}	73 ± 3^{bc}	94 ± 6^{d}	61 ± 10^{b}	90 ± 11^{d}	91 ± 9^{d}
	Esters (4)												
26	methyl acrylate	96-33-3	575	1.3	25 ± 3^{a}	61 ± 23^{bc}	98 ± 3^{e}	36 ± 3^{a}	72 ± 12^{cd}	87 ± 11^{de}	42 ± 12^{ab}	63 ± 10^{bc}	92 ± 10^{de}
27	ethyl acetate monomer	141-78-6	641	1.1	39 ± 4^{a}	79 ± 15^{cd}	88 ± 15^{de}	53 ± 6^{ab}	89 ± 8^{de}	95 ± 5^{de}	65 ± 6^{bc}	89 ± 8^{de}	96 ± 5^{e}
28	isoamyl acetate	123-92-2	854	1.3	38 ± 14^{abc}	77 ± 25^{de}	54 ± 3^{bcd}	92 ± 6^{e}	98 ± 2^{e}	74 ± 4^{de}	13 ± 6^{a}	22 ± 8^{ab}	37 ± 13^{cd}
29	ethyl acetate dimer	141-78-6	868	1.1	68 ± 9^{c}	92 ± 9^{ef}	74 ± 6^{cd}	94 ± 2^{ef}	98 ± 2^{f}	85 ± 2^{de}	21 ± 4^{a}	28 ± 5^{ab}	38 ± 10^{b}
	Furans (5)												
30	2-ethylfuran monomer	3208-16-0	681	1.1	44 ± 6^{b}	62 ± 10^{c}	88 ± 12^{de}	98 ± 3^{e}	22 ± 3^{a}	31 ± 2^{a}	92 ± 9^{de}	85 ± 3^{de}	84 ± 4^{d}
31	2-ethylfuran dimer	3208-16-0	681	1.1	77 ± 3^{bc}	73 ± 10^{b}	84 ± 15^{bc}	70 ± 1^{b}	90 ± 4^{bc}	95 ± 5^{c}	14 ± 0^{a}	26 ± 11^{a}	80 ± 23^{bc}
32	furfural monomer	98-01-1	823	1.3	93 ± 8^{f}	41 ± 8^{d}	13 ± 1^{a}	98 ± 2^{f}	50 ± 3^{e}	31 ± 1^{c}	99 ± 1^{f}	31 ± 2^{c}	22 ± 1^{b}
33	furfural dimer	98-01-1	825	1.1	92 ± 8^{de}	85 ± 5^{cd}	59 ± 4^{a}	64 ± 3^{a}	87 ± 4^{cd}	96 ± 4^{e}	97 ± 3^{e}	83 ± 2^{c}	72 ± 2^{b}
34	5-methylfurfural	620-02-0	967	1.1	90 ± 10^{c}	71 ± 7^{b}	39 ± 2^{a}	99 ± 1^{c}	73 ± 2^{b}	45 ± 1^{a}	94 ± 6^{c}	64 ± 10^{b}	51 ± 9^{a}
	Others (3)												
35	dimethyl sulfide	75-18-3	517	1.0	50 ± 3^{c}	23 ± 7^{a}	29 ± 7^{ab}	92 ± 7^{e}	34 ± 1^{b}	21 ± 1^{a}	98 ± 2^{e}	64 ± 1^{d}	47 ± 1^{c}
36	propane 2-methoxy-2-methyl	96-33-3	575	1.3	50 ± 3^{a}	76 ± 9^{bc}	99 ± 1^{d}	84 ± 4^{cd}	94 ± 6^{cd}	51 ± 6^{a}	58 ± 16^{ab}	79 ± 15^{c}	89 ± 13^{cd}
37	ethyl pyrazine	13925-00-3	911	1.1	92 ± 9^{d}	91 ± 1^{d}	74 ± 6^{bc}	98 ± 2^{d}	67 ± 1^{ab}	61 ± 3^{a}	99 ± 1^{d}	79 ± 3^{c}	74 ± 4^{bc}

Rt, retention time. Dt, drift time. Different letters in the same row indicate statistical differences between samples at a $p = 0.05$ level.

In addition, the signal intensities of butanal, 2-pentanone, and methyl acylate were higher in coffee S; the 3-pentanone and isoamyl acetate levels in coffee P were higher than in other samples; and the concentrations of furfural and 5-methylfurfural in C were higher than in S and P. The signal intensities of butanal monomer, butanal dimer, propanal, butanone monomer, 2-pentanone monomer, 2-pentanone dimer, 3-pentanone, butanol, 2-ethyl-1-propanol, 2-methyl-1-butanol, acetic acid, propanoic acid, methyl acrylate, athyl acetate monomer, 2-pentanone, 3-pentanone acetic acid, 2-ethylfuran, 2-methyl-1-butanol, methyl acrylate, propanal, and propane 2-methoxy-2-methyl- in S were higher than in P and C for the bean, powder, and brew statuses. The majority of hydrocarbon compounds were reported to have high odor thresholds, so they rarely contributed to coffee flavor. Acid compounds usually produce an unpleasant odor. Acetic acid is the volatile compound in coffee, having a sharp pungent odor like vinegar [19,23]. The concentration of acids will reduce in the process of roasting. 2-methyl-1-butanol was detected in all samples, which had a malt flavor. 2-methylpropanol was also found in all samples with almost the same number of concentrations. Furan could be generated by the thermal treatment of Maillard reaction precursors or lipids [24,25]. The formation of furan may also depend on the content of sugar [24]. 2-ethylfuran contributes a pungent and fruity flavor to coffee. Methyl pyrazine has a nutty, roasted, and chocolate aroma [26].

Isoamyl acetate, methional, and ethyl pyrazine in P had higher concentrations than in S and C. The methional in the bean and powder forms were significantly higher than in the brews. Methional is the key coffee odorant [5], which is generated by a Strecker reaction between α-dicarbonyl and the amino acid methionine. In addition, furfural, 5-methylfurfural, 3-methyl-2-butanol, 2,3-pentanedione, and 2-ethylfuran levels in C were higher than in S and P. 5-methylfurfural presents with a caramel, spicy, and maple flavor in coffee [19,27]. 2,3-pentanedione is considered one source of the buttery aroma in coffee [6,19], with an odor threshold of 30 ppb.

2.3. Comparison between GC-IMS and E-Nose

The flavor responses from the E-nose were based on signal responses from 14 sensors (Table S1). In order to further investigate the correlation between the E-nose and GC-IMS data, the PLSR was used for correlation analysis, where the GC-IMS data was used as the independent variable (X), and the E-nose data was used as the dependent variable (Y). The circle represents 100% of the explained variance. Among the variables explained by F1, most of the E-nose data (except E-1, E-3, and E-14) are located on the right side of the relevant load diagram, while the GC-IMS data are also located in the relevant location. As shown in Figure 3, the reference substance for the E-nose gas sensor array, E3 response to hydrogen, which had a strong positive correlation with the VOCs, such as C4, C115, and C108, measured by GC-IMS. The reference substances of sensors E1 and E14 were ammonia, amines, methane, and fuel gas and had a positive correlation with the VOCs, such as C2 and C75, measured by GC-IMS. Other sensing substances were mainly positively correlated with C28, C34, C81, and C32, located on the right side along the X-axis (Figure 3).

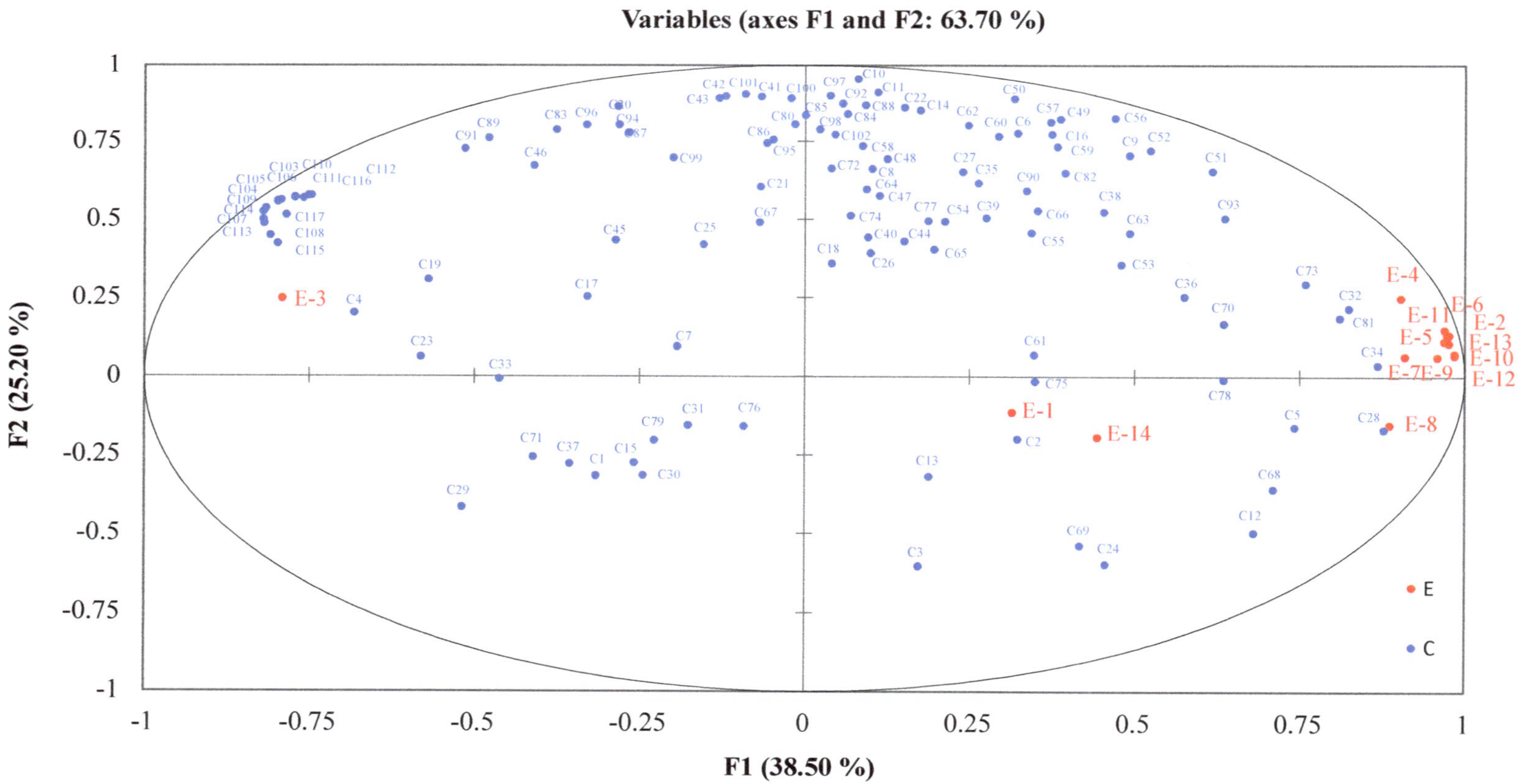

Figure 3. PLSR analysis to study the correlation between GC-IMS signals and E-nose sensor arrays. The VOCs identified by GC-IMS showed as C1 to C117 (blue dots). Data measured by E-nose showed as E1 to E 14 (red dots).

The correlation between the sample statuses, E-nose, and GC-IMS was then analyzed using multiple factor analysis (MFA) (Figure 4), where F1 and F2 accounted for 63.7% of variances. MFA is useful in studying relationships by simultaneously analyzing observations, variables, and tables [28]. In Figure 4, the centroid represents the resulting coordinates of the MFA. The points connected to the centroid represent the coordinates of the projections formed by the variations. The closer these projections are to the centroid, the greater the similarity between the descriptions. In Figure 4, S1, S2, and S3 represent the flavor characteristics of three commercial coffee samples (S, C, and P) in bean conditions; S4, S5, and S6 represent the flavor characteristics in powder conditions; S7, S8, and S9 represent the flavor characteristics in brewing conditions. Taking the average value of the E-nose and GC-IMS data as the sample value, from each sample value, it can be concluded that the same coffee sample dispersed further under different conditions, indicating that different conditions have a greater impact on coffee flavor characteristics. However, under the same conditions, different coffees are more closely distributed, indicating that even if the coffee samples are different, their flavor characteristics are similar under the same conditions. As can be seen in Figure 4, sample status and GC-IMS variables had a close distance to the centroid, indicating that GC-IMS is a good choice to discriminate coffee samples under different statuses. Our investigation showed GC-IMS could be a fast way to distinguish different coffee samples just like the E-noses do. Since E-noses only collect responses from 14 sensor arrays, GC-IMS could provide more VOCs, which could be served as biomarkers to indicate flavor differences in detail. The traditional GC-MS identified more VOCs, but a GC-IMS could show biomarkers in a time-saving and intuitive way.

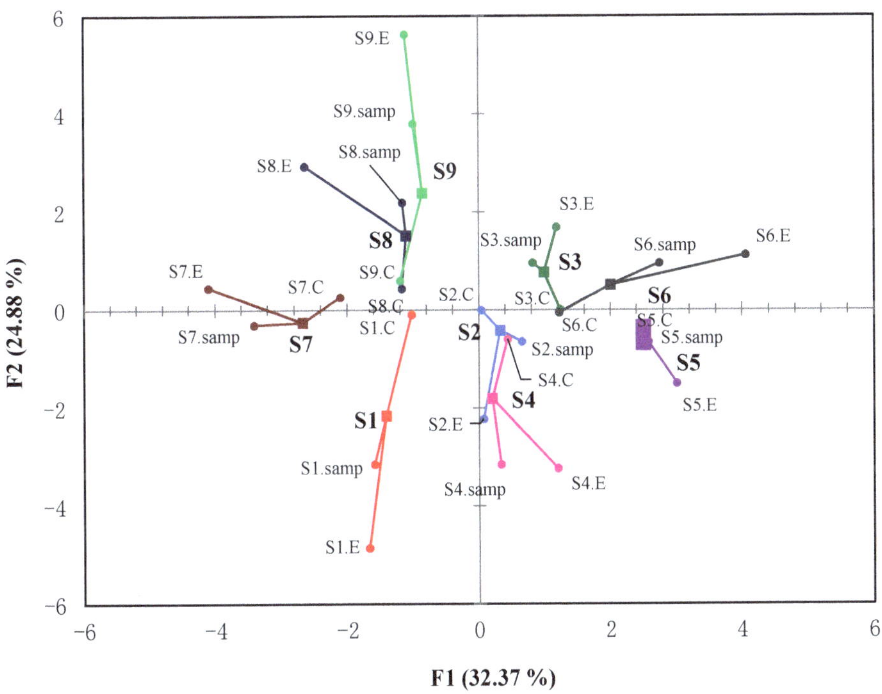

Figure 4. Comparison of the flavor profiles of three commercial coffee samples (S, C, and P) in three conditions (bean, powder, and brew) using multiple factor analysis (MFA). S1, S2, and S3 represent the flavor characteristics of three commercial coffee samples (S, C, and P) under bean conditions; S4, S5, and S6 represent the flavor characteristics under powder conditions; S7, S8, and S9 represent the flavor characteristics under brewing conditions.

3. Materials and Methods

3.1. Coffee Samples

Three brands of medium roasted coffee beans (S, C, and P) were purchased from local retailers and stored in a vacuum seal. They were *Arabica* beans from Colombia with moderately baking. The coffee powder was prepared by grinding the coffee with a manual grinder then passing it through a 700 μm sieve. Beverage was prepared as follows: 5.5 g of coffee powder was weighted into a coffee press beaker, added to 100 mL 92 °C hot water for brewing, immediately divided into a 20 mL headspace bottle, and sealed.

3.2. Gas Chromatography–Ion Mobility Spectrometry (GC-IMS)

Analyses of different coffee samples were performed on a commercial GC-IMS instrument (FlavourSpec®, Gesellschaft für Analytische Sensorsysteme, Dortmund, Germany) at Zhejiang University. The GC was connected to an Rtx-wax capillary column (30 m × 0.32 mm, 0.25 μm) and equipped with an auto-sampler unit (CTC Analytics AG, Zwingen, Switzerland), which used a 1 mL, air-tight, heated syringe to directly inject sample into the headspace.

The sample analysis method can be found in our previous report [16]. Samples (solid: 0.3 g, liquid: 1 mL) were transferred into a 20 mL vial and incubated at 40 °C for 20 min. Then, 100 μL headspace was automatically injected into the heated injector (80 °C, splitless mode) using a heated syringe (65 °C) at a speed of 500 μL/s. The carrier gas (Nitrogen, purity 99.99%) transferred the headspace of samples into GC column initially at a rate of 2 mL/min for 4 min, increased to 80 mL/min in 6 min, and, lastly, increased to 150 mL/min in 10 min. The analytes were eluted and separated at 45 °C then driven into an ionization chamber and ionized by a 3H ionization source with 300 MBq activity in positive ion mode. The resulting ions were driven to a drift tube (9.8 cm in length), which was operated at a constant temperature (45 °C) and voltage (5 kV). The flow rate of the drift gas (nitrogen gas) was set at 150 mL/min. Each spectrum was reported as the average of 12 scans.

The retention index (RI) of each compound was calculated by using C4–C9 n-ketone mixture standards (Sinopharm Chemical Reagent Co., Ltd., Beijing, China) as external references. Volatile compounds were identified based on RI and drift time compared with the GC-IMS library supplied by G.A.S. (Gesellschaft für analytische Sensor system mbH). The volatile compounds identified from IMS data were realized by the coupled software Laboratory Analytical Viewer (LAV, version 2.0.0), including three plug-ins and a GC-IMS library search to analyze samples from different perspectives.

3.3. E-Nose

The E-nose (Super Nose, Isenso Group Cooperation, Shanghai, China) used in this study was equipped with an array of 14 sensors. Each sensor was sensitive to hydrogen sulfide, sulfide, alcohols, ketones, aldehydes, and aromatic compounds.

The detection method can be found in our previous report [29]. Samples (solid: 3 g, liquid: 10 mL) were measured in a 100 mL beaker, covered tightly with a sealing film, and left at room temperature (25 °C) for 45 min until they reached equilibrium. Then, 20 mL of the headspace was extracted using a plastic, odorless free needle and injected into the sensor array chamber via sampling tubing. The sampling gas flowed at a constant speed rate of 600 mL/min and lasted for 160 s. The cleaning phase proceeded by pumping clean air for 100 s to normalize sensor signals between samples. Each sample replicated 12 parallels.

3.4. Statistical Analysis

The significant differences between coffee samples were found using SPSS statistical 23.0 software (IBM Watson Analytics, Beijing, China) at a $p = 0.05$ level with Duncan's multiple comparison method. PCA was analyzed using OriginPro 2016 (Northampton, MA, USA). Multiple factor analysis (MFA) was analyzed using XLSTAT 2016 (New York, NY, USA).

4. Conclusions

In this study, 117 peaks were identified in three coffee brands under three conditions (bean, powder, and brew) by GC-IMS. The PCA results clearly showed that three different brands of coffee under the same treatment could be divided into three parts and could be distinguished well using GC-IMS. In addition, 37 signal peaks from topographic plots were selected as biomarkers for the identification and classification of coffee samples. The comparison between E-nose and GC-IMS data using PLSR and MFA showed GC-IMS could present very close sample spaces. Compared with E-nose, GC–IMS could not only be used to classify coffee samples in a very short time but also provide VOC biomarkers to discriminate coffee samples.

Supplementary Materials: The following supporting information can be downloaded at: https://www.mdpi.com/article/10.3390/molecules27196262/s1, Table S1: The flavor responses from the E-nose based on signal responses from 14 sensors.

Author Contributions: Y.C.: data curation; visualization; writing–original draft; writing–review and editing; H.C.: investigation; D.C.: investigation; methodology; writing–original draft; X.F.: investigation; J.G.: investigation; Y.L.: project administration; supervision. All authors have read and agreed to the published version of the manuscript.

Funding: This research received no external funding.

Data Availability Statement: The data presented in this study are available on request from the corresponding author.

Acknowledgments: We are grateful to Isenso Group Cooperation for providing the E-nose and Wang Jun at Zhejiang University for providing the GC-IMS.

Conflicts of Interest: The authors declare no conflict of interest.

Sample Availability: Samples of the compounds are available or not from the authors.

References

1. Bressanello, D.; Liberto, E.; Cordero, C.; Sgorbini, B.; Rubiolo, P.; Pellegrino, G.; Ruosi, M.R.; Bicchi, C. Chemometric Modeling of Coffee Sensory Notes through Their Chemical Signatures: Potential and Limits in Defining an Analytical Tool for Quality Control. *J. Agric. Food Chem.* **2018**, *66*, 7096–7109. [CrossRef] [PubMed]
2. Baggenstoss, J.; Poisson, L.; Kaegi, R.; Perren, R.; Escher, F. Coffee Roasting and Aroma Formation: Application of Different Time-Temperature Conditions. *J. Agric. Food Chem.* **2008**, *56*, 5836–5846. [CrossRef] [PubMed]
3. Abdul Ghani, N.H.; Bingol, G.; Li, B.; Yu, W.; Young, B. Development of a novel 2D single coffee bean model and comparison with a 3D model under varying heating profiles. *J. Food Process Eng.* **2019**, *42*, e13063. [CrossRef]
4. Caporaso, N.; Whitworth, M.B.; Fisk, I.D. Prediction of coffee aroma from single roasted coffee beans by hyperspectral imaging. *Food Chem.* **2022**, *371*, 131159. [CrossRef] [PubMed]
5. Gigl, M.; Frank, O.; Barz, J.; Gabler, A.; Hegmanns, C.; Hofmann, T. Identification and Quantitation of Reaction Products from Quinic Acid, Quinic Acid Lactone, and Chlorogenic Acid with Strecker Aldehydes in Roasted Coffee. *J. Agric. Food Chem.* **2021**, *69*, 1027–1038. [CrossRef]
6. He, Y.; Zhang, H.; Wen, N.; Hu, R.; Wu, G.; Zeng, Y.; Li, X.; Miao, X. Effects of maltose and lysine treatment on coffee aroma by flash gas chromatography electronic nose and gas chromatography-mass spectrometry. *J. Sci. Food Agric.* **2018**, *98*, 154–165. [CrossRef]
7. Dong, W.; Hu, R.; Long, Y.; Li, H.; Zhang, Y.; Zhu, K.; Chu, Z. Comparative evaluation of the volatile profiles and taste properties of roasted coffee beans as affected by drying method and detected by electronic nose, electronic tongue, and HS-SPME-GC-MS. *Food Chem.* **2019**, *272*, 723–731. [CrossRef]
8. Cohen, M.J.; Karasek, F.W. Plasma Chromatography™—A New Dimension for Gas Chromatography and Mass Spectrometry. *J. Chromatogr. Sci.* **1970**, *8*, 330–337. [CrossRef]
9. Li, W.; Chen, Y.P.; Blank, I.; Li, F.; Li, C.; Liu, Y. GC x GC-ToF-MS and GC-IMS based volatile profile characterization of the Chinese dry-cured hams from different regions. *Food Res. Int.* **2021**, *142*, 110222. [CrossRef]
10. Chen, Y.P.; Cai, D.; Li, W.; Blank, I.; Liu, Y. Application of gas chromatography-ion mobility spectrometry (GC-IMS) and ultrafast gas chromatography electronic-nose (uf-GC E-nose) to distinguish four Chinese freshwater fishes at both raw and cooked status. *J. Food Biochem.* **2022**, *46*, e13840. [CrossRef]
11. Arroyo-Manzanares, N.; Martin-Gomez, A.; Jurado-Campos, N.; Garrido-Delgado, R.; Arce, C.; Arce, L. Target vs spectral fingerprint data analysis of Iberian ham samples for avoiding labelling fraud using headspace—Gas chromatography-ion mobility spectrometry. *Food Chem.* **2018**, *246*, 65–73. [CrossRef] [PubMed]

12. Garrido-Delgado, R.; Dobao-Prieto Mdel, M.; Arce, L.; Valcarcel, M. Determination of volatile compounds by GC-IMS to assign the quality of virgin olive oil. *Food Chem.* **2015**, *187*, 572–579. [CrossRef]
13. Franca, A.S.; Oliveira, L.S.; Oliveira, R.C.S.; Agresti, P.C.M.; Augusti, R. A preliminary evaluation of the effect of processing temperature on coffee roasting degree assessment. *J. Food Eng.* **2009**, *92*, 345–352. [CrossRef]
14. Schnabel, F.; de Melo Virginio Filho, E.; Xu, S.; Fisk, I.D.; Roupsard, O.; Haggar, J. Shade trees: A determinant to the relative success of organic versus conventional coffee production. *Agrofor. Syst.* **2017**, *92*, 1535–1549. [CrossRef]
15. Hendon, C.H.; Colonna-Dashwood, L.; Colonna-Dashwood, M. The role of dissolved cations in coffee extraction. *J. Agric. Food Chem.* **2014**, *62*, 4947–4950. [CrossRef]
16. Liu, D.; Bai, L.; Feng, X.; Chen, Y.P.; Zhang, D.; Yao, W.; Zhang, H.; Chen, G.; Liu, Y. Characterization of Jinhua ham aroma profiles in specific to aging time by gas chromatography-ion mobility spectrometry (GC-IMS). *Meat Sci.* **2020**, *168*, 108178. [CrossRef]
17. Li, M.; Yang, R.; Zhang, H.; Wang, S.; Chen, D.; Lin, S. Development of a flavor fingerprint by HS-GC-IMS with PCA for volatile compounds of Tricholoma matsutake Singer. *Food Chem.* **2019**, *290*, 32–39. [CrossRef]
18. Yuan, Z.-Y.; Qu, H.-Y.; Xie, M.-Z.; Zeng, G.; Huang, H.-Y.; Ren, F.; Chen, N.-H. Direct authentication of three Chinese materia medica species of the Lilii Bulbus family in terms of volatile components by headspace-gas chromatography-ion mobility spectrometry. *Anal. Methods* **2019**, *11*, 530–536. [CrossRef]
19. Bressanello, D.; Liberto, E.; Cordero, C.; Rubiolo, P.; Pellegrino, G.; Ruosi, M.R.; Bicchi, C. Coffee aroma: Chemometric comparison of the chemical information provided by three different samplings combined with GC-MS to describe the sensory properties in cup. *Food Chem.* **2017**, *214*, 218–226. [CrossRef]
20. ILópez-Galilea, I.; Fournier, N.; Cid, C.; Guichard, E. Changes in headspace volatile concentrations of coffee brews caused by the roasting process and the brewing procedure. *J. Agric. Food Chem.* **2006**, *54*, 8560–8566. [CrossRef] [PubMed]
21. Zhang, M.; Zhang, M.; Li, Y.; Wang, B. Research Progress in Flavor Components of Coffee. *Food Res. Dev.* **2016**, *37*, 213–218. [CrossRef]
22. Liu, C.; Yang, Q.; Linforth, R.; Fisk, I.D.; Yang, N. Modifying Robusta coffee aroma by green bean chemical pre-treatment. *Food Chem.* **2019**, *272*, 251–257. [CrossRef] [PubMed]
23. Cheong, M.W.; Tan, A.A.; Liu, S.Q.; Curran, P.; Yu, B. Pressurised liquid extraction of volatile compounds in coffee bean. *Talanta* **2013**, *115*, 300–307. [CrossRef] [PubMed]
24. Nie, S.; Huang, J.; Hu, J.; Zhang, Y.; Wang, S.; Li, C.; Marcone, M.; Xie, M. Effect of pH, temperature and heating time on the formation of furan in sugar–glycine model systems. *Food Sci. Hum. Wellness* **2013**, *2*, 87–92. [CrossRef]
25. Märk, J.; Pollien, P.; Lindinger, C.; Blank, I.; Märk, T. Quantitation of furan and methylfuran formed in different precursor systems by proton transfer reaction mass spectrometry. *J. Agric. Food Chem.* **2006**, *54*, 2786–2793. [CrossRef]
26. Liu, C.; Yang, N.; Yang, Q.; Ayed, C.; Linforth, R.; Fisk, I.D. Enhancing Robusta coffee aroma by modifying flavour precursors in the green coffee bean. *Food Chem.* **2019**, *281*, 8–17. [CrossRef]
27. Caporaso, N.; Whitworth, M.B.; Cui, C.; Fisk, I.D. Variability of single bean coffee volatile compounds of Arabica and robusta roasted coffees analysed by SPME-GC-MS. *Food Res. Int.* **2018**, *108*, 628–640. [CrossRef]
28. Sebald, K.; Dunkel, A.; Schafer, J.; Hinrichs, J.; Hofmann, T. Sensoproteomics: A New Approach for the Identification of Taste-Active Peptides in Fermented Foods. *J. Agric. Food Chem.* **2018**, *66*, 11092–11104. [CrossRef]
29. Xu, Y.; Chen, Y.P.; Deng, S.; Li, C.; Xu, X.; Zhou, G.; Liu, Y. Application of sensory evaluation, GC-ToF-MS, and E-nose to discriminate the flavor differences among five distinct parts of the Chinese blanched chicken. *Food Res. Int.* **2020**, *137*, 109669. [CrossRef]

molecules

Article

Untargeted Lipidomics Method for the Discrimination of Five Crab Species by Ultra-High-Performance Liquid Chromatography High-Resolution Mass Spectrometry Combined with Chemometrics

Jiaxu Yao [1], Jinrui Zhu [1], Minjie Zhao [2], Li Zhou [1,*] and Eric Marchioni [2]

1 National Demonstration Center for Experimental Ethnopharmacology Education, School of Pharmaceutical Sciences, South-Central Minzu University, Wuhan 430074, China
2 Equipe de Chimie Analytique des Molécules Bioactives et Pharmacognoise, Institut Pluridisciplinaire Hubert Curien (UMR 7178, CNRS/UDS), 74 Route du Rhin, 67400 Illkirch, France
* Correspondence: zhou2018@scuec.edu.cn

Abstract: In this study, ultra-high-performance liquid chromatography high-resolution accurate mass-mass spectrometry (UHPLC-HRAM/MS) was applied to characterize the lipid profiles of five crab species. A total of 203 lipid molecular species in muscle tissue and 176 in edible viscera were quantified. The results indicate that *Cancer pagurus* contained high levels of lipids with a docosahexaenoic acid (DHA) and eicosapntemacnioc acid (EPA) structure in the muscle tissue and edible viscera. A partial least squares discriminant analysis (PLS-DA) showed that PE 16:0/22:6, PE P-18:0/20:5, PA 16:0/22:6 and PC 16:0/16:1 could be used as potential biomarkers to discriminate the five kinds of crabs. In addition, some lipids, such as PE 18:0/20:5, PC 16:0/16:1, PE P-18:0/22:6 and SM 12:1;2O/20:0, could be used as characteristic molecules to distinguish between *Cancer magister* and *Cancer pagurus*, which are similar in appearance. This study provides a new perspective on discriminating crab species from MS-based lipidomics.

Keywords: seafood; molecular species; mass spectrum; quantification; identification; lipid

Citation: Yao, J.; Zhu, J.; Zhao, M.; Zhou, L.; Marchioni, E. Untargeted Lipidomics Method for the Discrimination of Five Crab Species by Ultra-High-Performance Liquid Chromatography High-Resolution Mass Spectrometry Combined with Chemometrics. *Molecules* **2023**, *28*, 3653. https://doi.org/10.3390/molecules28093653

Academic Editor: Susy Piovesana

Received: 22 February 2023
Revised: 14 April 2023
Accepted: 19 April 2023
Published: 22 April 2023

1. Introduction

Crab, as one of the most famous delicacies in the world, is greatly adored by people for its unique taste and distinct flavor [1]. Due to the different consumption habits, crab muscle and edible viscera are individually appreciated. The former is mainly from the claws, legs and abdomen and the latter from the hepatopancreas and gonads [2]. Crab is an excellent source of numerous nutrients essential for human health, including highly unsaturated fatty acids, proteins, minerals and vitamins [3], which bring benefits to humans, such as anti-inflammatory and immunity and cognitive enhancing properties [4].

The 2021 Power of Seafood report published by the Food Industry Association demonstrated that the overall seafood category was up 28.4% in retail around the world. In China, the demand for crab is also constantly increasing, as evidenced by the growth of crab farming from 821,000 tons in 2010 to 1,063,000 tons in 2020. Crab ranks third in global seafood production after shrimp and lobsters and has significant commercial value [5,6]. Mainly, crab is one of the most diverse of the decapod crustaceans, which can be found in both fresh and ocean waters [7]. The nutritional value of different crabs might be different due to the fact of their different nutritional components.

It has been documented that approximately 7000 species of crab survive throughout the world, either in freshwater lakes or in the ocean [8]. However, the majority of crabs are used for food, whether they are raised or wild. Commercially valuable crabs can be easily purchased for consumption, such as red king crab (*Paralithodes camtschaticus*), swimming crab (*Portunus trituberculatus*), Chinese mitten crab (*Eriocheir sinensis*), Dungeness crab (*Cancer magister*) and brown edible crab (*Cancer pagurus*) [9].

Lipids, a major constituent of various foods, play vital roles in many cellular processes, such as energy storage, signal-mediated processes and tumor suppression [10–12]. Lipids contribute to the quality features of food products, including flavor and nutritional value. Numerous studies have focused on lipids, resulting in the emergence of lipidomics. Lipidomics, a branch of metabolomics, has superiority in providing lipid profiles of biological samples, especially those rich in lipids. In nutritional research, lipidomics has been routinely employed to elucidate the interactions among diet, nutrients and human metabolism to optimize food processing and to evaluate the nutrition of foods [13]. Consumers can establish guidelines for personalized nutrition based on lipid information [14].

Statistically, the number of publications on the lipidomics of marine animals showed an explosive increase over the past few years [15], and numerous studies have been conducted concerning the lipids of crabs. The research objectives are also focused on improving the level of beneficial lipids or survival rates by changing the dietary structure of crabs [16,17] and analyzing the lipid profile of crabs to determine their nutritional information [18]. Nevertheless, few studies have been performed comparing the lipids of multiple crab species. Thus, it is meaningful to comprehensively characterize and compare the lipid compositions among crab species from a lipidomics perspective.

In recent years, one of the most valuable technologies for lipid identification has been ultra-high-performance liquid chromatography coupled to high-resolution mass spectrometry [19], which possesses the advantages of accurate quality and excellent sensitivity. Quadrupole-Exactive high-resolution accurate mass spectrometry (HRAM) is a developed technique that can provide complete lipid molecular information without derivatization [20]. Thus, this study aimed to identify the lipid classes and lipid molecular species of the muscles and edible viscera tissue from five kinds of edible crabs by a UHPLC-HRAM/MS approach. Subsequently, the different compositions of the lipid profiles were analyzed by lipidomics combined with chemometrics. This is the first investigation involving an in-depth and comprehensive identification and comparison of the lipid profiles of different crabs. This study may help us better understand the nutritional values of edible crabs.

2. Results and Discussion

2.1. Total Lipid Content of Crab Muscle and Edible Viscera

The total lipid content of crab muscle and edible viscera are shown in Table S2. No significant difference ($p > 0.05$) was exhibited in the content of the crab muscle lipids. As for edible viscera, the five crabs exhibited a significant difference ($p < 0.05$). The total lipid content of the edible viscera from *P. camtschaticus* (766.3 mg/g) and *E. sinensis* (761.0 mg/g) showed a higher level than the other three crab lipids. Overall, the total lipid content of the edible viscera was higher than that in the muscles, which is consistent with the results of previous reports [2,21]. The reason for this is probably that crab muscles mainly store protein, while the viscera stores fat and cholesterol. The high level of total lipids in the crab samples results in biological weight gain, which is associated with consuming crab-containing diets [22]. In addition, many factors might affect the lipid content of different crab species, such as temperature [23], maturity [24] and diets [25]. The variation of the lipid content in crabs under different physiological conditions requires further experimental analysis.

2.2. Validation of the UHPLC-HRAM/MS Method

As described in this method, all lipids were quantified by the internal standard method. The internal standards were mixed with each test sample to quantify the lipid molecules, so the relative standard deviation (RSD) values of the peak areas of the internal standards could be used as a measure of the stability of the UHPLC-HRAM/MS instrument during the measurement of the sample [26]. The RSD values of the peak area of the internal standard are displayed in Figure S1. It was found that the RSD values of 84.3% of features were less than 20%, which demonstrated that the signal was stable during the sample detection.

Further validation was carried out using seven internal standards. The linear regression equations derived from the different concentrations of the seven lipid standards are shown in Table S3, which reveals an excellent linearity ($R^2 \geq 0.990$). The limit of detection (LOD) value was less than 2.28 ng/mL, and the limit of quantification (LOQ) values were between 0.99 ng/mL and 7.52 ng/mL.

2.3. Lipid Identification

The muscle lipids and edible viscera lipids of the five crabs were identified by UHPLC-HRAM/MS, with the electrospray ionization (ESI) source in positive and negative ion modes. Different lipid classes were detected in specific ESI source patterns due to the fact of their polarities and electric charges [27]. The PC, LPC, PE, LPE, TAG and DAG were analyzed under the positive ion mode ($[M + H]^+$ and $[M + NH_4]^+$), while the remaining lipids were examined under the negative ion mode ($[M - H]^-$, $[M + HCOO^-]^-$ and $[M - 2H]^{2-}$). The lipid profiles of the muscle (Figure 1a) and edible viscera (Figure 1b) from five crab species could be detected by ESI sources in the positive ion mode within 20 min. From the total ion chromatograms plot in the positive ion mode, the weakly polar lipid such as TAG were retained for a short time, and the retention times of the molecular species in the same lipid classes were similar. However, the different peak shapes indicate that that the contents of common lipid classes could be different in these crab samples. Figure 1c presents an example of extracted ion chromatograms (EICs) of *m/z* 876.8015, 750.5432, 832.5851 and 522.3554 from *C. magister* muscle lipids.

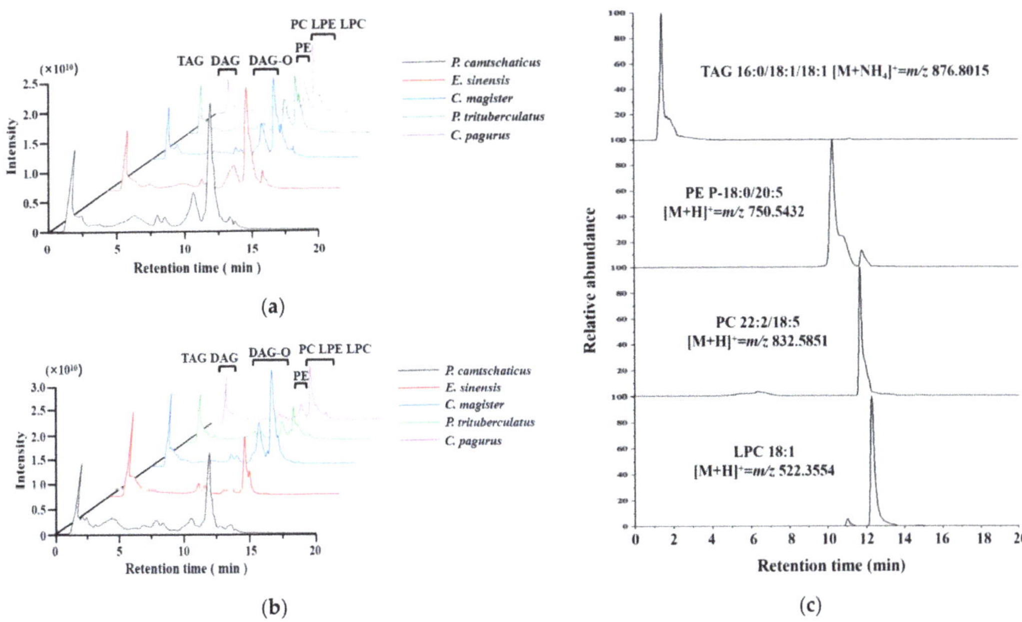

Figure 1. Total ion chromatograms of lipids in the positive ion mode from the (**a**) muscle and (**b**) edible viscera of five crabs. (**c**) Extracted individual ion chromatograms of the muscle lipids of *C. magister*, showing the species TAGs of 16:0/18:1/18:1, PE P-18:0/20:5, PC 22:2/18:5 and LPC 18:1.

The lipid molecular species were identified by MS/MS spectrometry on the basis of their characteristic mass values and fragment ions. Lipid classes can be classified by the types of polar headgroups, which exhibit characteristic fragments in the MS/MS spectrum.

For example, m/z 184.0730 $[C_5H_{15}NO_4P]^+$ is the typical fragment of the choline headgroup of PC, and m/z 241.0130 $[C_6H_{10}O_8P]^-$ is the polar head of PI. In addition, other MS/MS fragments can provide information to infer the fatty acyl group of the lipids and the total relative molecular masses.

A total of 14 lipid subclasses and 203 molecular species were determined in the muscle lipids (Table S4) and 13 lipid subclasses and 176 molecular species in the edible viscera lipids (Table S5). Seventy-one and fifty-five common molecular species were detected, respectively, in the muscles and edible viscera of the five crabs (Figure 2a,b). The concentrations of each lipid subclass in the crab muscles and edible viscera are displayed in Tables S6 and S7, respectively. The proportion of each lipid subclass is calculated and shown in Figure S2. PE, PC and TAG were the main lipid subclasses in the crab muscle samples, while TAG was the predominant one in the edible viscera of *E. sinensis*, which accounted for more than 89%. These results are similar to that published by Wang et al. [28], who illustrated that crabs stored great numbers of TAGs in visceral organs for energy expenditure during starvation, molting or reproduction [29,30]. Especially, a relatively high level of PA was found in the crab muscles, which was not encountered in previous studies. According to research, more than 70% of PA is consumed at the time crabs mature from larvae [31]. The absence of PA might be one of the causes of the high mortality of crab larvae, so diets adding relevant lipids would be used to increase the survival rate of crabs. In comparison, phospholipids have a crucial effect on cell membranes, mainly by maintaining endogenous systems, serving as binding sites for proteins and participating in signaling [32], which play a vital role in the growth and metabolism of crabs [33,34]. Moreover, they are precursors of essential steroids [35].

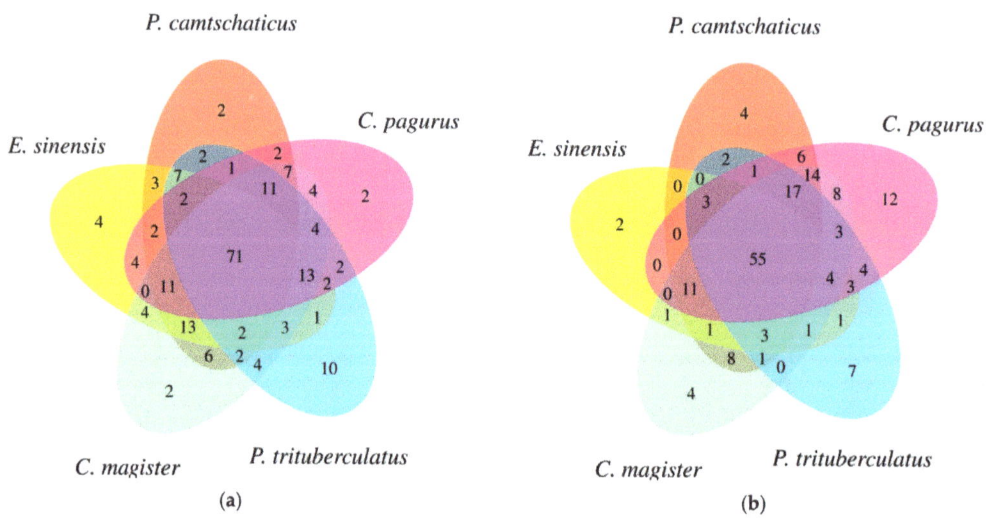

Figure 2. Distribution of lipid molecular species in the (**a**) muscle and (**b**) edible viscera of five crabs.

It is well known that the lipids containing long-chain polyunsaturated fatty acyl chains (LC-PUFAs), such as EPA and DHA, play an irreplaceable role in brain development [36] and have properties such as antitumor [37] and anti-inflammatory [5]. In addition, LC-PUFAs exhibit better bio-efficacy and bioavailability than other lipids [38]. In this study, some DHA/EPA-TAG and DHA/EPA-PL molecules were detected in the muscle and edible viscera of the five crab species. As shown in Figure S3, the contents of DHA/EPA-PLs and DHA/EPA-TAGs (DHA/EPA-PLs + DHA/EPA-TAGs) were the highest (72.26%) in *C. pagurus* muscle, while they appeared to have the lowest levels in *E. sinensis* edible visceral tissue. The total contents of DHA/EPA-PLs and DHA/EPA-TAGs were the highest in *C. pagurus*, with significant differences ($p < 0.05$) compared to the other groups. The DHA/EPA distribution in PLs and TAGs was consistent with the overall fatty acid composition of the plentiful PUFAs in marine animals and plant plankton [39]. Herein, our results indicate that crab muscles and edible viscera are an excellent source of DHA/EPA.

Furthermore, ether-linked phospholipids were found in the lipids of the muscle and edible viscera of the crabs, which are essential signaling molecules needed for the execution of complex life activity [40,41]. A representative MS/MS fragmentation spectrum in the positive ion mode of PE P-20:1/20:5 (*m/z* 776.5596) from *P. camtschaticus* muscle is shown in Figure 3a. This molecular species only existed in the *P. camtschaticus* muscle and was most abundant (3183 nmol/g) in plasmalogen PE. The product ion at *m/z* 141.0698 ($C_2H_8NO_4P$) was the characteristic fragment of the PE molecule. The product ion at *m/z* 635.5086 corresponded to the loss of the ethanolamine headgroup. The oxygen at the sn-1 position attacked the phosphorus atom, which led to the formation of a new oxygen–phosphorous bond. At the same time, hydrogen was extracted from C-2 of the glycerol backbone, forming a double bond between C-1 and C-2 of the glycerol backbone [42,43]. Therefore, the ion fragment at *m/z* 418.3072 had the structure of 20:1 ether and $C_2H_8NO_3P$, and the ion fragment at *m/z* 320.3316 was a neutral loss corresponding to H_3PO_4 on this basis.

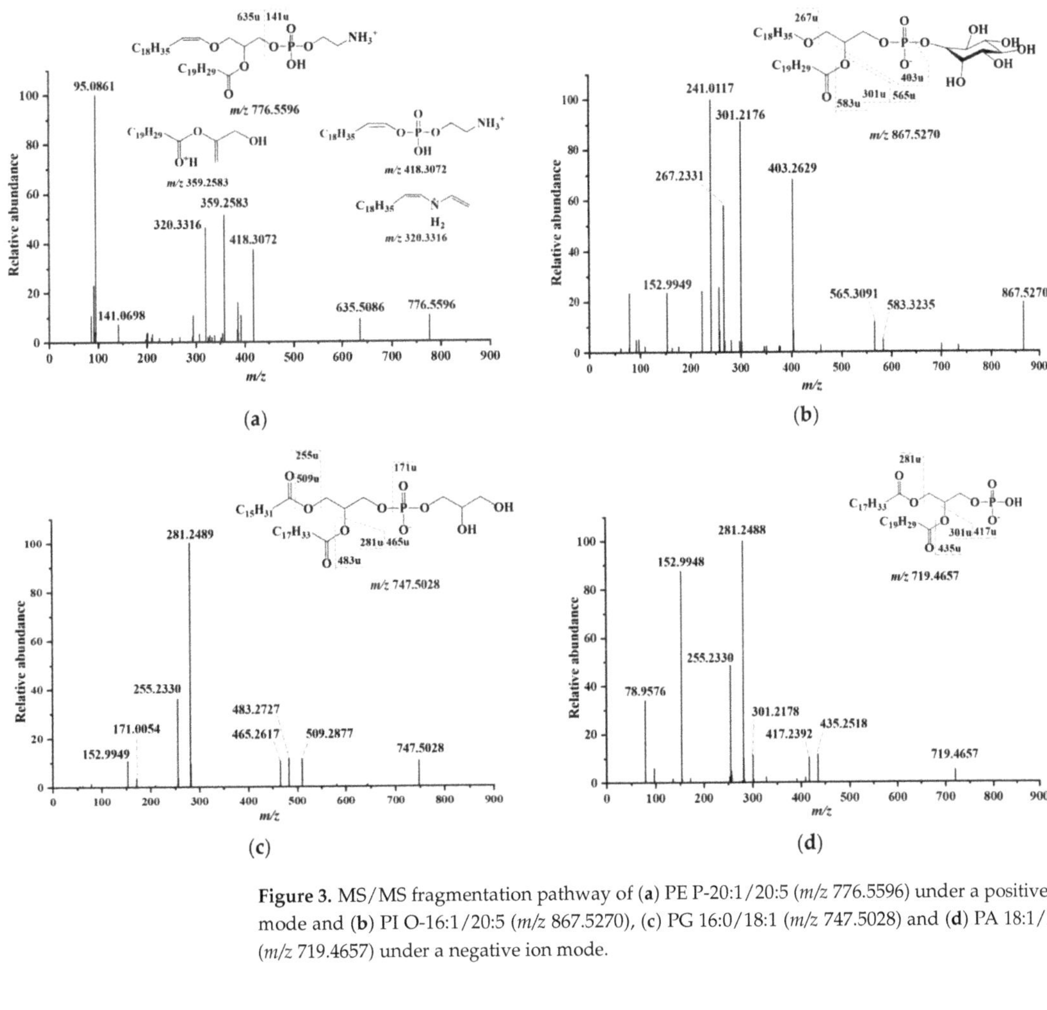

Figure 3. MS/MS fragmentation pathway of (**a**) PE P-20:1/20:5 (*m/z* 776.5596) under a positive ion mode and (**b**) PI O-16:1/20:5 (*m/z* 867.5270), (**c**) PG 16:0/18:1 (*m/z* 747.5028) and (**d**) PA 18:1/20:5 (*m/z* 719.4657) under a negative ion mode.

The representative MS/MS fragmentation spectrum in the negative ion mode of PI O-16:1/20:5 (*m/z* 841.5156) from *C. magister* edible viscera is provided in Figure 3b. The product ion at *m/z* 301.2182 corresponds to the ion of EPA, and *m/z* 539.2991 corresponds to the loss of EPA [44]. The ion fragment at *m/z* 377.2467 was identified as the residue fragment, which was generated by the loss of the inositol headgroup of the ion fragment at *m/z* 539.2991. The ion fragment at *m/z* 557.3096 corresponded to the loss of C20:5 acyl group from the precursor ion. The ion fragment at *m/z* 241.0117 and 152.9948 were characteristic signals of the PI molecules, which were generated from the inositol headgroup. The former was a structure formed by dehydrating the head bond, while the latter was a five-membered ring structure formed by disconnecting two fatty acid chains and a six-membered ring.

PG 16:0/18:1 was detected in the muscles of all five kinds of crab, and the fragmentation pathways of PG 16:0/18:1 are displayed in Figure 3c. The ion fragment at *m/z* 747.5028 corresponds to the empirical formula $C_{40}H_{76}O_{10}P$. The ion fragment at *m/z* 171.0054 was the $C_3H_7O_6P$ structure, which was the head group of the PG molecule, and the ion fragment at *m/z* 152.9949 was a five-membered ring structure formed by dehydration on this basis. As seen in Figure 3d, the PA 18:1/20:5 from *P. camtschaticus*, which had the highest level (750.6 ± 41.4 nmol/g) among all species, displayed signals at *m/z* 719.4567 in the MS^- spectrum, which corresponded to the empirical formula $C_{41}H_{69}O_8P^-$. The ion fragments at *m/z* 281.2490 and *m/z* 301.2176 were oleic acid and EPA, respectively. The ion fragments at *m/z* 417.2411 and 4435.2516 were derived from the loss of the EPA and C20:5 acyl groups, respectively, from the precursor molecule.

2.4. Multivariate Analysis

As MS-based lipidomics generates a large amount of data in the analysis process [45], principal component analysis (PCA), a multivariate statistical method, was used to examine the correlations among the lipidomics data with multiple variables and to compare the lipid molecules and contents of the different crabs to facilitate the analysis and visualization. Figure 4a presents the PCA score plots ($R^2X = 0.994$, $Q^2X = 0.982$) of the crab lipids in the different muscles, displaying the clustering of each sample in the first two principal component (t1 and t2) score plots, accounting for 0.367 and 0.251 of the total variance, respectively. The PCA score plots ($R^2X = 0.997$, $Q^2X = 0.985$) of the edible visceral lipids in the different crabs indicate that the first two principal components (t1 and t2) accounted for 0.356 and 0.242 of the total variance. From the PCA score plot (Figure 4a), the five crab types were nonoverlapping with each other, which indicates that the metabolites of each crab were discrepant. Therefore, the PCA model had good separation and clustering results and can be used as a method to effectively distinguish different samples.

Furthermore, the separations were tested in the PLS-DA score plot. R^2 represents the degree of fitting between the model and crab lipid data, and Q^2 (cumulative) represents the prediction ability of the model for new data [46]. For the crab muscle samples, the PLS-DA score plot is displayed in Figure 4b ($R^2X = 0.998$, $R^2Y = 0.999$, $Q^2 = 0.997$), which demonstrates a good separation effect and prediction ability. After 999 permutations (Figure 4c), the values of $R^2 = (0.0, 0.062)$ and $Q^2 = (0.0, -0.823)$ of this model were more prominent than the random values of all samples, which is similar to the PLS-DA score plot of the edible visceral samples. All these data underlie the classification of the five crab species and show a good predictive ability for a new data set.

Subsequently, the value of VIP (variable importance in projection) was applied to screen for significant differences in the lipids of the five crab species, which was available from the PLS-DA model. In the crab muscle model, a total of 57 lipid molecules with VIP > 1 were analyzed by cluster analysis to assess the similarity of the crab samples (Figure 4d). According to the dendritic diagram, *C. magister* and *C. pagurus* was in one group. At the same time, 51 lipid molecules met VIP > 1, and *C. pagurus* and *P. camtschaticus* belonged to one group based on the similarities in the composition of the edible viscera. The results were also well reflected in the PCA diagram. Thus, studying the pedigree relationship of lipidomics characteristics can provide a reference for species classification.

Additionally, one-way analysis of variance (ANOVA) was conducted to determine the final statistically significant ($p < 0.05$) lipid species. In general, markers need to meet $p < 0.05$ and VIP > 1 [47], which could be identified as potentially characteristic metabolites, thus achieving the ability to identify species accurately. In the crab muscle model, 26 lipid molecules were selected as markers (Table S8), such as PE 16:0/22:6, PE P-18:0/20:5, PA 16:0/22:6 and PC 16:0/16:1. The distribution of 26 lipids in the muscle content of the five species of crab is shown in Figure S4a. For example, when the detected range of PE 16:0/22:6 was 8000 nmol/g, the crab species can be judged as *E. sinensis*. In the samples of edible viscera, 17 lipid molecules were selected as markers (Table S8), including TAG 16:0/20:1/18:2, SM 14:1;2O/22:0, PE 18:1/20:5 and TAG 16:0/18:1/20:1, whose distributions are shown in Figure S4b.

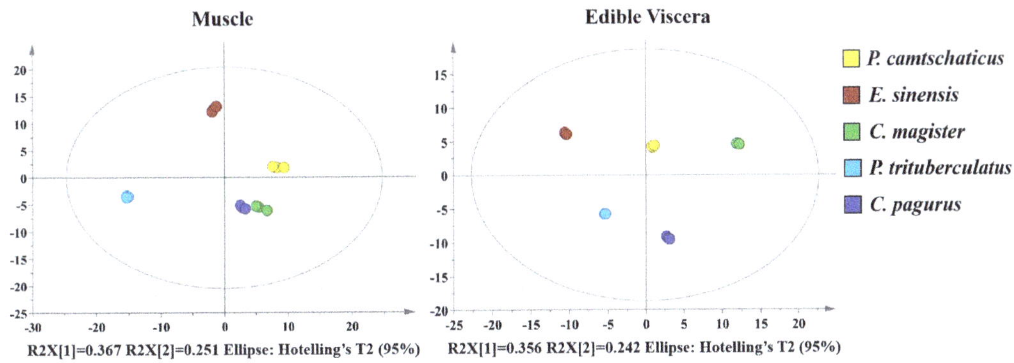

(**a**)

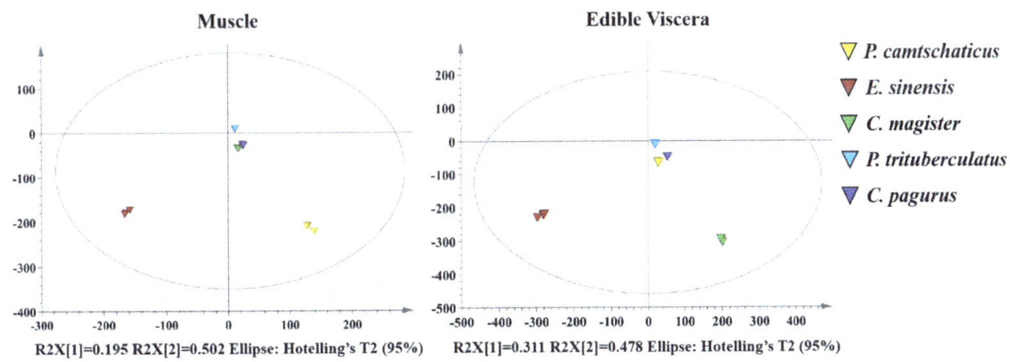

(**b**)

Figure 4. *Cont.*

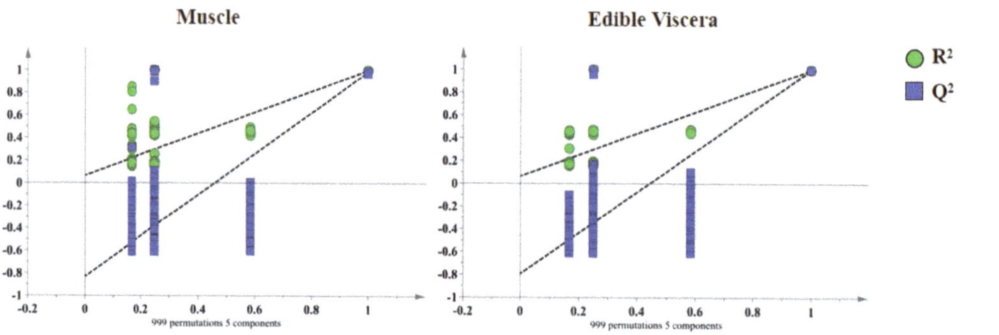

(**c**)

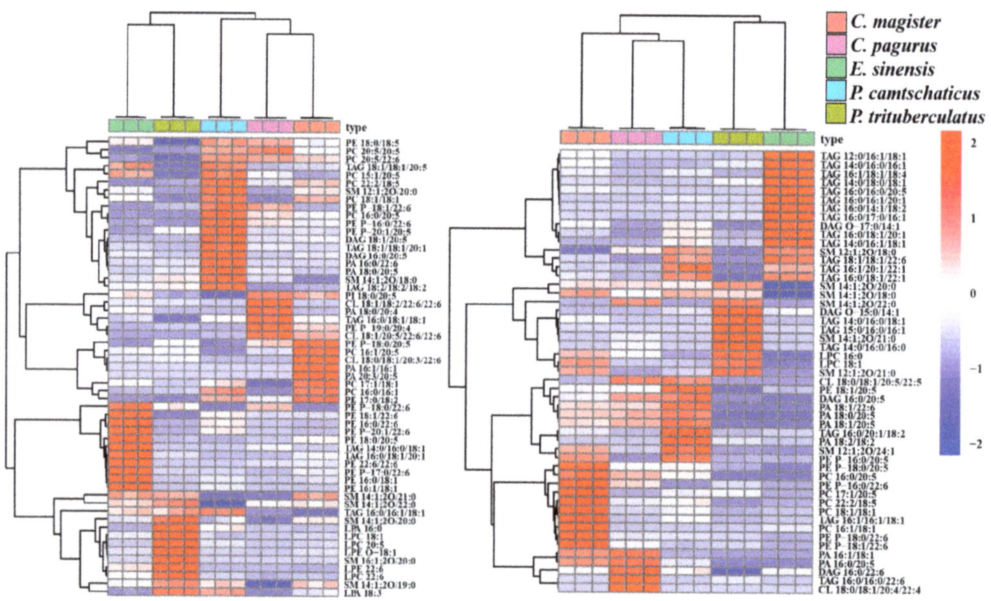

(**d**)

Figure 4. (**a**) PCA score plot of identified lipids in muscle with $R^2X = 0.994$ and $Q^2X = 0.982$ and edible viscera with $R^2X = 0.997$ and $Q^2X = 0.985$. (**b**) PLS-DA score plot of identified lipids in muscle with $R^2X = 0.998$, $R^2Y = 0.999$ and $Q^2 = 0.997$ and edible viscera with $R^2X = 0.998$, $R^2Y = 0.999$ and $Q^2 = 0.998$. (**c**) Permutation constructed on the basis of the PLS-DA of the contribution molecular species in muscle and edible viscera, and (**d**) Hierarchical cluster dendritic diagram of muscle and edible viscera and hierarchical cluster analysis based on molecular species in muscle and edible viscera with VIP > 1. Colors represent different concentrations indicated by the color bar.

2.5. Orthogonal Partial Least Squares Discrimination Analysis

Due to the highly similar appearance of *C. magister* and *C. pagurus*, the two species were often confused without relevant books and professionals. Here, the differences between the two crabs were analyzed from a lipidomics perspective combined with a metrological approach.

In this study, the principle for the selection of significantly different lipid species between *C. magister* and *C. pagurus* is a sufficiently high variable importance for the projection (VIP > 1.3) and the univariate statistical analysis, including a fold change ($\log_2(FC) \geq 1$ or ≤ -1) and T-test criteria ($p < 0.05$). Overall, 174 lipids from muscles were calculated and are shown in volcano plots in Figure 5a, among which 78 lipids were upregulated and 68 were downregulated in *C. magister* compared with *C. pagurus*. Analogously, Figure 5b demonstrates that 76 lipids from edible viscera were upregulated and 43 were downregulated. In addition, the VIP values of each lipid molecule were obtained from orthogonal partial least squares discrimination analysis (OPLS-DA). Figure 5c,d represent the VIP value map of some quantified metabolites from the muscles and edible viscera, respectively. Finally, 20 lipid molecules from muscles and 17 lipid molecules from edible viscera were selected as the characteristic molecules to distinguish *C. magister* and *C. pagurus* (Table 1).

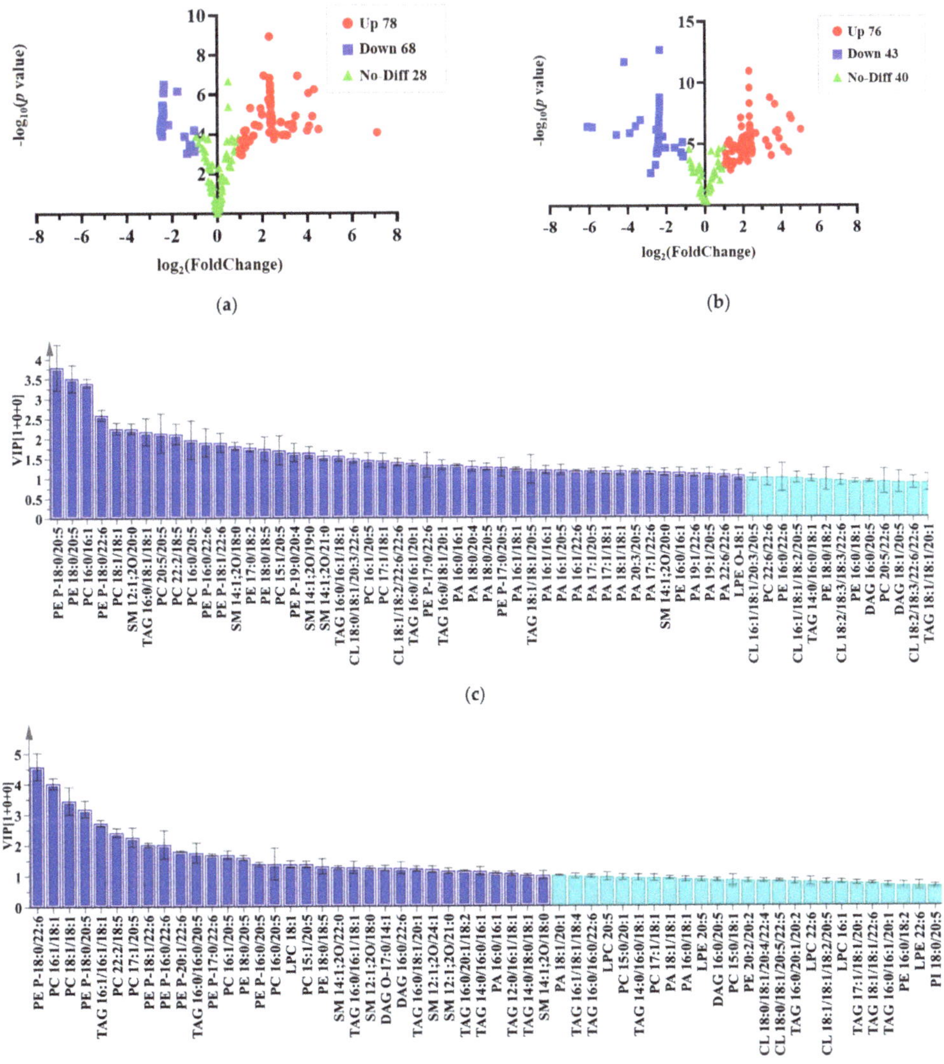

Figure 5. An orthogonal partial least squares discrimination analysis was performed for *C. magister* and *C. pagurus*. Volcano plots of lipids from (**a**) muscle and (**b**) edible viscera. VIP maps of lipid molecules in (**c**) crab muscle and (**d**) edible viscera. Dark blue indicates VIP > 1.

Table 1. The significantly different lipids in the muscles and edible viscera of *C. magister* and *C. pagurus*.

Compound	Muscle VIP	log_2(FC)	*p*-Value	Compounds	Edible Viscera VIP	log_2(FC)	*p*-Value
PE 18:0/20:5	3.28	2.41	<0.01	PE P-18:0/22:6	4.43	2.49	<0.01
PC 16:0/16:1	3.28	4.06	<0.01	PC 16:1/18:1	3.97	2.39	<0.01
PE P-18:0/22:6	2.37	−1.75	<0.01	PC 18:1/18:1	3.37	3.49	<0.01
SM 12:1;2O/20:0	2.21	2.39	<0.01	PE P-18:0/20:5	3.16	1.20	<0.01
PC 18:1/18:1	2.08	1.94	<0.01	TAG 16:1/16:1/18:1	2.68	5.02	<0.01
PC 20:5/20:5	1.96	−1.16	<0.01	PC 22:2/18:5	2.33	2.14	<0.01
PC 22:2/18:5	1.93	1.20	<0.01	PC 17:1/20:5	2.23	4.38	<0.01
PE P-18:1/22:6	1.78	−2.46	<0.01	PE P-16:0/22:6	2.04	1.14	<0.01
SM 14:1;2O/18:0	1.78	−2.40	<0.01	PE P-18:1/22:6	1.95	1.88	<0.01
PE 17:0/18:2	1.61	1.48	<0.01	PE P-20:1/22:6	1.76	3.68	<0.01
PC 15:1/20:5	1.56	−1.22	<0.01	TAG 16:0/16:0/20:5	1.70	1.81	<0.01
SM 14:1;2O/19:0	1.56	2.85	<0.01	PE P-17:0/22:6	1.65	3.42	<0.01
PE P-19:0/20:4	1.52	−1.03	<0.01	PC 16:1/20:5	1.65	3.36	<0.01
SM 14:1;2O/21:0	1.51	4.24	<0.01	PE 18:0/20:5	1.54	1.92	<0.01
CL 18:0/18:1/20:3/22:6	1.46	2.44	<0.01	PE P-16:0/20:5	1.36	3.73	<0.01
TAG 16:0/16:1/18:1	1.45	−2.36	<0.01	LPC 18:1	1.33	1.91	<0.01
PC 17:1/18:1	1.41	7.12	<0.01	PC 15:1/20:5	1.33	2.21	<0.01
CL 18:1/18:2/22:6/22:6	1.37	−2.39	<0.01				
PC 16:1/20:5	1.36	2.33	<0.01				
TAG 16:0/16:1/20:1	1.35	2.40	<0.01				

VIP: variable importance in projection; FC: fold change.

3. Materials and Methods

3.1. Ethical Statement

In this study, all experimental animals were adequately cared for, their pain minimized, and killed painlessly. In addition, the number of all animals was controlled to the minimum required to obtain scientific results.

3.2. Sample Preparation

Five living female farmed crab species (Figure S5), including *P. camtschaticus* (*Paralithodes camtschaticus*), *E. sinensis* (*Eriocheir sinensis*), *C. magister* (*Cancer magister*), *P. trituberculatus* (*Portunus trituberculatus*) and *C. pagurus* (*Cancer pagurus*), were purchased from a local seafood market in Wuhan, China, in October 2020, with the help of professionals (12 fresh crabs of each species, individually weighed, *P. camtschaticus*: 1500–1700 g; *E. sinensis*: 150–200 g; *C. magister*: 750–800 g; *P. trituberculatus*: 250–300 g; and *C. pagurus*: 900–1000 g). The ages of each species of crab shell were similar, and all crab limbs were free of damage. According to a previous study [48], all crabs were electrocuted through 110 volts and 2 amps of current for 10 s with a Crustastun machine (Studham Technologies, Scotland, UK) and killed immediately. The edible viscera and muscle were manually separated from the crabs, immediately preserved in liquid nitrogen and then lyophilized using a vacuum freeze-dryer (Labconco Corporation, Kansas, MO, USA) with a vacuum of 0.06 MPa. The freeze-dried crab tissues were stored at −80 °C until analysis.

3.3. Standards and Reagents

The PL standards were purchased from Avanti Polar Lipids, including phosphatidylcholine (PC 17:0/17:0), phosphatidylethanolamine (PE 17:0/17:0), phosphatidylglycerol (PG 17:0/17:0), phosphatidic acid (PA 17:0/17:0), phosphatidylinositol (PI 8:0/8:0), sphingomyelin (SM d18:1/12:0) and SPLASH LIPIDOMIX (isotopic internal standard mixture). Glyceryl trilinoleate (TAG) was purchased from Sigma Aldrich. Other chemicals, including methanol, chloroform, acetonitrile and ammonium formate, were purchased from Macklin Biochemical Technology Co., Ltd. (Shanghai, China) (high-performance liquid chromatography or analytical reagent grade).

3.4. Lipid Extraction

All biological replicates were sourced from 12 crabs of the counterpart species, and a triplicate analysis was performed for each crab species. Total lipids were extracted using the methods described by Yang et al. with minor modifications [49]. Briefly, the freeze-dried edible viscera and muscle tissues were weighed and homogenized in 10 mL of methanol in a homogenizer (Bead Mill 4, Fisherbrand, Tokyo, Japan) for 30 s (×2 cycles). Subsequently, 10 mL of the homogenate (1 g) was transferred to a beaker, and 20 mL of chloroform was added. The mixture was placed in an ultrasonic bath and processed for 30 min at 20 °C. Then, the mixture was centrifuged at 4000× *g* for 10 min. The supernatant was transferred to a clean bottle, and the lower phase was re-extracted twice with chloroform:methanol (2:1, *v/v*). Finally, the organic solvents were combined and evaporated under a vacuum rotary evaporator (40 °C, 350 kPa). The residue was further dried using a gentle stream of nitrogen. Dried lipid extracts were weighted and stored at −20 °C for further analysis by UHPLC-HRAM/MS. All analyses were completed within two weeks after extraction.

3.5. Lipid Identification and Semi-Quantification

Lipid molecular species were identified with MS-DIAL software (version 4.48) and Xcalibur 4.0 software (ThermoFisher, Waltham, MA, USA), as well as the LipidMaps and LipidBank websites. False positive data were manually excluded based on database matching and fragmentation information.

Seven PL standards, including PC 17:0/17:0 (standard for PC and lysophophatidylcholine (LPC)), PE 17:0/17:0 (standard for PE and lysophosphatidylethanolamine (LPE)), PG 17:0/17:0 (standard for PG, lysophosphatidylglycerol (LPG) and cardiolipin (CL)), PA 17:0/17:0 (standard for PA and lysophosphatidic acid (LPA)), SM d18:1/12:0 (standard for SM), PI 8:0/8:0 (standard for PI and lysophosphatidylinositol (LPI)) and TAG 18:2/18:2/18:2 (standard for TAG and diacylglycerol (DAG)), were used for the construction of the calibration curves. These standards were quantitatively dissolved in a chloroform:methanol (2:1, *v/v*) mixture as stock solutions and then diluted into a series of concentrations, with a constant concentration of internal standards from SPLASH LIPIDOMIX (as displayed in Table S1 of the Supplementary Materials), including PC 15:0/18:1(d7), PG 15:0/18:1(d7), PE 15:0/18:1(d7), TAG 15:0/18:1(d7)/15:0, PI 15:0/18:1(d7), SM d18:1/18:1(d9) and PA 15:0/18:1(d7). The internal standard method was applied to quantify the lipid concentrations of crab samples [50]. Briefly, the same concentration of internal standards was added to each crab lipid, and the samples were injected for detection. The crab lipids were quantified according to the calibration curves of the peak area ratio of the standard and internal standard within the same class so the concentration (mg/mL) of lipids could be obtained. Subsequently, the concentration (nmol/g) of each lipid molecule in a dried biological sample was available by calculation. Furthermore, all molecular quantifications were based on the EICs of individual lipids, with *m/z* expansion at ±5 ppm, and all crab samples were analyzed with three duplicate samples.

3.6. UHPLC-HRAM/MS Analysis

3.6.1. UHPLC Conditions

The total lipids of the different crabs were identified using a UHPLC system (Dionex, UltiMate, 3000 RSLC) coupled with QE Orbitrap/MS (ThermoFisher Scientific, Bremen, Germany). A BEH-HILIC column (100 × 1.0 mm, 1.7 μm, Sigma–Aldrich/Supelco, Bellefonte, PA, USA) was used for chromatographic separation. The flow rate was 0.1 mL /min, and the column temperature was 40 °C. Eluent A and eluent B were 5 mM ammonium formate in water and in acetonitrile, respectively. The chromatographic gradient elution mode was as follows: 0 min, 5% B; 4 min, 5% B; 10 min, 40% B; 15 min, 40% B; 16 min, 5% B; and 20 min, 5% B. The injection volume was 2 μL.

3.6.2. Quadrupole-Exactive High-Resolution Accurate Mass Spectrometry

The desolvation ESI source parameters were set as follows—electrospray voltage: 3.2 kV; a sheath gas flow rate: 35 arbitrary units (arb unit); auxiliary gas flow rate: 10 arbs; capillary temperature: 325 °C; heater temperature: 350 °C; collision energy: 35 eV; dynamic exclusion: 10 s; and isolation window: 3.0 *m/z*. The data were acquired in the positive ionization mode (*m/z* 120–1800) and negative ionization mode (*m/z* 120–1800) with dependent MS/MS acquisition. The resolutions of the full-scan spectra and the fragment spectra were 140,000 and 70,000, respectively. All lipid identifications were determined by MS and MS/MS, with an MS mass error of <5 ppm and MS/MS mass error of <8 ppm.

To investigate the background contamination and the method's validation, quality control procedures were applied to this experiment, which contained solvent blanks, solvent spiked with internal standards and matrix blanks without spiking the internal standards [51]. To test the stability, the RSD values of the peak areas of the internal standards in three parallel samples were calculated. The LOD and LOQ were the concentrations of the standards corresponding to the signal-to-noise (S/N) values of 3 and 10 [52].

3.7. *Statistical Analysis*

The data are expressed as the mean ± SD. SPSS software (version 24.0) was used for the one-way analysis of variance, and $p < 0.05$ was considered statistically significant. Lipid molecule species were identified and quantified with MS-DIAL software (version 4.48) and Xcalibur 4.0 software (ThermoFisher, Waltham, MA, USA), as well as the LipidMaps and LipidBank websites. MetaboAnalyst 5.0 was used for the cluster analysis, and SIMCA 14.1 was used for the PCA, PLS-DA and OPLS-DA.

4. Conclusions

This study applied the UHPLC-HRAM/MS approach to discriminate the lipid molecules in the muscles and edible viscera of five kinds of crabs, including *P. camtschaticus*, *E. sinensis*, *C. magister*, *P. trituberculatus* and *C. pagurus*. Combined with chemometric analysis, 14 lipid subclasses in muscle tissue and 13 lipid subclasses in edible viscera were detected and quantified. *C. pagurus* had the highest DHA/EPA-PL molecule content in the muscle and edible viscera tissue of five crab species. Our work can help researchers better understand the lipid composition of crab muscle and edible visceral tissues and better select or assess their nutritional values.

In addition, we developed a new way to distinguish two kinds of crabs with similar appearances (*C. magister* and *C. pagurus*) based on lipidomics. In the muscle samples, twenty characteristic lipids were screened, such as PE 18:0/20:5, PC 16:0/16:1, PE P-18:0/22:6 and SM 12:1; 2O/20:0, as the basis for distinguishing *C. magister* and *C. pagurus*. Seventeen characteristic lipids, such as PE P-18:0/22:6, PC 16:1/18:1, PC 18:1/18:1 and PE P-18:0/20:5, were screened from the edible visceral samples. However, our work focused on the differences in the crab varieties, while some factors that might affect the lipid composition, such as maturity and living environment, were not considered. Therefore, the effects of various factors on the lipid composition of crabs will be investigated in further study.

Supplementary Materials: The following supporting information can be downloaded at: https://www.mdpi.com/article/10.3390/molecules28093653/s1, Figure S1: RSD distribution of the peak area of seven internal standards in 10 groups of different samples; Figure S2: Lipid composition (%) in the muscles and edible viscera of five crab species; Figure S3: The percentage contents of DHA/EPA-PLs and DHA/EPA-TAGs in the total lipids extracted from crab muscle and edible viscera. The data are expressed as the mean ± SD. Different letters indicate a significant difference ($p < 0.05$) among the different samples; Figure S4: Boxplots of the significantly different lipids in the muscles (A) and edible viscera (B) of five crab species; Figure S5: Optical images of the crab species; Table S1: Concentration (ng/mL) of 7 lipid standards and corresponding internal standard (IS) for the construction of calibration curves; Table S2: Total lipids content in five kinds of crabs; Table S3: The linear regression equations, linear regression coefficients (R^2), LOD and LOQ of lipid standards; Table S4: Molecular species of lipids extracted from the muscles of five kinds of crabs;

Table S5: Molecular species of lipids extracted from the edible viscera of five kinds of crabs; Table S6: Concentrations of the different lipid subclasses in the five kinds of crab muscles (nmol/g); Table S7: The concentrations of different lipid subclasses of the five kinds of edible viscera (nmol/g); Table S8: The significantly different lipids in the muscle and edible viscera of five crab species.

Author Contributions: J.Y. and J.Z. were responsible for the experimental study, data analysis and manuscript writing; M.Z. was responsible for the revision of the manuscript; L.Z. and E.M. were responsible for the overall design, planning of the project and revision of the manuscript. All authors have read and agreed to the published version of the manuscript.

Funding: This work was supported by grants from the National Natural Science Foundation of China (31501521) and the Natural Science Foundation of Hubei Province (2022CFB258). The authors thank the Analytical and Measuring Center, School of Pharmaceutical Sciences, South-Central Minzu University, for the spectral analyses.

Institutional Review Board Statement: Animal use and care were confirmed by the IAEC of South-Central Minzu University (SYXK (Wuhan) 2016-0089, no: 2021-SCUEC-AEC-033), Wuhan, China.

Informed Consent Statement: Not applicable.

Data Availability Statement: The data that support the findings of this study are available on request from the corresponding author.

Conflicts of Interest: The authors declare no conflict of interest.

Sample Availability: Not applicable.

References

1. Yang, F.; Guo, H.; Gao, P.; Yu, D.; Xu, Y.; Jiang, Q.; Yu, P.; Xia, W. Comparison of methodological proposal in sensory evaluation for Chinese mitten crab (*Eriocheir sinensis*) by data mining and sensory panel. *Food Chem.* **2021**, *356*, 129698. [CrossRef] [PubMed]
2. Lu, T.; Shen, Y.; Cui, G.; Yin, F.; Yu, Z.; Zhou, D. Detailed Analysis of Lipids in Edible Viscera and Muscles of Cooked Crabs *Portunus trituberculatus* and *Portunus pelagicus*. *J. Aquat. Food Prod. Technol.* **2020**, *29*, 391–406. [CrossRef]
3. Barrento, S.; Marques, A.; Teixeira, B.; Anacleto, P.; Vaz-Pires, P.; Nunes, M.L. Effect of season on the chemical composition and nutritional quality of the edible crab Cancer pagurus. *J. Agric. Food Chem.* **2009**, *57*, 10814–10824. [CrossRef]
4. Nanda, P.K.; Das, A.K.; Dandapat, P.; Dhar, P.; Bandyopadhyay, S.; Dib, A.L.; Lorenzo, J.M.; Gagaoua, M. Nutritional aspects, flavour profile and health benefits of crab meat based novel food products and valorisation of processing waste to wealth: A review. *Trends Food Sci. Technol.* **2021**, *112*, 252–267. [CrossRef]
5. Narayanasamy, A.; Balde, A.; Raghavender, P.; Shashanth, D.; Abraham, J.; Joshi, I.; Nazeer, R.A. Isolation of marine crab (*Charybdis natator*) leg muscle peptide and its anti-inflammatory effects on macrophage cells. *Biocatal. Agric. Biotechnol.* **2020**, *25*, 101577. [CrossRef]
6. Venugopal, V.; Gopakumar, K. Shellfish: Nutritive Value, Health Benefits, and Consumer Safety. *Compr. Rev. Food Sci. Food Saf.* **2017**, *16*, 1219–1242. [CrossRef] [PubMed]
7. Yeo, D.C.J.; Ng, P.K.L.; Cumberlidge, N.; Magalhães, C.; Daniels, S.R.; Campos, M.R. Global diversity of crabs (Crustacea: Decapoda: Brachyura) in freshwater. *Hydrobiologia* **2007**, *595*, 275–286. [CrossRef]
8. Ng, P.K.L. Collecting and processing freshwater shrimps and crabs. *J. Crust. Biol.* **2017**, *37*, 115–122. [CrossRef]
9. Ferdoushi, Z.; Zhang, X.G.; Hasan, M.R. Mud crab (*Scylla* sp.) marketing system in Bangladesh. *Asian J. Food Agro-Ind.* **2010**, *3*, 248–265.
10. Yao, Y.; Ding, L.; Huang, X. Diverse Functions of Lipids and Lipid Metabolism in Development. *Small Methods* **2020**, *4*, 1900564. [CrossRef]
11. Welte, M.A.; Gould, A.P. Lipid droplet functions beyond energy storage. *Biochim. Biophys. Acta Mol. Cell. Biol. Lipids* **2017**, *1862 Pt B*, 1260–1272. [CrossRef]
12. Casas-Agustench, P.; Cherubini, A.; Andres-Lacueva, C. Lipids and physical function in older adults. *Curr. Opin. Clin. Nutr. Metab. Care* **2017**, *20*, 16–25. [CrossRef] [PubMed]
13. Hyötyläinen, T.; Bondia-Pons, I.; Orešič, M. Lipidomics in nutrition and food research. *Mol. Nutr. Food Res.* **2013**, *57*, 1306–1318. [CrossRef] [PubMed]
14. Castro-Alves, V.; Orešič, M.; Hyötyläinen, T. Lipidomics in nutrition research. *Curr. Opin. Clin. Nutr. Metab. Care* **2022**, *25*, 311–318. [CrossRef] [PubMed]
15. Imbs, A.B.; Ermolenko, E.V.; Grigorchuk, V.P.; Sikorskaya, T.V.; Velansky, P.V. Current Progress in Lipidomics of Marine Invertebrates. *Mar. Drugs* **2021**, *19*, 660. [CrossRef] [PubMed]
16. Yuan, Y.; Xu, F.; Jin, M.; Wang, X.; Hu, X.; Zhao, M.; Cheng, X.; Luo, J.; Jiao, L.; Betancor, M.; et al. Untargeted lipidomics reveals metabolic responses to different dietary n-3 PUFA in juvenile swimming crab (*Portunus trituberculatus*). *Food Chem.* **2021**, *354*, 129570. [CrossRef]

17. Ding, Z. Lipid metabolism disorders contribute to the pathogenesis of *Hepatospora eriocheir* in the crab *Eriocheir sinensis*. *J. Fish Dis.* **2021**, *44*, 305–313. [CrossRef]
18. Cherif, S.; Frikha, F.; Gargouri, Y.; Miled, N. Fatty acid composition of green crab (*Carcinus mediterraneus*) from the Tunisian mediterranean coasts. *Food Chem.* **2008**, *111*, 930–933. [CrossRef]
19. Nguyen, T.P.L.; Nguyen, V.T.A.; Do, T.T.T.; Nguyen Quang, T.; Pham, Q.L.; Le, T.T. Fatty Acid Composition, Phospholipid Molecules, and Bioactivities of Lipids of the Mud Crab *Scylla paramamosain*. *J. Chem.* **2020**, *2020*, 1–9. [CrossRef]
20. Zhang, J.; Wu, X.; Qiu, J.; Zhang, L.; Zhang, Y.; Qiu, X.; Huang, Z.; Xu, W. Comprehensive Comparison on the Chemical Profile of Guang Chen Pi at Different Ripeness Stages Using Untargeted and Pseudotargeted Metabolomics. *J. Agric. Food Chem.* **2020**, *68*, 8483–8495. [CrossRef]
21. Dvoretsky, A.G.; Bichkaeva, F.A.; Baranova, N.F.; Dvoretsky, V.G. Fatty acid composition of the Barents Sea red king crab (*Paralithodes camtschaticus*) leg meat. *J. Food Compost. Anal.* **2021**, *98*, 103826. [CrossRef]
22. Wang, T.; Xiao, X.; Regenstein, J.M.; Wu, W.; Zhou, Y.; Wang, S.; Cheng, Y.; Wu, X.; Bao, B. Effect on lipid metabolism of mice continuously fed a crab-containing diet. *Food Biosci.* **2019**, *30*, 100422. [CrossRef]
23. Stoner, A.W.; Ottmar, M.L.; Copeman, L.A. Temperature effects on the molting, growth, and lipid composition of newly-settled red king crab. *J. Exp. Mar. Biol. Ecol.* **2010**, *393*, 138–147. [CrossRef]
24. Wu, H.; Ge, M.; Zhou, X.; Jiang, S.; Lin, L.; Lu, J. Nutritional qualities of normal and precocious adult male Chinese mitten crabs (*Eriocheir sinensis*). *Aquac. Res.* **2019**, *50*, 2267–2275. [CrossRef]
25. Han, T.; Wang, J.; Hu, S.; Li, X.; Jiang, Y.; Wang, C. Effects of different dietary lipid sources on growth performance and tissue fatty acid composition of juvenile swimming crab *Portunus trituberculatus*. *Chin. J. Oceanol. Limnol.* **2015**, *33*, 957–965. [CrossRef]
26. Liu, F.; Ren, D.; Guo, D.A.; Pan, Y.; Zhang, H.; Hu, P. Method Development for Gypenosides Fingerprint by High Performance Liquid Chromatography with Diode-Array Detection and the Addition of Internal Standard. *Chem. Pharm. Bull.* **2008**, *56*, 389–393. [CrossRef]
27. Huang, C.; Li, Y.; Wang, K.; Xi, J.; Xu, Y.; Si, X.; Pei, D.; Lyu, S.; Xia, G.; Wang, J.; et al. Analysis of lipidomics profile of Carya cathayensis nuts and lipid dynamic changes during embryonic development. *Food Chem.* **2022**, *370*, 130975. [CrossRef]
28. Wang, F.; Lin, W.; Lv, S.; Jiang, S.; Lin, L.; Lu, J. Comparison of Lipids Extracted by Different Methods from Chinese Mitten Crab (*Eriocheir sinensis*) Hepatopancreas. *J. Food Sci.* **2019**, *84*, 3594–3600. [CrossRef]
29. Jobling, M.; Johansen, S.; Foshaug, H.; Burkow, I.C.; JøRgensen, E.H. Lipid dynamics in anadromous Arctic charr, Salvelinus alpinus (L.): Seasonal variations in lipid storage depots and lipid class composition. *Fish Physiol. Biochem.* **1998**, *18*, 225–240. [CrossRef]
30. Wang, W.; Wu, X.; Liu, Z.; Zheng, H.; Cheng, Y. Insights into hepatopancreatic functions for nutrition metabolism and ovarian development in the crab *Portunus trituberculatus*: Gene discovery in the comparative transcriptome of different hepatopancreas stages. *PLoS ONE* **2014**, *9*, e84921. [CrossRef]
31. Lu, Z.; Shi, C.; Liu, L.; Mu, C.; Ye, Y.; Wang, C. Phospholipid Compositions in *Portunus trituberculatus* Larvae at Different Developmental Stages. *J. Ocean. Univ. China* **2022**, *21*, 152–162. [CrossRef]
32. Zhang, Y.; Zhang, M.; Dong, L.; Chang, J.; Wang, H.; Shen, Q. Lipidomics Screening of Polyunsaturated Phospholipid Molecular Species in Crab (*Portunus trituberculatus*) Muscular Tissue: A Nontarget Approach by HILIC-MS. *Eur. J. Lipid Sci. Technol.* **2021**, *124*, 2100097. [CrossRef]
33. Suprayudi, M.A.; Takeuchi, T.; Hamasaki, K. Phospholipids Effect on Survival and Molting Synchronicity of Larvae Mud Crab Scylla serrata. *HAYATI J. Biosci.* **2012**, *19*, 163–168. [CrossRef]
34. Xu, H.; Wang, J.; Han, T.; Li, X.; Zheng, P.; Yin, F.; Wang, C. Effects of dietary phospholipids levels on growth performance, lipid metabolism, and antioxidant capacity of the early juvenile green mud crab, *Scylla paramamosain* (Estampador). *Aquac. Res.* **2018**, *50*, 513–520. [CrossRef]
35. Fitzpatrick, J. Subjectivity and objectivity: Polanyi and Lonergan. *High. Educ. Q.* **1982**, *36*, 183–195. [CrossRef]
36. Weiser, M.J.; Butt, C.M.; Mohajeri, M.H. Docosahexaenoic Acid and Cognition throughout the Lifespan. *Nutrients* **2016**, *8*, 99. [CrossRef] [PubMed]
37. Hossain, Z.; Hosokawa, M.; Takahashi, K. Growth inhibition and induction of apoptosis of colon cancer cell lines by applying marine phospholipid. *Nutr. Cancer* **2009**, *61*, 123–130. [CrossRef]
38. Zhang, L.Y.; Ding, L.; Shi, H.H.; Xu, J.; Xue, C.H.; Zhang, T.T.; Wang, Y.M. Eicosapentaenoic acid in the form of phospholipids exerts superior anti-atherosclerosis effects to its triglyceride form in ApoE(-/-) mice. *Food Funct.* **2019**, *10*, 4177–4188. [CrossRef]
39. Zhang, T.T.; Xu, J.; Wang, Y.M.; Xue, C.H. Health benefits of dietary marine DHA/EPA-enriched glycerophospholipids. *Prog. Lipid Res.* **2019**, *75*, 100997. [CrossRef]
40. Donovan, E.L.; Pettine, S.M.; Hickey, M.S.; Hamilton, K.L.; Miller, B.F. Lipidomic analysis of human plasma reveals ether-linked lipids that are elevated in morbidly obese humans compared to lean. *Diabetol. Metab. Syndr.* **2013**, *5*, 24. [CrossRef]
41. Rangholia, N.; Leisner, T.M.; Holly, S.P. Bioactive Ether Lipids: Primordial Modulators of Cellular Signaling. *Metabolites* **2021**, *11*, 41. [CrossRef] [PubMed]
42. Aboshi, T.; Nishida, R.; Mori, N. Identification of plasmalogen in the gut of silkworm (*Bombyx mori*). *Insect Biochem. Mol. Biol.* **2012**, *42*, 596–601. [CrossRef] [PubMed]
43. Zemski Berry, K.A.; Murphy, R.C. Electrospray ionization tandem mass spectrometry of glycerophosphoethanolamine plasmalogen phospholipids. *J. Am. Soc. Mass. Spectrom.* **2004**, *15*, 1499–1508. [CrossRef] [PubMed]

44. Liu, Z.Y.; Zhou, D.Y.; Wu, Z.X.; Yin, F.W.; Zhao, Q.; Xie, H.K.; Zhang, J.R.; Qin, L.; Shahidi, F. Extraction and detailed characterization of phospholipid-enriched oils from six species of edible clams. *Food Chem.* **2018**, *239*, 1175–1181. [CrossRef]
45. Wiklund, S.; Johansson, E.; Sjöström, L.; Mellerowicz, E.J.; Edlund, U.; Shockcor, J.P.; Gottfries, J.; Moritz, T.; Trygg, J. Visualization of GC/TOF-MS-Based Metabolomics Data for Identification of Biochemically Interesting Compounds Using OPLS Class Models. *Anal. Chem.* **2008**, *80*, 115–122. [CrossRef]
46. Liu, H.; Hui, T.; Zheng, X.; Li, S.; Wei, X.; Li, P.; Zhang, D.; Wang, Z. Characterization of key lipids for binding and generating aroma compounds in roasted mutton by UPLC-ESI-MS/MS and Orbitrap Exploris GC. *Food Chem.* **2022**, *374*, 131723. [CrossRef]
47. He, C.; Cao, J.; Bao, Y.; Sun, Z.; Liu, Z.; Li, C. Characterization of lipid profiling in three parts (muscle, head and viscera) of tilapia (*Oreochromis niloticus*) using lipidomics with UPLC-ESI-Q-TOF-MS. *Food Chem.* **2021**, *347*, 129057. [CrossRef]
48. Roth, B.; Ines, S. Stunning and killing of edible crabs (*Cancer pagurus*). *Anim. Welf.* **2010**, *19*, 287–294. [CrossRef]
49. Yang, F.; Zhao, M.; Zhou, L.; Zhang, M.; Liu, J.; Marchioni, E. Identification and Differentiation of Wide Edible Mushrooms Based on Lipidomics Profiling Combined with Principal Component Analysis. *J. Agric. Food Chem.* **2021**, *69*, 9991–10001. [CrossRef]
50. Song, S.; Cheong, L.Z.; Wang, H.; Man, Q.Q.; Pang, S.J.; Li, Y.Q.; Ren, B.; Wang, Z.; Zhang, J. Characterization of phospholipid profiles in six kinds of nut using HILIC-ESI-IT-TOF-MS system. *Food Chem.* **2018**, *240*, 1171–1178. [CrossRef]
51. Sun, F.; Chen, H.; Chen, D.; Tan, H.; Huang, Y.; Cozzolino, D. Lipidomic Changes in Banana (*Musa cavendish*) during Ripening and Comparison of Extraction by Folch and Bligh-Dyer Methods. *J. Agric. Food Chem.* **2020**, *68*, 11309–11316. [CrossRef] [PubMed]
52. Shi, C.; Guo, H.; Wu, T.; Tao, N.; Wang, X.; Zhong, J. Effect of three types of thermal processing methods on the lipidomics profile of tilapia fillets by UPLC-Q-Extractive Orbitrap mass spectrometry. *Food Chem.* **2019**, *298*, 125029. [CrossRef] [PubMed]

molecules

Article

Histamine and Tyramine in Chihuahua Cheeses during Shelf Life: Association with the Presence of tdc and hdc Genes

Eduardo Campos-Góngora *, María Teresa González-Martínez †, Abad Arturo López-Hernández, Gerardo Ismael Arredondo-Mendoza, Ana Sofía Ortega-Villarreal and Blanca Edelia González-Martínez *

Universidad Autónoma de Nuevo León, Centro de Investigación en Nutrición y Salud Pública, Facultad de Salud Pública y Nutrición, Monterrey 64460, Nuevo León, México
* Correspondence: eduardo.camposg@uanl.mx (E.C.-G.); blanca.gonzalezma@uanl.mx (B.E.G.-M.)
† This paper is written in loving memory of Ma. Teresa González Martínez, a great human being and skillful researcher.

Abstract: Cheese is a product of animal origin with a high nutritional value, and it is one of the most consumed dairy foods in Mexico. In addition, Chihuahua cheese is the most consumed matured cheese in Mexico. In the production process of Chihuahua cheese, maturation is carried out by adding acid lactic microorganisms, mainly of the *Lactococcus* genus and, in some cases, also the *Streptococcus* and *Lactobacillus* genus. As part of the metabolism of fermenting microorganisms, biogenic amines can develop in matured foods, which result from the activity of amino decarboxylase enzymes. In cheeses, histamine and tyramine are the main amines that are formed, and the consumption of these represents a great risk to the health of consumers. In this work, the presence of biogenic amines (histamine and tyramine) was determined by HPLC at different times of the shelf life of Chihuahua cheeses. In addition, the presence of genes hdc and tdc that code for the enzymes responsible for the synthesis of these compounds (histidine and tyrosine decarboxylase, or HDC and TDC) was determined by molecular techniques. A significant correlation was observed between the presence of both histamine and tyramine at the end of shelf life with the presence of genes that code for the enzymes responsible for their synthesis.

Keywords: Chihuahua cheese; histamine; tyramine; hdc and tdc genes; HDC and TDC proteins

Citation: Campos-Góngora, E.; González-Martínez, M.T.; López-Hernández, A.A.; Arredondo-Mendoza, G.I.; Ortega-Villarreal, A.S.; González-Martínez, B.E. Histamine and Tyramine in Chihuahua Cheeses during Shelf Life: Association with the Presence of tdc and hdc Genes. *Molecules* **2023**, *28*, 3007. https://doi.org/10.3390/molecules28073007

Academic Editor: Andrea Salvo

Received: 25 February 2023
Revised: 14 March 2023
Accepted: 22 March 2023
Published: 28 March 2023

1. Introduction

Cheese is a dairy product with a high nutritional value. Since it is one of the most consumed foods of animal origin, it is part of the basic food basket in several Latin American countries. In Mexico, a wide variety of cheeses from different regions of the country (Panela, Oaxaca, Cotija, Asadero, Chihuahua, Sierra, etc.) are consumed, with fresh cheeses being the most consumed. Within the group of matured cheeses, the most consumed is Chihuahua or Mennonite cheese, so-called because it was originally made by this resident community of the state of Chihuahua. Although this type of cheese originated in Chihuahua, it is currently prepared by different companies throughout the country, and it is also exported to the United States to meet the demands of the Hispanic community [1,2].

In the dairy industry, the starter cultures used to produce fermented products (cheese, cottage cheese, butter, kefir, yogurt, etc.) are microorganisms that are used to ferment glucose, transforming it into lactic acid. The microorganisms that are most used in these processes are lactic acid bacteria (LAB), whose main characteristic is that they cause a decrease in pH, which inhibits the growth of other microorganisms in fermented food, also promoting the coagulation of casein [3]. Starter cultures grow from the beginning of inoculation until reaching very high cell densities (10^8–10^9 CFU/mL) in the first hours of fermentation. Subsequently, there is a gradual decrease throughout the cheese maturation process. Although starter cultures can be made up of different types of microorganisms, mesophilic bacteria cultures of the genera *Lactococcus*, *Lactobacillus*, and *Leuconostoc*, or thermophilic species of *Lactobacillus*, such as *L. delbrueckii*, *L. helveticus*, or *Streptococcus thermophilus*, are generally used for the maturation of cheeses [4,5].

During cheese maturation, LABs are not only responsible for the production of lactic acid, but they also participate in proteolysis by hydrolyzing polypeptide chains and releasing amino acids that serve as precursors for the formation of volatile compounds largely responsible for the organoleptic properties of cheese [6]. The amino acids produced are catabolized by an initial pathway in which an aminotransferase enzyme acts, followed by two catabolic pathways where deamination or decarboxylation occurs. While deamination generates aromatic products due to the action of dehydrogenases (α-ketoacid and ammonia) or oxidases (aldehyde and ammonia), by-products of decarboxylation of biogenic amines can be generated (histamine, tyramine, putrescine, cadaverine, etc.). Some of these compounds can cause adverse physiological effects in susceptible consumers [7–9].

The activity of decarboxylase enzymes depends on the genetic potential, as well as other factors such as the pH of food (slightly acidic), availability of free amino acids, presence of fermentable carbohydrates (glucose), presence of pyridoxal phosphate (cofactor of the decarboxylation reaction), maturation temperatures, humidity, and a low concentration of salts, which favors the action of bacteria with amino decarboxylase enzymes [10–12]. Although temperatures above 15 °C favor the development of microorganisms and, therefore, of biogenic amines, the activity of the amino decarboxylase enzyme has been reported at temperatures of 4 °C [13–16].

The formation of histidine and tyramine (the biogenic amines that present a greater health risk) during the metabolism of LABs used in the elaboration of cheese is due to the presence of histidine decarboxylase and tyrosine decarboxylase, the enzymes responsible for their synthesis encoded by the hdc and tdc genes, respectively. Under normal conditions, the concentrations of biogenic amines in food do not present a health risk because the digestive system of the human organism has an efficient mechanism to eliminate these molecules through the enzymes monoamine oxidase (MAO) and diamine oxidase (DAO). However, under specific conditions, such as high concentrations of amines or in individuals that present deficiency or inhibition of these enzymes, there may be some risk of adverse reactions [17]. Symptoms of histamine poisoning occur in manifestations of the digestive system, such as nausea, vomiting, and diarrhea, in addition to hypotension, edema, tachycardia, headaches, and migraines [18,19].

Fermented cheeses have also been linked to severe hypertensive crises in patients who take medications that inhibit the enzyme monoamine oxidase (MAO). The amine related these events is tyramine, therefore, these clinical manifestations are called "cheese reactions" [20,21]. Although the symptoms appear from within a few minutes to three hours of ingestion of the contaminated product, it has been described that its duration depends, in large part on: (1) The physiological condition of the patient, reestablishing in hours or up to several days; and (2) The concentrations of histamine and other amines ingested with food [22]. On the other hand, it has been reported that an intake of 5–10 mg of histamine can be considered a risk factor for sensitive people, from 10 to 100 mg is considered a tolerable limit, from 100 to 1000 mg produces a medium intoxication, and amounts greater than 1000 mg produce severe intoxication [23]. Despite the toxic effects related to the biogenic amines, histidine and tyramine, in Mexico, there is no regulation of the allowable limits of these amines in cheeses. It is noteworthy that the production of biogenic amines is carried out mainly during the maturation process of Chihuahua cheese, which covers an approximate period of two months at temperatures between 4 and 27 °C. The production of these compounds continues during the storage of the cheese, usually at refrigeration, between 10 and 20 °C [24,25].

The aim of this study was to evaluate the concentration of histamine and tyramine using High-Performance Liquid Chromatography (HPLC) in Chihuahua cheeses at different times of their shelf life, and to determine the presence of the genes that code for the enzymes responsible for the synthesis of these compounds in cheese samples using molecular techniques (Polymerase Chain Reaction: PCR).

2. Results

2.1. Microbiological Quality of the Cheese Samples

Results corresponding to the microbiological quality of cheeses are presented in Table 1. As expected in matured products, all the cheeses have a high LAB content (ranging from 7.6 to 9 Log CFU/g). The total count of aerobic mesophilic bacteria was found between 3.1 and 3.8 Log CFU/g, and the total coliform bacteria was between 2.6 and 3.4 Log CFU/g. Statistical analysis showed a significant difference between the different brands.

Table 1. Microbiological quality of Chihuahua cheeses.

Cheeses	Aerobic Mesophilic Log CFU/g	Total Coliforms Log CFU/g	LAB Log UFC/g
A	3.7 ± 0.03^{a}	3.3 ± 0.01^{c}	9.0 ± 0.02^{a}
B	3.1 ± 0.02^{b}	3.1 ± 0.02^{a}	8.3 ± 0.04^{c}
C	3.1 ± 0.01^{b}	3.3 ± 0.02^{c}	8.3 ± 0.04^{c}
D	3.6 ± 0.01^{a}	3.3 ± 0.01^{c}	8.6 ± 0.04^{c}
E	3.7 ± 0.02^{a}	2.6 ± 0.01^{b}	7.6 ± 0.03^{e}
F	3.4 ± 0.03^{c}	3.3 ± 0.01^{cd}	8.8 ± 0.04^{b}
G	3.8 ± 0.02^{e}	3.4 ± 0.02^{d}	7.8 ± 0.03^{d}
H	3.8 ± 0.02^{e}	3.3 ± 0.03^{cd}	7.8 ± 0.03^{d}

Capital letters represent each one of included cheeses. Values correspond to average ± standard deviation of Log CFU/g in three independent experiments. Different superscripts $^{a–e}$ indicate significant differences between cheese brands ($p < 0.05$). Log CFU/g: Logarithm of Colony Forming Unit per gram. LAB: Lactic Acid Bacteria.

At the end of the shelf life, the microorganisms recognized as histamine and tyramine producers were identified by API biochemical system. The identification of bacteria was performed up to the species level; the results are presented in Table 2. *L. pentosus* was found in three different brands, while *L. rhamnosus* was found in two brands.

Table 2. Biochemical identification of strains recognized as histamine and tyramine producers in the Chihuahua cheeses.

Cheeses	HDC	TDC
A	*L. pentosus*	*L. pentosus*, *L. plantarum*
B	ND	ND
C	ND	*L. pentosus*
D	ND	*L. lactis sub lactis*
E	ND	ND
F	ND	*L. rhamnosus*
G	*L. pentosus*	*L. rhamnosus*
H	ND	ND

Identification of bacteria based on API biochemical tests at the end of the shelf life. Capital letters represent each different included cheese. HDC: histidine decarboxylase; TDC: tyrosine decarboxylase; ND: bacteria were not detected.

2.2. Histamine and Tyramine Quantification

The biogenic amine quantification was carried out by the HPLC method with a fluorescence detector (FLD). The retention times for histamine and tyramine were 15 and 30 min, respectively. Calibration curves carried out with the standards showed linearity with different concentrations (25, 50, 100, 250, and 400 mg/L) of compounds. A correlation coefficient of 0.993 was determined for histamine and 0.987 for tyramine. Based on these values, a limit of detection (LOD) of 25 mg/L was determined for both amines under the experimental conditions of this study.

HPLC analysis was performed at three moments in the shelf life of cheeses (Table 3). In the first week of its acquisition (beginning of the shelf's life), the presence of histamine was not detected in any of the cheeses, while tyramine was present in 37.5% of the cheeses (Cheeses B, C, and F). Tyramine presence reached concentrations ranging from 34–122 mg/kg. At half of shelf life, the presence of histamine was detected only in one of the cheeses (12.5% of cheeses), while tyramine was detected in 62.5% of the cheeses (five out of eight). Interestingly, it was observed that the concentration of tyramine in Cheeses B, C, and F (in which tyramine was detected from the first analysis) increased significantly to reach values of 160, 160, and 139 mg/kg, respectively. At this stage, in Cheese A, both histamine and tyramine were detected at concentrations between 45 mg/kg and 59 mg/kg, respectively.

Table 3. Detection of biogenic amines in Chihuahua cheeses.

Cheese	Histamine (mg/kg) Shelf Life			Tyramine (mg/kg) Shelf Life		
	Start	Half	End	Start	Half	End
A	ND	45 ± 4	192 ± 20	ND	59 ± 15	115 ± 10
B	ND	ND	ND	122 ± 12	160 ± 18	209 ± 2
C	ND	ND	ND	34 ± 2	160 ± 17	166 ± 6
D	ND	ND	64 ± 11	ND	147 ± 4	194 ± 24
E	ND	ND	ND	ND	ND	ND
F	ND	ND	ND	42 ± 8	139 ± 9	150 ± 1
G	ND	ND	46 ± 6	ND	ND	205 ± 4
H	ND	ND	ND	ND	ND	ND

Biogenic amine detection was performed by HPLC at three moments in the shelf life of cheeses. Capital letters represent each one of included cheese. Values correspond to average ± standard deviation of amines concentration (mg/kg) determined in three independent experiments. mg/kg: correspond to amines quantity by kilogram of cheese.

At the end of the shelf life (approximately four months after its acquisition) histamine was detected in three of the eight cheeses (A, D, and G), which corresponds to 37.5% of them. In Cheese A, histamine concentration increased by more than 400% (from 45 to 192 mg/kg). On the other hand, tyramine was detected in 75% of cheeses (six out of eight) in concentrations from 115 mg/kg to 209 mg/kg.

It is noteworthy that while the concentration of tyramine increased in Samples B, C, D, and F over time, in Cheeses E and H, the presence of histamine and tyramine was not detected at the different storage times analyzed.

2.3. *Detection of hdc and tdc Genes in Cheese Samples*

For the determination of bacterial DNA in the cheese samples, the first-stage PCR amplification was performed using the specific oligonucleotides PO (GAGAGTTTGATC-CTGGCTCAG) and 338-F (GCTGCCTCCCGTAGGAGT), which were reported by Ventura, et al., (2001) [26]. These oligonucleotides allow for the amplification of a 332 base-pair fragment corresponding to the gene that encodes the 16S ribosomal subunit of eubacteria. In all the cheeses analyzed, the expected PCR products were obtained (data not shown), indicating the presence of bacterial DNA in all cheese samples.

For the specific analysis of hdc gene presence, specific oligonucleotides (HDC1: TTGACCGTATCTCAGTGAGTCCAT and HDC2: ACGGTCATACGAAACAATACCATC), which were designed and reported by Fernandez et al., (2006) [27], were used. In addition, the obtained PCR products (bands of 174 bp) were found only in three of the samples (Cheeses A, D, and G), showing that 37.5% of cheeses analyzed contain bacteria able to produce HDC (Figure 1A).

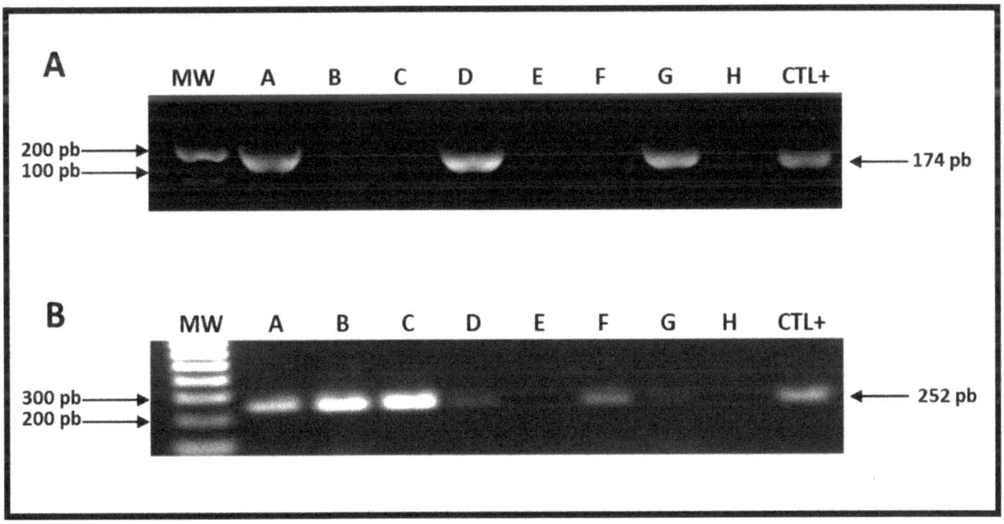

Figure 1. PCR amplification of histidine decarboxylase (**A**) and tyrosine decarboxylase (**B**) genes in LAB present in Chihuahua-type cheeses. (**A**) Amplification of histidine decarboxylase gene. Oligonucleotides HDC1 and HDC2 were used to amplify a 174-bp fragment of hdc gene in the bacterial DNA present in the different cheese samples (line A-H); genomic DNA corresponding to *Lactobacillus* sp. 30a strain was used as positive PCR-control (lane CTL+). (**B**) Amplification of tyrosine decarboxylase gene. Degenerated oligonucleotides TER-F and TER-R were used to amplify a 252-bp fragment of tdc gene from bacterial DNA present in the different cheese samples (line A-H) genomic DNA corresponding to *L. brevis* strain was used as positive PCR-control (lane CTL+). DNA molecular weight marker was included in the electrophoresis analyses (line MW).

On the other hand, to test the usefulness of degenerated primers designed to specifically amplify the tdc gene in LAB, DNA obtained from the cells of *L. brevis* (reference strain) was used as a template (Figure 1B, Line CTL+). PCR reactions using the degenerate oligonucleotides TER-F (GCWAAYYTDGARGGDYTHTGGTATGC) and TER-R (CCAWGAATARTGYTTHGTTTGTGG) were performed at different alignment temperatures (50–60 °C). As PCR products, DNA bands of the expected size (252 bp) were obtained. In six of eight cheeses (A, B, C, D, F, and G), the corresponding band to the tdc gene was observed, suggesting the presence of bacteria carrying the tdc gene, in 75% of the cheeses analyzed (Figure 1B).

3. Discussion

Chihuahua cheese is the matured cheese more consumed in Mexico and, in general, in several Latin American countries. Lactic acid bacteria are used in the Chihuahua cheese elaboration. However, LAB can produce biogenic amines during the process, causing adverse physiological effects in cheese consumers [9]. In Mexico, there is no regulation about the permitted levels of these compounds. In addition, there is no standardized process allowing the biogenic amines identification or species capable of producing them in this type of food.

Biogenic amine concentration can vary between cheeses since it depends on different factors during its elaboration: the quality of the raw material, the presence of strains used as starters, or the presence of contaminating microorganisms with amino decarboxylase activity. It has been reported that certain bacteria produce biogenic amines: *L. buchneri*, *L. fermentum*, *L. helveticus*, *L. rhamnosus L. brevis* [28], *L. fermentum*, *L. plantarum*, *L. helveticus* [29], *L. sakei*, and *L. pentosus* [30]. These species are commonly found in cheeses and are used as starter lactic acid bacteria cultures. On the other hand, both histamine and tyramine formations in cheeses have also been related to the presence of different bacterial species known as non-initiating lactic acid bacteria, mainly *Lactobacillus curvatus* and *L. lactis lactis*, respectively [28,30,31].

In this work, it has been demonstrated that the presence of biogenic amines in samples (histamine and tyramine) of Chihuahua cheese at different storage times. During the first week of shelf life, the histamine concentration was low or undetectable by the HPLC methodology. On the other hand, tyramine was found in concentrations ranging from 34–122 mg/kg in three of the eight samples. During the mid-shelf-life analysis, histamine and tyramine were detected in 12.5% and 62.5% of the samples, respectively. In the final tests (end shelf life), 37.5% of the cheeses were positive for the presence of histamine, and 75% for tyramine.

These results suggest that biogenic amine production takes place not only during the maturation process of the cheeses, but that such compound production continues during their shelf life. This phenomenon has already been reported by Diaz–Cinco et al., (1992) [32], who stated that cheese samples in their study were stored between 5–25 °C for 12 days, and the concentrations of histamine and tyramine increased both in higher temperatures and longer storage times. It is noteworthy that samples in our study were always kept at 1 °C during the storage period. This could indicate that even when cheeses are stored at low temperatures, the production of amines continues. The effect of low temperatures is well-known in delaying the growth rate of several bacterial microorganisms. However, such a condition is probably not adequate for the inactivation of enzymes responsible for amine production.

Histamine and tyramine concentrations in the different cheese samples showed marked variations despite belonging to the same type of cheese. These variations could be because cheeses were made by different manufacturers who probably used different starter cultures (or different strains), variations in the production process, and conservation of raw material.

In previous studies on matured cheeses, tyramine has been highlighted as the most frequent of the biogenic amines [33–35]. Similarly, in this study, tyramine was detected in six of the eight analyzed samples at the end of shelf life, while histamine was detected only in three of eight cheese samples. This biogenic amine is considered of particular interest due to the vasoactive effect it produces on susceptible consumers who have inhibition of MAO (monoamine oxidase enzyme). It has been described that the intake of 6 mg of tyramine showed mild effects, while the intake of 10–25 mg increased the risk of hypertension when ingested in combination with MAO inhibitors [36].

Despite the well-documented toxic effects of biogenic amines, current regulations in Mexico related to the maximum concentration of histamine only exist for fresh and processed fish, and there are no regulations for tyramine. According to the Norma Oficial Mexicana (NOM-242-SSA1-2009) [37], the limit of histamine is 100 mg/kg, which is similar to the value established by the European Union for fermented foods [38]. Within this margin, only a sample of Chihuahua cheeses analyzed in this study did not comply with this regulation. The Food and Drug Administration (FDA, Silver Spring, MD, USA) sets a maximum limit of 50 mg/kg for histamine in foods and considers of risk the concentrations between 50–200 mg/kg [39]. According to this, it could be stated that at the end of shelf life, 37% of Chihuahua cheeses may be a health risk (since three cheese samples presented higher concentration levels).

Using degenerated primers allowed for the detection of bacterial species capable of producing biogenic amines along the shelf life of cheeses. The presence of these LAB species at early shelf life could not be detected by routine microbiological methods, probably because the number of bacteria was not enough for its detection by these methods. However, such bacteria quantity could be detected by PCR due to the detection limit and sensibility this technique possesses. In our results, the use of degenerated primers agrees with the detection of tyramine by the HPLC technique (six of eight cheese samples), but it is interesting to note that the HPLC detection in six samples was performed only at the end of the shelf life.

4. Materials and Methods

4.1. Sample Origin

Eight Chihuahua kinds of cheese of different brands marketed in the Monterrey, N.L., Mexico metropolitan area were selected and analyzed. Cheese samples were kept at refrigeration temperature (2–8 °C) during transportation, and they were kept refrigerated at 1 °C throughout the study. A one-letter code was assigned to each sample. Relevant information was recorded (the elaboration and expiration date and the production batch). Three samples of each cheese were selected, and the analyses were carried out in triplicate.

4.2. Microbiological Analyses

Microbiological analyses were carried out during the first week of the acquisition of the cheeses. The analyses were performed according to the criteria of the Norma Oficial Mexicana (NOM-121-SSA1-1994) [40], which establishes sanitary specifications for fresh, matured, and processed cheeses, and the Norma Mexicana (NMX-F-209-1985) [41], which establishes specifications of the product named "Chihuahua-type cheese".

The aerobic mesophilic bacteria count was performed by seeding in trypticase soybean agar, and samples were incubated at 37 °C for 24 h. For the coliform bacteria cultures, the specific culture medium Rida® COUNT coliform (R-Biopharm AG Darmstadt, Germany) was used; the cultures were incubated for 24 h at 35 °C, and the presence and count of *Escherichia coli* were determined using Petrifilm *E. coli*/Coliform Count Plate (3M, Minnesota, USA) following the procedure recommended by the manufacturer. The count of LAB was made using cheese samples of 10 g and were homogenized with 90 mL of saline solution (0.85%) in sterile jars. Next, each bacterial culture was serially diluted (10^{-1}–10^{-10}) and pour-plated onto MRS agar and incubated at 37 °C for 48 h under anaerobic conditions using Gas-Pack-System (BD, Ontario, Canada), as recommended by [42]. The determination of LAB was performed in triplicate for each sample, and the identification was performed using the API® (BioMérieux, Marcy-l'Étoile, France) biochemical system.

4.3. Histamine and Tyramine Analysis

The determination and quantification of the biogenic amines (histamine and tyramine) were carried out in three stages of the cheeses' shelf life: start (during the first three days after the acquisition); half (in the middle of its shelf life, considering the product label statement); and end (at the end of the shelf life, regarding the expiration date declared on the label). For the amine extraction from cheese samples, the methodology described by Elsanhoty, Mahrous, and Ghanaimy (2009) [43] was followed. Briefly, samples (10 g of cheese) were mixed with 10 mL of 10% trichloroacetic acid solution and homogenized for 15 min (Ultra-turrax homogenizer®, Daigger, IL, USA); the products obtained were centrifuged 10 min at 3000 rpm at 4 °C (Eppendorf, model 5804 R, Hamburg, Germany). The supernatant was filtered (Whatman Paper No. 1), transferred to 15 mL polypropylene tubes (Corning, NY, USA), and kept at −20 °C until use.

The quantification of histamine and tyramine was performed following the official method 977.13 of AOAC (Association of Official Analytical Chemists). This method is sensitive for the identification and quantification of histamine in seafood and consists of the extraction of the compound, the formation of a derivative with *o*-phthaldialdehyde (OPA), and the detection of the fluorescence developed [44]. For the formation of the fluorescent derivatives, 300 µL of cheese sample diluted in 120 µL of 0.4 M borate buffer solution were added 5 min before to the HPLC analysis with 120 µL of o-phthaldialdehyde reagent (200 mg of OPA mixed in 9 mL of methanol, 1 mL of 0.4 M sodium borate, pH 10, and 160 µL of 2-mercaptoethanol) (Sigma–Aldrich, Saint Louis, MO, USA).

The fluorescent derivatives were separated by HPLC (high-performance liquid chromatography) for amine quantification (Thermo Scientific, Spectra System, Waltham, MA,

USA). The HPLC System was equipped with a fluorescence detector fixed to 338 nm and 430 nm (absorption and emission wavelength, respectively). A Waters Ultrasphere ODS 5 μm (4.6 × 250 mm) C18 column was set, and a circumvention gradient formed by 12.5 mM of phosphate buffer, pH 6.5 (Eluent A), and acetonitrile (Eluent B) was used. The gradient started with an 85:15% ratio of Solvents A and B, respectively, and a flow of 0.9 mL/min during 3 min; subsequently, Solvent B was increased to 60% with a final flow rate of 1.3 mL/min (3–24 min) and then returned to 15% of Solvent B (24–50 min) and remained isocratic for five more minutes.

Histamine dihydrochloride and tyramine hydrochloride (Sigma–Aldrich, Saint Louis, MO, USA) were used as standards for the identification and quantification of these compounds in the cheese samples. Standard solutions were prepared at different concentrations (25, 50, 100, 250, and 400 mg/L), and a standard calibration curve was constructed. For the HPLC analysis, 10 μL of standards and samples were injected.

4.4. *Molecular Biology Techniques Standardization*

Standard techniques were used for the molecular analysis (DNA extraction and PCR analysis). To establish the molecular protocols that would allow the identification of bacteria-producing biogenic amines, the reference strains ATCC 33222 of *Lactobacillus* 30a and ATCC 367 of *Lactobacillus brevis*, were used as controls. For preservation and experimental conditions, both strains were grown in MRS broth for 48 h and were kept stirring at 200 rpm. The *Lactobacillus* 30a culture was performed at 37 °C under anaerobic conditions, while the *L. brevis* culture was performed under aerobic conditions at 30 °C.

4.4.1. DNA Extraction

DNA extraction from reference strain cultures was carried out following the technique described by Hoffman and Winston, (1999) [45], in which the cell wall is broken both by mechanical fractionation with glass beads and by chemical lysis using a buffer solution (TSNT buffer: 2% Triton, 1% SDS, 100 mM NaCl, 10 mM Tris-HCl, pH 8.0) followed by a DNA purification step with a phenol–chloroform mixture (50:50) and subsequent precipitation of DNA with 3 mM of sodium acetate and ethanol (96%).

For the direct extraction of bacterial DNA present in cheeses, an adaptation of the technique described by Randazzo et al., (2002) [46] was used. DNA purification was performed using the SV Total RNA Isolation System kit (Promega, Madison, WI, USA). Briefly, cheese samples (5 g) were placed in sterile dilution jars (130 mL with screw cap) containing 40 mL of a sterile solution of sodium citrate (2%) and were incubated at 45 °C for 30 min. Subsequently, 200 μL of sterile glass beads (0.1 mm diameter) were added, and the mixture was stirred manually for 5 min and left to stand for 10 min at room temperature. Next, 1 mL of the supernatant was transferred to a sterile 1.5 mL polypropylene tube and centrifuged for 3 min at 12,000 rpm (Eppendorf mini spin centrifuge; Eppendorf; Hamburg, Germany); the supernatant was eliminated. This process was repeated until the pellet product of all volumes (10 mL) of the homogenized cheese was obtained. From this product, the DNA was obtained using the SV Total RNA Isolation System kit, following the instructions from the manufacturer; the obtained DNA was diluted with sterile water (50 μL) and stored at −20 °C until use.

4.4.2. Amplification of Genes Encoding hdc and tdc

The presence of the genes encoding decarboxylase enzymes (histidine and tyrosine) on DNA obtained from cheese was determined by PCR amplification. PCR results obtained from DNA amplification from different cheese samples were considered positive when PCR reactions were positive in the three representative cheese samples. For the identification of hdc and tdc genes in cheese samples, the conditions described by Fernandez et al., (2006) [27] were used. For the hdc gene amplification, the specific oligonucleotides HDC1 and HDC described by Ventura et al., (2001) [26] were used. For the tdc gene identification, a pair of degenerate oligonucleotides (TER-F and TER-R) was designed on the sequences from the tdc gene corresponding to different bacterial species.

4.4.3. Design of Degenerated Oligonucleotides for the tdc Gene

Nucleotide sequences corresponding to genes that code for the enzyme tyrosine decarboxylase from different species of bacteria (*Enterococcus hirae, E. durans, E. faecium, Lactobacillus curvatus, L. brevis and Streptococcus* sp.) were obtained from databases. Alignments of these sequences (which allowed identifying the conserved regions) were made (both the amino acids and nucleotides levels) using the BLAST (http://blast.ncbi.nlm.nih.gov/Blast.cgi, accessed on 10 July 2020) and CLUSTAL-W (http://www.genome.jp/tools/clustalw/, accessed on 10 July 2020) programs. The regions that presented higher conservation were identified and used to design the degenerate oligonucleotides TER-F (forward primer) and TER-R (reverse primer).

The standardization of the experimental conditions of PCR amplification and determination of the specificity of both the specific oligonucleotides for hdc and the degenerate ones for the tdc genes was carried out using DNA obtained from the reference strains ATCC 33222 of *Lactobacillus* 30a and ATCC 367 of *L. brevis*, respectively. PCR reactions were performed in a Thermocycler (Axygen, Therm-1000, Maxygene) using a temperature gradient (50–60 °C) as an alignment/hybridization temperature. Final PCR conditions were 95 °C for 5 min (a cycle), followed by 95 °C for 45 s, 50–60 °C for 45 s, and 72 °C for 2 min (35 cycles), with a final extension cycle (72 °C for 5 min). The products obtained from PCR reactions were analyzed by electrophoresis in 1.5% agarose gels, stained with ethidium bromide [47], and were visualized under UV light with a Gel-Doc Photo Documentation System (Vision Works, UVP, Upland, CA, USA).

4.5. Statistical Analysis

Statistical analyses were performed using the statistical software package IBM SPSS Statistics V21.0. Differences between treatments were analyzed using the analysis of variance (ANOVA) test. The statistical significance was determined at $p < 0.05$. Data corresponding to each variable are expressed as means ± SD, unless otherwise indicated. All determinations were performed by triplicate.

5. Conclusions

In this study, the presence of biogenic amines, histamine and tyramine, was detected in 37% and 75%, respectively, at the end of the shelf life of the Chihuahua cheese tested. It was observed that at this point, the number of samples positive for the presence of histamine and tyramine increased, suggesting that even at cooling temperature, the production of these compounds continues. The presence of genes encoding for hdc and tdc was detected by molecular techniques using degenerated primers. The detected tyramine levels (<600 mg/kg) do not present a risk to healthy consumers (without MAO inhibitors prescription); in the other hand, the histamine content at the end of the shelf of cheeses (>50 mg/kg) represent a health risk, according to EFSA and FDA regulations. The presence of biogenic amines in foods, such as Chihuahua cheese, could be controlled by the strict use of good manufacturing hygiene practices and raw material selection.

In Mexico, there is no regulation about the maximum concentration of biogenic amines (histamine or tyramine) in cheeses. Although there are standardized protocols for amine detection, these processes are time-consuming and labor-intensive. In this work, we presented evidence of the usefulness of degenerated primers to be employed as tools for amplification by PCR techniques to achieve a direct detection of bacterial species that produce biogenic amines.

Moreover, we suggest the appliance of molecular techniques as a detection tool for bacteria that produce biogenic amines in food as it can prevent the early development of these bacteria and, therefore, biogenic amines in food. In addition, maximum regulatory limits on biogenic amines must be declared to protect consumers and avoid possible public health problems.

Author Contributions: Conceptualization, B.E.G.-M. and E.C.-G.; investigation and data curation M.T.G.-M., A.A.L.-H., A.S.O.-V. and G.I.A.-M.; methodology, M.T.G.-M., A.A.L.-H., A.S.O.-V. and G.I.A.-M.; validation, formal analysis, and resources, M.T.G.-M., A.A.L.-H., A.S.O.-V. and G.I.A.-M.; writing—original draft preparation, A.A.L.-H., A.S.O.-V., G.I.A.-M. and E.C.-G.; writing—review and editing, E.C.-G., A.A.L.-H., A.S.O.-V. and G.I.A.-M.; supervision, B.E.G.-M. and E.C.-G.; project administration and funding acquisition, B.E.G.-M. and E.C.-G. All authors have read and agreed to the published version of the manuscript.

Funding: This research was supported in part by the PROMEP-SEP Program with a fellowship to M.T.G.-M. and by the UANL-PAICYT Program (No. CE 1025-11 and CN 354-14). The APC was funded by UANL-Facultad de Salud Pública y Nutrición.

Institutional Review Board Statement: Not applicable.

Informed Consent Statement: Not applicable.

Data Availability Statement: Not applicable.

Conflicts of Interest: The authors declare no conflict of interest. The funders had no role in the design of the study, in the collection, analyses, or interpretation of data, in the writing of the manuscript, or in the decision to publish the results.

References

1. Renye, J.A.; Somkuti, G.A.; Van Hekken, D.L.; Guerrero Prieto, V.M. Characterization of Microflora in Mexican Chihuahua Cheese. *J. Dairy Sci.* **2011**, *94*, 3311–3315. [CrossRef] [PubMed]
2. Van Hekken, D.L.; Tunick, M.H.; Renye, J.A. The Character of Queso Chihuahua. *ACS Symp. Ser.* **2020**, *1347*, 142–160. [CrossRef]
3. Sharma, R.; Garg, P.; Kumar, P.; Bhatia, S.K.; Kulshrestha, S. Microbial Fermentation and Its Role in Quality Improvement of Fermented Foods. *Fermentation* **2020**, *6*, 106. [CrossRef]
4. Beresford, T.P.; Fitzsimons, N.A.; Brennan, N.L.; Cogan, T.M. Recent Advances in Cheese Microbiology. *Int. Dairy J.* **2001**, *11*, 259–274. [CrossRef]
5. Özer, E.; Kesenkaş, H. The Effect of Using Different Starter Culture Combinations on Ripening Parameters, Microbiological and Sensory Properties of Mihaliç Cheese. *J. Food Sci. Technol.* **2019**, *56*, 1202. [CrossRef]
6. Blaya, J.; Barzideh, Z.; LaPointe, G. Symposium Review: Interaction of Starter Cultures and Nonstarter Lactic Acid Bacteria in the Cheese Environment. *J. Dairy Sci.* **2018**, *101*, 3611–3629. [CrossRef]
7. Dimou, A.; Tsimihodimos, V.; Bairaktari, E. The Critical Role of the Branched Chain Amino Acids (BCAAs) Catabolism-Regulating Enzymes, Branched-Chain Aminotransferase (BCAT) and Branched-Chain & alpha-Keto Acid Dehydrogenase (BCKD), in Human Pathophysiology. *Int. J. Mol. Sci.* **2022**, *23*, 4022.
8. Ganesan, B.; Weimer, B.C. Amino Acid Catabolism and Its Relationship to Cheese Flavor Outcomes. *Cheese Chem. Phys. Microbiol. Fourth Ed.* **2017**, *1*, 483–516. [CrossRef]
9. Linares, D.M.; del Río, B.; Ladero, V.; Martínez, N.; Fernández, M.; Martín, M.C.; Alvarez, M.A. Factors Influencing Biogenic Amines Accumulation in Dairy Products. *Front. Microbiol.* **2012**, *3*, 180. [CrossRef]
10. Aliakbarlu, J.; Alizadeh, M.; Razavi-Rohani, S.M.; Vahabzade, Z.; Saei, S.S.; Agh, N. Effects of Processing Factors on Biogenic Amines Production in Iranian White Brine Cheese. *Res. J. Biol. Sci.* **2009**, *4*, 23–28.
11. Fernández, M.; Linares, D.M.; Del Río, B.; Ladero, V.; Alvarez, M.A. HPLC Quantification of Biogenic Amines in Cheeses: Correlation with PCR-Detection of Tyramine-Producing Microorganisms. *J. Dairy Res.* **2007**, *74*, 276–282. [CrossRef] [PubMed]
12. Pinho, O.; Pintado, A.I.E.; Gomes, A.M.P.; Pintado, M.M.E.; Malcata, F.X.; Ferreira, I.M.P.L.V. Interrelationships among Microbiological, Physicochemical, and Biochemical Properties of Terrincho Cheese, with Emphasis on Biogenic Amines. *J. Food Prot.* **2004**, *67*, 2779–2785. [CrossRef] [PubMed]
13. Standarová, E.; Vorlová, L.; Kordiovská, P.; Janštová, B.; Dračková, M.; Borkovcová, I. Biogenic Amine Production in Olomouc Curd Cheese (Olomoucké Tvarůžky) at Various Storage Conditions. *Acta Vet. Brno* **2010**, *79*, 147–156. [CrossRef]
14. Lavizzari, T.; Veciana-Nogués, M.T.; Weingart, O.; Bover-Cid, S.; Mariné-Font, A.; Vidal-Carou, M.C. Occurrence of Biogenic Amines and Polyamines in Spinach and Changes during Storage under Refrigeration. *J. Agric. Food Chem.* **2007**, *55*, 9514–9519. [CrossRef]
15. Mariné i Font, A. *Les Amines Biògenes En Els Aliments: Història i Recerca En El Marc de Les Ciències de L'alimentació*; Institut d'Estudis Catalans: Barcelona, Spain, 2005; ISBN 8472837882.
16. Arena, M.E.; Manca de Nadra, M.C. Biogenic Amine Production by Lactobacillus. *J. Appl. Microbiol.* **2001**, *90*, 158–162. [CrossRef]

17. Roig-Sagues, A.X.; Hernandez-Herrero, M.; Lopez-Sabater, E.I.; Rodriguez-Jerez, J.J.; Mora-Ventura, M.T. Histidine Decarboxylase Activity of Bacteria Isolated from Raw and Ripened Salchichon, a Spanish Cured Sausage. *J. Food Prot.* **1996**, *59*, 516–520. [CrossRef] [PubMed]
18. Maintz, L.; Novak, N. Histamine and Histamine Intolerance. *Am. J. Clin. Nutr.* **2007**, *85*, 1185–1196. [CrossRef] [PubMed]
19. Barbieri, F.; Montanari, C.; Gardini, F.; Tabanelli, G. Biogenic Amine Production by Lactic Acid Bacteria: A Review. *Foods* **2019**, *8*, 17. [CrossRef] [PubMed]
20. McCabe-Sellers, B.J.; Staggs, C.G.; Bogle, M.L. Tyramine in Foods and Monoamine Oxidase Inhibitor Drugs: A Crossroad Where Medicine, Nutrition, Pharmacy, and Food Industry Converge. *J. Food Compos. Anal.* **2006**, *19*, S58–S65. [CrossRef]
21. Rao, T.S.S.; Yeragani, V.K. Hypertensive Crisis and Cheese. *Indian J. Psychiatry* **2009**, *51*, 65–66. [CrossRef]
22. Food and Agriculture Organization of the United Nations/World Health Organization. *Public Health Risks of Histamine and Other Biogenic Amines from Fish and Fishery Products*; Food and Agriculture Organization/World Health Organization: Rome, Italy, 2012; ISBN 9789251078495.
23. Karovičová, J.; Kohajdová, Z. Biogenic Amines in Food. *Chem. Pap.* **2005**, *59*, 70–79.
24. Van Hekken, D.L.; Tunick, M.H.; Tomasula, P.M.; Corral, F.J.M.; Gardea, A.A. Mexican Queso Chihuahua: Rheology of Fresh Cheese. *Int. J. Dairy Technol.* **2007**, *60*, 5–12. [CrossRef]
25. Tunick, M.H.; Van Hekken, D.L.; Molina-Corral, F.J.; Tomasula, P.M.; Call, J.; Luchansky, J.; Gardea, A.A. Queso Chihuahua: Manufacturing Procedures, Composition, Protein Profiles, and Microbiology. *Int. J. Dairy Technol.* **2008**, *61*, 62–69. [CrossRef]
26. Ventura, M.; Reniero, R.; Zink, R. Specific Identification and Targeted Characterization of *Bifidobacterium lactis* from Different Environmental Isolates by a Combined Multiplex-PCR Approach. *Appl. Environ. Microbiol.* **2001**, *67*, 2760–2765. [CrossRef] [PubMed]
27. Fernández, M.; Del Río, B.; Linares, D.M.; Martín, M.C.; Alvarez, M.A. Real-Time Polymerase Chain Reaction for Quantitative Detection of Histamine-Producing Bacteria: Use in Cheese Production. *J. Dairy Sci.* **2006**, *89*, 3763–3769. [CrossRef]
28. Izquierdo, P.; Allara, M.; Torres, G.; García, A.; Barboza, Y.; Piñero, M.I. Histamina En Quesos Madurados: Manchego, Parmesano y de Año. *Rev. Cient.* **2003**, *13*, 431–435.
29. Burdychova, R.; Komprda, T. Biogenic Amine-Forming Microbial Communities in Cheese. *FEMS Microbiol. Lett.* **2007**, *276*, 149–155. [CrossRef]
30. Tittarelli, F.; Perpetuini, G.; Di Gianvito, P.; Tofalo, R. Biogenic Amines Producing and Degrading Bacteria: A Snapshot from Raw Ewes' Cheese. *LWT* **2019**, *101*, 1–9. [CrossRef]
31. Novella-Rodríguez, S.; Veciana-Nogués, M.T.; Roig-Sagués, A.X.; Trujillo-Mesa, A.J.; Vidal-Carou, M.C. Evaluation of Biogenic Amines and Microbial Counts throughout the Ripening of Goat Cheeses from Pasteurized and Raw Milk. *J. Dairy Res.* **2004**, *71*, 245–252. [CrossRef]
32. Diaz-Cinco, M.E.; Fraijo, O.; Grajeda, P.; Lozano-Taylor, J.; De Mejía, E.G. Microbial and Chemical Analysis of Chihuahua Cheese and Relationship to Histamine and Tyramine. *J. Food Sci.* **1992**, *57*, 355–356. [CrossRef]
33. Standarová, E.; Borkovcová, I.; Dušková, M.; Přidalová, H.; Dračková, M. Effect of Some Factors on the Biogenic Amines and Polyamines Content in Blue-Veined Cheese Niva. *Spec. Issue Czech J. Food Sci* **2009**, *27*, S410–S413. [CrossRef]
34. Ibrahim, E.M.A.; Amer, A.A.M. Comparison of Biogenic Amines Levels in Different Processed Cheese Varieties with Regulatory Specifications. *World J. Dairy Food Sci.* **2010**, *5*, 127–133.
35. Roig-Sagués, A.X.; Molina, A.P.; Hernández-Herrero, M. Histamine and Tyramine-Forming Microorganisms in Spanish Traditional Cheeses. *Eur. Food Res. Technol.* **2002**, *215*, 96–100. [CrossRef]
36. McCabe, B.J. Dietary Tyramine and Other Pressor Amines in MAOI Regimens: A Review. *J. Am. Diet. Assoc.* **1986**, *86*, 1059–1064. [CrossRef] [PubMed]
37. Secretaría de Salud. NORMA Oficial Mexicana NOM-242-SSA1-2009, Productos y Servicios. Productos de la Pesca Frescos, Refrigerados, Congelados y Procesados. Especificaciones Sanitarias y Métodos de Prueba. Diario Oficial de la Federación. 2009. Available online: https://www.dof.gob.mx/normasOficiales/4295/salud2a/salud2a.htm (accessed on 24 February 2023).
38. FDA (Food and Drug Administration). *Hazards and Controls Guidance*, 3rd ed.; Center of Food Safety and Nutrition: Washington, DC, USA, 2001.
39. FDA (Food and Drug Administration). *Fish and Fishery Products Hazards and Controls Guidance*, 4th ed.; FDA: Silver Spring, MD, USA, 2020; pp. 113–152.
40. Secretaría de Salud. PROYECTO de Norma Oficial Mexicana NOM-121-SSA1-1994, Bienes y Servicios. Quesos: Frescos, Madurados y Procesados. Especificaciones Sanitarias. Diario Oficial de la Federación. 1994. Available online: https://dof.gob.mx/nota_detalle.php?codigo=4729075&fecha=15/08/1994#gsc.tab=0 (accessed on 24 February 2023).
41. Secretaría de Salud. NORMA Mexicana NMX-F-209-1985. Alimentos-Lácteos-Queso Tipo Chihuahua. Dirección General de Normas. 1985. Available online: http://cide.uach.mx/pdf/NORMAS%20MEXICANAS%20NMX/PRODUCTOS%20ALIMENTICIOS/ALIMENTOS.LACTEOS.QUESO%20TIPO%20CHIHUAHUA.pdf (accessed on 24 February 2023).
42. Hamilton-Miller, J.; Shah, S.; Winkler, J. Public Health Issues Arising from Microbiological and Labelling Quality of Foods and Supplements Containing Probiotic Microorganisms. *Public Health Nutr.* **1999**, *2*, 223–229. [CrossRef] [PubMed]
43. Elsanhoty, R.M.; Mahrous, H.; Ghanaimy, G.A. Chemical, Microbial Counts and Evaluation of Biogenic Amines during the Ripening of Egyptian Soft Domiati Cheese Made from Raw and Pasteurized Buffaloes Milk. *Int. J. Dairy Sci.* **2009**, *4*, 80–90. [CrossRef]

44. AOAC. Histamine in seafood: Fluorometric method Sec. 35.1.32, Method 977.13. In *Official Methods of Analysis of AOAC International*; Cunniff, P.A., Ed.; AOAC International: Gaithersburg, MD, USA, 1995; pp. 6–17.
45. Hoffman, C.; Winston, F. A Ten-Minute DNA Preparation from Yeast Efficiently Releases Autonomous Plasmids for Transformation of *Escherichia coli*. *Gene* **1987**, *57*, 267–272. [CrossRef]
46. Randazzo, C.L.; Torriani, S.; Akkermans, A.D.L.; De Vos, W.M.; Vaughan, E.E. Diversity, Dynamics, and Activity of Bacterial Communities during Production of an Artisanal Sicilian Cheese as Evaluated by 16S RRNA Analysis. *Appl. Environ. Microbiol.* **2002**, *68*, 1882–1892. [CrossRef] [PubMed]
47. Sambrock, J.; Fritsch, E.F.; Maniatis, T. *Molecular Cloning, A Laboratory Manual*, 2nd ed.; Volumes 1, 2 and 3; Cold Spring Harbor Laboratory Press: Cold Spring Harbor, NY, USA, 1989; ISBN 0-87969-309-6.

molecules

MDPI

Article

Variation of Aroma Components of Pasteurized Yogurt with Different Process Combination before and after Aging by DHS/GC-O-MS

Mu Zhao [1,†], Hongliang Li [2,†], Dongjie Zhang [2], Jie Li [1], Rong Wen [2], Hairan Ma [2], Tingting Zou [1,*], Yaqiong Hou [2,*] and Huanlu Song [1,*]

1 Laboratory of Molecular Sensory Science, Beijing Technology and Business University, No. 11, Fucheng Road, Haidian District, Beijing 100048, China
2 Inner Mongolia Mengniu Dairy (Group) Co., Ltd., Hohhot 011500, China
* Correspondence: zoutingting@th.btbu.edu.cn (T.Z.); houyaqiong@mengniu.cn (Y.H.); songhl@th.btbu.edu.cn (H.S.)
† These authors contributed equally to this work.

Abstract: Pasteurized yogurt is a healthy yogurt that can be stored in ambient temperature conditions. Dynamic headspace sampling (DHS) combined with gas chromatography-olfactory mass spectrometry (GC-O-MS), sensory evaluation, electronic nose (E-nose), and partial least squares discriminant analysis (PLS-DA) were used to analyze the flavor changes of pasteurized yogurt with different process combinations before and after aging. The results of odor profiles showed that the sensory descriptors of fermented, sweet, and sour were greatly affected by different process combinations. The results of odor-active compounds and relative odor activity value (r-OAV) showed that the combination of the production process affected the overall odor profile of pasteurized yogurt, which was consistent with the sensory evaluation results. A total of 15 odor-active compounds of 38 volatile compounds were detected in pasteurized yogurt samples. r-OAV results revealed that hexanal, *(E)*-2-octenal, 2-heptanone, and butanoic acid may be important odor-active compounds responsible for off-odor in aged, pasteurized yogurt samples. PLS-DA and variable importance of projection (VIP) results showed that butanoic acid, hexanal, acetoin, decanoic acid, 1-pentanol, 1-nonanal, and hexanoic acid were differential compounds that distinguish pasteurized yogurt before and after aging.

Keywords: pasteurized yogurt; dynamic headspace sampling (DHS); GC-O-MS; odor-active compounds; r-OAV; PLS-DA

Citation: Zhao, M.; Li, H.; Zhang, D.; Li, J.; Wen, R.; Ma, H.; Zou, T.; Hou, Y.; Song, H. Variation of Aroma Components of Pasteurized Yogurt with Different Process Combination before and after Aging by DHS/GC-O-MS. *Molecules* **2023**, *28*, 1975. https://doi.org/10.3390/molecules28041975

Academic Editor: Igor Jerković

Received: 20 January 2023
Revised: 1 February 2023
Accepted: 17 February 2023
Published: 19 February 2023

1. Introduction

Pasteurized yogurt, also known as room-temperature yogurt, is made by adding a second pasteurization process to the original process of traditional yogurt. O'Brien et al. found no significant differences in immunological parameters and gastrointestinal comfort between pasteurized yogurt and low-temperature yogurt [1]. Additionally, the shelf life of pasteurized yogurt can be as long as 4–6 months under ambient temperature conditions, breaking the limitation of low-temperature yogurt requiring full cold chain transportation and storage and short shelf life. Therefore, pasteurized yogurt has gradually entered the life of consumers in recent years and has become a popular yogurt.

At present, there have been some studies on the effect of the yogurt production process on yogurt. Tian et al. [2] studied the effect of lactic acid bacteria species on the aroma profile of yogurt. Miyaji et al. [3] studied the effect of pasteurization conditions on pasteurized drinking yogurt. Ding et al. [4] studied the characteristics of metabolomics and physicochemical properties of yogurt under aerobic and strict anaerobic fermentation conditions. Yang et al. [5] studied the effect of fermentation temperature on yogurt quality.

Xia et al. [6] studied the effect of fermentation time on yogurt quality. Summarizing the above studies, it was found that most of the studies involved fermentation strains and fermentation conditions setting, but there were very few studies on the influence of other processes. Sfakianakis et al. analyzed the odor, flavor, taste, sensory, and volatile components profile of yogurt derived from milk homogenized by ultrasound or pressure, respectively [7]. Therefore, the homogenization process of raw milk will affect the quality of yogurt. In the process of yogurt production, both the homogenization and smooth pump can make the fat particles smaller and the texture smoother. However, the influence of the above two processes on the yogurt flavor has not been reported.

With the presence of large amounts of live bacteria in yogurt, the volatile compounds in yogurt change all the time during storage [8]. Most studies have analyzed the changes of volatile compounds or odor-active compounds during the storage of low-temperature yogurt [9–11]. Compared with low-temperature yogurt, pasteurized yogurt has a longer storage time and unlimited storage temperature, so it is more prone to processes such as fat degradation and lipid oxidation during its storage. Milk powder and milk often occur fat oxidation and produce off-odor [12,13]. Although the odor-active off-flavor compounds in aged, pasteurized yogurt have been identified in our previous studies [14], there are still some shortcomings. The previous research results were only discussed and studied based on a group of samples, and it is unclear how the research results would change when the sample range was expanded.

In this study, the aging process was used to simulate the state of pasteurized yogurt before and after storage. In this study, sensory evaluation combined with a variety of instrument analysis was used to comprehensively analyze the flavor characteristics of pasteurized yogurt, and data were further explored by combining with multivariate statistical analysis, so as to realize the purpose of analyzing the changes in aroma components of different process combinations of pasteurized yogurt before and after aging. This study can analyze the effect of the production process on pasteurized yogurt flavor and flavor changes during storage from the perspective of small molecular compounds, which is helpful in improving the production process, prolong shelf life and improve the quality of pasteurized yogurt.

2. Results and Discussion

2.1. Establishment of Overall Odor Profile

Firstly, the perception of the overall odor of the six pasteurized yogurt samples was evaluated, and a radar chart was drawn (Figure 1). The fermented odor appeared to be the main characteristic odor of the pasteurized yogurt, as evidenced by its higher score by the sensory panelists. Among the samples before aging, bhF sample showed strong characteristics of fermented odor, sour odor, and sweet odor, which have been proven to be beneficial to the overall odor profile of yogurt in previous studies [15,16]. There was no significant difference between creamy odor and milky odor. After aging, all samples showed obvious oxidized and fishy odor and showed a trend of weakening in fermented odor, sweet odor, sour odor, and milky odor compared with the samples before aging, resulting in the decline of the overall odor quality of pasteurized yogurt. There was a similar phenomenon in other dairy products, with the extension of storage time will also appear oxidized odor and other off-odor [12,13]. However, this change does not occur in the later period of low-temperature yogurt storage, which is due to the short shelf life and its strict low-temperature storage conditions, avoiding the production of lipid oxidation and other possible off-odor sources [7]. It was noteworthy that the bhF sample had good flavor before aging but had the strongest oxidized and fishy odor after aging (bhA sample). In summary, pasteurized yogurt samples with different process combinations had different odor profiles and had off-odor after aging. The overall odor profile is formed by the interaction of volatile compounds in the sample, so it is very necessary to study the volatile compounds in pasteurized yogurt samples, which is conducive to clarifying the reasons for the odor changes of pasteurized yogurt from the perspective of small molecular compounds.

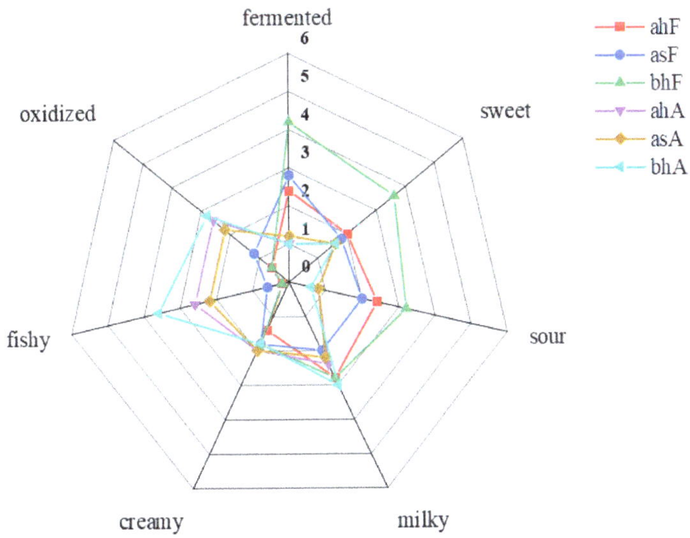

Figure 1. Odor profile evaluation of the six samples.

2.2. Volatiles in Pasteurized Yogurt Analyzed via E-Nose

The electronic nose is equipped with ten different receptors (Table 1), it does not accurately identify or describe a certain odor, only certain compounds, so it is often used to distinguish between different samples. This study was mainly used to investigate differences in volatile components of different pasteurized yogurts. Figure 2A shows the radar chart drawn from the E-nose receptors' response values. The six samples were similar in shape. However, the ten receptors differed in intensity. Among them, the W5S receptor, which is sensitive to oxynitride, responded most strongly, which may indicate that these compounds were present at high levels and that the W5S receptor was the most important indicator to distinguish the six samples. Specifically, the bhA sample had the highest receptor response value. Figure 2B was the result of PCA, the horizontal axis indicates the contribution rate of PC1 (98.07%), whereas the vertical axis represents the contribution rate of PC2 (1.85%), and the total contribution rate reaches 99.92%. The greater the total contribution rate, the more the data can reflect the original information of the samples [17]. According to the sum of the contribution rates, the two principal components retained the main characteristics and information of the volatiles of the samples, and the contribution of the horizontal axis was larger. The distance between the bhA sample and other samples was large in the PC1 direction, indicating that this sample was quite different from other samples. The other samples showed little difference in the PC1 direction, and there were some differences in the PC2 direction. The volatile components of the three samples, ahF, asF, and bhF, overlapped and intersected, and the volatile components of the three samples were similar to a certain extent. In summary, these results supported the overall odor profile results.

Table 1. The description of receptor on E-nose.

No.	Receptor	Specification
R1	W1C	aromatic hydrocarbon
R2	W5S	broad range
R3	W3C	aromatic ammonia
R4	W6S	hydrogen
R5	W5C	arom-aliph

Table 1. *Cont.*

No.	Receptor	Specification
R6	W1S	broad-methane
R7	W1W	sulfur-organic
R8	W2S	broad-alcohol
R9	W2W	sulfur-chlor
R10	W3S	methane-aliph

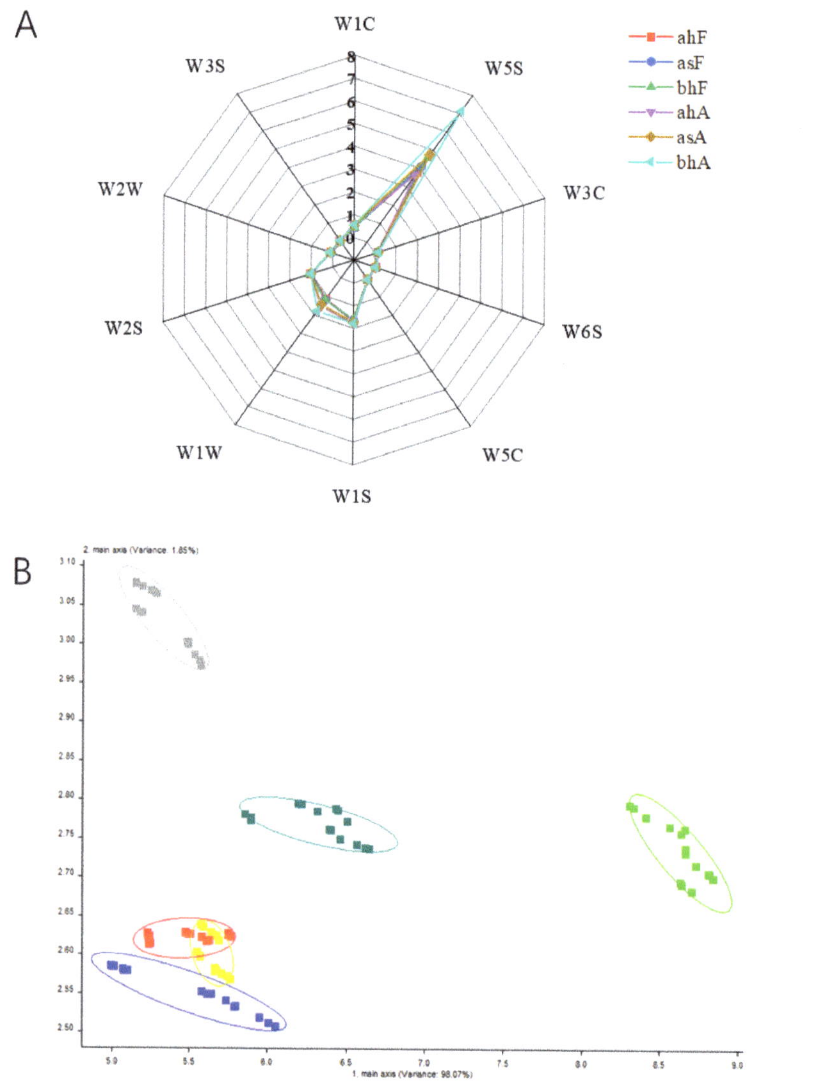

Figure 2. Radar chart and PCA analysis by E-nose. (**A**) Radar chart of the E-nose response of different types of volatiles for different pasteurized yogurt samples, (**B**) PCA plot of the E-nose for different pasteurized yogurt samples.

2.3. Volatile Compounds in Pasteurized Yogurt Analyzed via GC-MS

According to the content distribution of volatile compounds in six pasteurized yogurt samples, a heat map was drawn, as shown in Figure 3. The contents of 38 volatile compounds in six samples were displayed, and the types and contents of volatile compounds were different, including aldehydes (8), ketones (8), alcohols (7), acids (7), esters (4), and others (4). This result was consistent with other studies [6,15]. Among the three samples before aging, volatile compounds were the most abundant in bhF samples, followed by ahF and asF samples. The contents of acetaldehyde, 2,3-butanedione, 2,3-pentanedione, acetoin, 3-methyl-2-buten-1-ol, etc., in bhF samples were higher than those in ahF and asF samples. These compounds were representative of fine volatile odor compounds in yogurt [15]. The difference in volatile compound content between ahF and asF samples may be caused by the existence of temperature and pressure during homogenization. Specifically, studies have indicated that during milk processing, the degradation of lipids and proteins during hydrostatic pressure treatment can increase the formation of flavor compounds such as aldehydes [18], and the rise of processing temperature can increase the formation of ketones [19]. Therefore, it is possible to explain the higher content of some aldehydes and ketones in the samples produced by the homogenizer (ahF and bhF), which uses pressure and temperature to break up fat particles. After pasteurized yogurt samples were aged, the contents of some volatile compounds showed an obvious increase trend, including aldehydes, ketones, acids, etc. These compounds were proven products of thermal degradation and oxidation of unsaturated fatty acids [8]. It is not difficult to know that a series of reactions, such as fatty acid degradation and lipid oxidation, may occur in the aging process of pasteurized yogurt and ultimately lead to an increase in the content of these compounds. Similarly, the phenomenon has been seen in other foods, such as powdered milk, roasted mutton, bacon, etc. [13,16,20]. Among all the samples, bhA sample had the most abundant volatile compounds, which was consistent with the response value of the E-nose receptor.

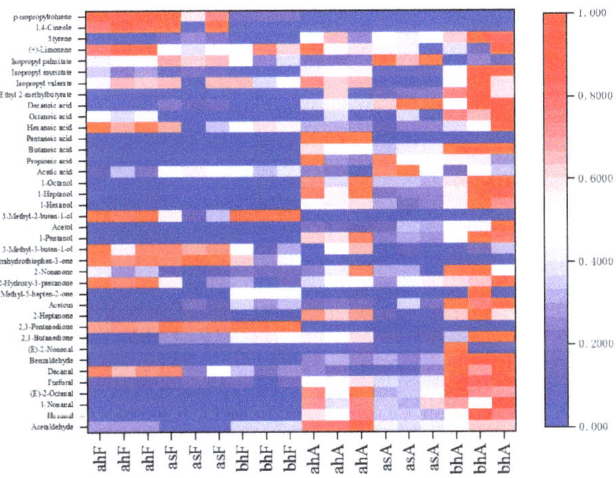

Figure 3. The heat map analysis of the six samples. Every square represents a volatile compound. If a compound had a lower concentration, the color of the square is close to blue, otherwise, the color is close to red.

2.4. Odor-Active Compounds in Pasteurized Yogurt Analyzed via GC-O-MS

The results of the identification of odor-active compounds in all samples are shown in Table 2. Aldehydes are the key carbonyl compound of yogurt, and four aldehyde odor-active compounds were identified, including acetaldehyde, hexanal, *(E)*-2-octenal, and benzaldehyde. In the samples before aging, acetaldehyde (grass-like) and benzaldehyde (almond-like) have higher content. Acetaldehyde is formed in several ways, the most important of which is the breakdown of threonine into acetaldehyde and glycine. As one of the most important odor-active compounds in yogurt, the content of acetaldehyde was significantly different among the three samples of ahF, asF, and bhF. Specifically, bhF was the most abundant sample, followed by ahF and asF. The fermentation process and homogenization process of pasteurized yogurt affect the generation of acetaldehyde. In the aged samples, the content of benzaldehyde and hexanal increased significantly, on the contrary, the content of acetaldehyde did not change significantly. Benzaldehyde may be derived from the degradation of phenylalanine and the oxidation of α-linolenic acid [21]. The content of hexanal was even increased by a factor of 100–500. It was formed in milk and yogurt by β-oxidation of the *n*-6 fatty acids linoleic and arachidonic acids [13], which may be facilitated by the aging process leading to a large increase in hexanal content. When the concentration of hexanal is low, it can give food a certain fragrance and fruity odor, but when the concentration reaches 4.5 μg/kg, it shows a grass-like odor, which seriously affects the flavor of the food [22]. Additionally, it was worth noting that *(E)*-2-octenal was newly formed after the aging of the samples. *(E)*-2-Octenal was metabolized by lipoxygenase using linoleic acid as a precursor, which was one of the products of lipid oxidation during aging.

Ketones are mainly formed by thermal degradation of amino acids and oxidation of unsaturated fatty acids. A total of five odor-active compounds of ketones were identified, including 2,3-butanedione, 2,3-pentanedione, 2-heptanone, acetoin, and 2-nonanone. They were buttery, creamy, and sweet, and could provide a positive odor characteristic for pasteurized yogurt. They were formed as follows: 2,3-butanedione and 2,3-pentanedione were produced by the chemical decarboxylation of their precursors, 2-acetolactate and 2-aceto-2-hydroxybutyrate, respectively. Acetoin was readily converted from 2,3-butanedione by diacetyl reductase [9,23,24]. This formation pathway could explain the phenomenon that the content of acetoin was the most abundant and the content of 2,3-butanedione was lower in this study. Acetoin, 2,3-butanedione and 2,3-pentanedione, three of the positive odor compounds recognized as important in yogurt, all showed the highest levels in the bhF samples. Combined with previous studies, it can be known that the heating process of homogenizer promoted the formation of ketones in raw milk [19], kept the promoting effect, and finally made the bhF sample has a high content of ketones. Similarly, Reis et al. also demonstrated that temperature and pressure can significantly affect the formation of ketones in milk [25]. However, the detailed influence process is still unknown, which requires further in-depth research. Such research is very beneficial to improve the quality of yogurt from the production equipment itself. After aging, the contents of acetoin and 2,3-butanedione tended to increase, and the content of 2-heptanone increased significantly. The variation trend of acetoin and 2,3-butanedione was consistent with the results of previous studies [10]. 2-Heptanone is usually formed by β-oxidation of free fatty acids or β-ketoacid decarboxylation, and aging may promote this process and lead to a large increase in its content. In previous studies, a significant increase in 2-heptanone content in other dairy products, such as milk, was found during storage [12].

Table 2. Odor-active compounds determined in six pasteurized yogurt samples by GC-O-MS.

No.	CAS	Compounds	Odor	Identification	RI [a]		Concentration (μg/g) [b]					
					DB-WAX	DB-5	ahF	asF	bhF	ahA	asA	bhA
1	75-07-0	Acetaldehyde	grass-like	MS/RI/O	<800	-	1.71 ± 0.00145 ^{b}	0.545 ± 0.0237 ^{c}	2.16 ± 0.0357 ^{a}	4.13 ± 0.469 ^{a}	2.28 ± 0.0834 ^{a}	2.99 ± 0.264 ^{a}
2	431-03-8	2,3-Butanedione	pungent buttery	MS/RI/O	980	<800	2.45 ± 0.0105 ^{b}	3.28 ± 0.0248 ^{a}	3.88 ± 0.136 ^{a}	4.03 ± 0.117 ^{a}	2.75 ± 0.0350 ^{b}	4.63 ± 0.861 ^{a}
3	600-14-6	2,3-Pentanedione	buttery	MS/RI/O	1053	<800	2.12 ± 0.0507 ^{a}	2.43 ± 0.116 ^{a}	2.84 ± 0.0757 ^{a}	-	-	-
4	66-25-1	Hexanal	grass-like	MS/RI/O	1079	803	0.423 ± 0.0427 ^{c}	0.124 ± 0.0408 ^{c}	0.125 ± 0.00338 ^{c}	53.6 ± 9.62 ^{ab}	30.3 ± 1.32 ^{b}	64.5 ± 19.8 ^{a}
5	110-43-0	2-Heptanone	sweet	MS/RI/O	1179	888	2.56 ± 0.350 ^{b}	2.94 ± 0.391 ^{b}	1.43 ± 0.189 ^{b}	13.1 ± 0.990 ^{a}	3.18 ± 0.631 ^{b}	13.8 ± 3.10 ^{a}
6	100-42-5	Styrene	plastic-like	MS/RI/O	1246	-	0.184 ± 0.00218 ^{c}	0.475 ± 0.0538 ^{b}	0.180 ± 0.00937 ^{c}	0.892 ± 0.247 ^{a}	0.880 ± 0.0618 ^{a}	1.44 ± 0.251 ^{a}
7	513-86-0	Acetoin	mild creamy	MS/RI/O	1281	<800	68.1 ± 1.53 ^{c}	76.7 ± 7.03 ^{bc}	93.7 ± 4.60 ^{b}	82.4 ± 2.03 ^{b}	84.6 ± 11.1 ^{b}	138 ± 8.23 ^{a}
8	821-55-6	2-Nonanone	sweet	MS/RI/O	1381	-	0.443 ± 0.0730 ^{b}	0.156 ± 0.00409 ^{c}	0.235 ± 0.0148 ^{bc}	0.569 ± 0.194 ^{ab}	0.384 ± 0.00406 ^{b}	0.913 ± 0.236 ^{a}
9	2548-87-0	(*E*)-2-Octenal	fresh	MS/RI/O	1420	-	-	-	-	0.834 ± 0.246 ^{ab}	0.423 ± 0.0638 ^{b}	0.911 ± 0.174 ^{a}
10	64-19-7	Acetic acid	vinegar-like	MS/RI/O	1461	<800	1.35 ± 0.194 ^{a}	1.79 ± 0.0537 ^{a}	1.63 ± 0.0837 ^{a}	1.23 ± 0.163 ^{a}	2.12 ± 0.331 ^{a}	1.54 ± 0.213 ^{a}
11	100-52-7	Benzaldehyde	almond-like	MS/RI/O	1509	956	1.65 ± 0.248 ^{c}	1.49 ± 0.0684 ^{c}	1.41 ± 0.213 ^{c}	4.58 ± 0.280 ^{b}	4.83 ± 0.508 ^{b}	12.3 ± 0.830 ^{a}
12	107-92-6	Butyric acid	sweaty	MS/RI/O	1630	<800	8.05 ± 0.772 ^{c}	5.54 ± 1.38 ^{cd}	3.63 ± 0.157 ^{d}	80.0 ± 4.75 ^{b}	59.7 ± 3.96 ^{b}	143 ± 4.18 ^{a}
13	109-52-4	Pentanoic acid	sweaty	MS/RI/O	1734	912	-	-	-	1.42 ± 0.0243	-	-
14	142-62-1	Hexanoic acid	cheesy	MS/RI/O	1851	984	17.9 ± 2.58 ^{a}	10.3 ± 5.44 ^{ab}	13.4 ± 0.526 ^{a}	9.49 ± 0.327 ^{ab}	8.11 ± 0.966 ^{b}	13.9 ± 3.36 ^{a}
15	124-07-2	Octanoic acid	cheesy	MS/RI/O	2083	1174	11.8 ± 0.255 ^{a}	9.59 ± 0.0197 ^{b}	9.51 ± 0.00387 ^{b}	10.3 ± 0.824 ^{ab}	7.69 ± 0.371 ^{b}	13.7 ± 0.763 ^{a}

[a] The retention index (RI) on capillaries DB-WAX and DB-5. [b] Mean values of triplicates with standard deviations. Different lower letters in the same row (such as a,b and c) indicate significant differences ($p \leq 0.05$).

Acids are essential compounds in yogurt, which can enrich the sour odor of yogurt. In this study, a total of 5 acid compounds showed odorant activity, including acetic acid, butanoic acid, pentanoic acid, hexanoic acid, and octanoic acid. Their odor characteristics were not exactly the same, they could present vinegar-like, sweaty, and cheesy odor. Different scholars hold different opinions on the formation of acids in yogurt. Some researchers believe that C_2–C_4 is produced by lactic acid bacteria and lactic acid metabolism, C_4–C_{20} is mainly formed by fat decomposition, and some researchers believe that the key precursors of most volatile fatty acids are amino acids [15]. Hexanoic acid and octanoic acid were the most abundant acids in pasteurized yogurt samples, which was consistent with the results of Moineau-Jean et al. [10] and Tian et al. [2] but inconsistent with those of Rychlik et al. [11]. It was noted that the content of butanoic acid increased significantly after aging, and new pentanoic acid was produced in ahA sample. It may be caused by the oxidative decomposition of fatty acids in the aging process of pasteurized yogurt, and the higher content of butanoic acid can bring cheese-like off-odor to yogurt.

Styrene (plastic-like) was detected in the samples, and its content increased significantly after aging. It was speculated that styrene may have migrated from the packaging material to the pasteurized yogurt during aging [26]. Moreover, alcohols were also important volatile compounds in pasteurized yogurt (Figure 3), but they have not been identified as odor-active compounds, possibly because they have a higher threshold than other aldehydes and ketones [27] and therefore cannot be detected on the olfactory detection port.

2.5. The Contribution Degree of Odor-Active Compounds to the Overall Odor Profile of Pasteurized Yogurt Evaluated by r-OAV

The compounds with r-OAV > 1 are generally considered to be of greater importance to the overall odor profile of the sample [28]. Specifically, the larger the value, the greater the contribution to the overall odor profile of the sample. The r-OAV calculation results are summarized in Table 3. Of all odor-active compounds of samples ahF, asF, and bhF, 2,3-butanedione had the highest r-OAV of 408, 547, and 647, respectively, attributing to the buttery odor of fresh pasteurized yogurt samples. This result was consistent with the previous study results [29]. Secondly, 2,3-pentanedione had a higher r-OAV of between 106 to 142, and its odor attribute was also buttery. Both 2,3-butanedione and 2,3-pentanedione had the largest r-OAV in bhF sample, which may have led to the higher score in the evaluation of overall odor profile (Figure 1). The other compounds, such as hexanal, acetaldehyde, 2-heptanone, and 2-nonanone had the slightly larger r-OAV, which was attributed to the grass-like and sweet odor in fresh pasteurized yogurt samples. Acetoin was the most abundant compound in samples, but its r-OAV was only 10, which was due to its high threshold, so its contribution to the formation of the overall odor profile of fresh pasteurized yogurt was relatively weak.

As shown in Table 3, after pasteurized yogurt aging, there were some odor-active compounds with significantly increased r-OAV. Among all compounds, the most significant change was hexanal, which replaced 2,3-butanedione to become the odor-active compound with the largest r-OAV in the aged, pasteurized yogurt samples. Combining with the production pathway of hexanal, it was speculated that this change was the reason for the oxidized odor of aged, pasteurized yogurt samples. Previous studies have shown that aldehydes were the main contributing compounds to the oxidative flavor of UHT milk [30]. Secondly, the r-OAV values of *(E)*-2-octenal, 2-heptanone, and butanoic acid also increased significantly. *(E)*-2-Octenal had a great influence on the overall odor profile of aged, pasteurized yogurt samples because of its low threshold. The results also confirmed that enals produced by fat oxidation have a great influence on food flavor [31]. Hexanal, *(E)*-2-octenal, 2-heptanone, and butanoic acid all had the highest r-OAV values in the bhA sample, which may have resulted in the strongest oxidized odor of the sample (Figure 1). Overall, the changes in these compounds resulted in changes in the overall odor profile of pasteurized yogurt samples after aging.

Table 3. Odor thresholds (OT) and relative odor activity values (r-OAV) of odor-active compounds in six pasteurized yogurt samples.

CAS	Compounds	OT (mg/kg)	r-OAV					
			ahF	asF	bhF	ahA	asA	bhA
431-03-8	2,3-Butanedione	0.006	408	547	647	672	458	772
600-14-6	2,3-Pentanedione	0.02	106	122	142	-	-	-
66-25-1	Hexanal	0.005	84	24	24	10,726	6062	12,896
75-07-0	Acetaldehyde	0.063	27	9	34	66	36	47
110 43 0	2-Heptanone	0.14	18	21	10	93	23	99
821-55-6	2-Nonanone	0.041	10	4	6	14	9	22
513-86-0	Acetoin	8	9	10	12	10	11	17
100-52-7	Benzaldehyde	0.35	5	4	4	13	14	35
124-07-2	Octanoic acid	3	4	3	3	3	3	5
100-42-5	Styrene	0.065	3	7	3	14	14	22
107-92-6	Butanoic acid	2.4	3	2	2	33	25	60
142-62-1	Hexanoic acid	18	1	1	1	1	<1	1
64-19-7	Acetic acid	22	<1	<1	<1	<1	<1	<1
2548-87-0	(*E*)-2-Octenal	0.003	-	-	-	277	140	303
109-52-4	Pentanoic acid	1.207	-	-	-	1	-	-

2.6. Differential Compounds of Pasteurized Yogurt before and after Aging Identified by PLS-DA

PLS-DA is a supervised pattern recognition method, which emphasizes the differences between groups to minimize the differences within groups and better grasp the overall characteristics and change rules of multidimensional data [32]. As shown in Figure 4A,B, a clear separation of pasteurized yogurt samples before and after aging could be observed through a dependable PLS-DA model. The PLS-DA model was established based on the semi-quantitative results of volatile compounds in all samples. The parameters R^2 and Q^2 represent the explanatory and predictive abilities of the model, respectively, and the values of R^2 and Q^2 should be greater than 0.5. The results are more accurate as R^2 and Q^2 approach 1 [33]. The R^2Y of the PLS-DA model was 0.98, and Q2 was 0.88, which indicated satisfactory explanatory and predictive effects of the PLS-DA model for the classification of pasteurized yogurt samples before and after aging.

Based on PLS-DA model, the variable importance of projection (VIP) diagram of the model was obtained (Figure 4C). The VIP is the vector, used to summarize the total importance of the variable in explaining the model [32]. Specifically, VIP value > 1 will contribute to the model. The abscissa of the VIP result graph in Figure 4C was the CAS number of the compound. The main compounds which contribute to the distinction between the two groups (compounds with VIP value > 1) were butanoic acid, hexanal, acetoin, decanoic acid, 1-pentanol, 1-nonanal and hexanoic acid (arranged in descending order of VIP value). In the previous analysis, butanoic acid and hexanal were found to play an important role in the formation of oxidized odor of aged, pasteurized yogurt samples. Fatty acids with more than four carbons are readily produced by fat degradation [15]. The change of 1-pentanol was consistent with the results of previous studies and was considered to be a product of lipid decomposition [34].

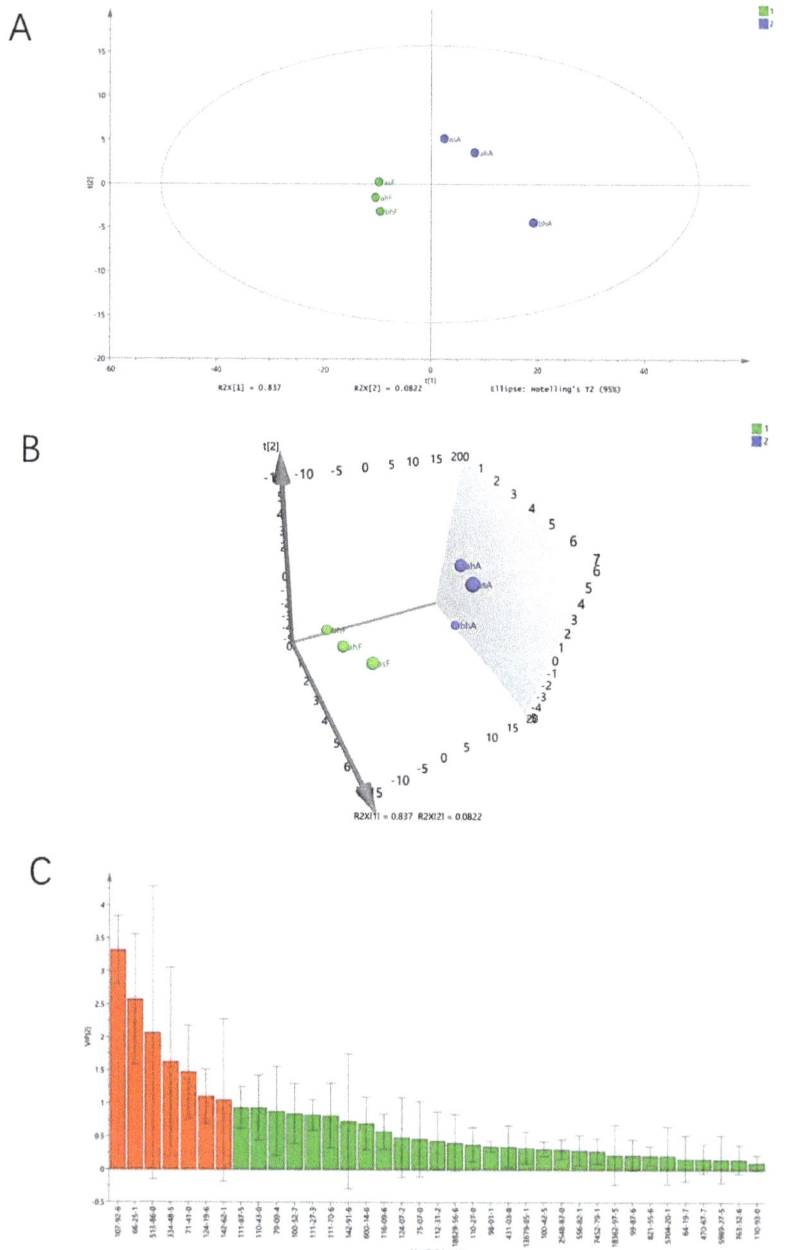

Figure 4. 2D (**A**) and 3D (**B**) score charts of PLS-DA and VIP diagram (**C**; red means VIP > 1, green means VIP < 1) of the six samples ($R^2X = 0.919$, $R^2Y = 0.98$, and $Q^2 = 0.88$).

3. Materials and Methods

3.1. Samples

The pasteurized yogurt sample used in this study was a factory pilot sample produced and provided by Inner Mongolia Mengniu Dairy (Group) Co., Ltd. (Inner Mongolia, China), which could not be purchased in the market, and no other volatile substances were added to the sample. The samples were all obtained in 2022, and the information is listed in Table 4. The basic index information of the sample is as follows: Water content of 78%, protein content of 3.2%, fat content of 3.5%, acidity of 75 °T. The homogenization temperature is 50–70 °C and the total pressure is 1×10^7 –1.8×10^7 Pa. Smooth pump speeds are 20–50 Hz. Aged pasteurized yogurt was obtained by fresh pasteurized yogurt accelerated aging for 13 d. This process is completed by Inner Mongolia Mengniu Dairy (Group) Co., Ltd., and it has been proved that this process can effectively simulate and reproduce the status of pasteurized yogurt after the shelf life when it is normally sold and stored. The specific operation and evaluation criteria are still in the confidential stage, belonging to the category of trade secrets. Samples canned and sealed directly after production are kept in a clean, odor-free refrigerator dedicated to the laboratory. All samples were analyzed within 10 d after production. There are six kinds of samples, 10 copies of each sample, 180 g each, a total of 10.8 kg.

Table 4. Detailed information of six pasteurized yogurt samples.

Sample Type	Process Combination	Abbreviation
Fresh pasteurized yogurt (F)	Fermentation process a & homogenizer treated (ah)	ahF
	Fermentation process a & smooth pump treated (as)	asF
	Fermentation process b & homogenizer treated (bh)	bhF
Aged pasteurized yogurt (A)	Fermentation process a & homogenizer treated (ah)	ahA
	Fermentation process a & smooth pump treated (as)	asA
	Fermentation process b & homogenizer treated (bh)	bhA

3.2. Reagents and Chemicals

Sodium chloride (NaCl, analytical reagent, purity ≥99.5%) was obtained from Sinopharm Chemical Reagent Co., Ltd. (Beijing, China). *n*-Alkanes standard solution (C_7–C_{30}) and 2-methyl-3-heptanone (99% purity) were obtained from Sigma-Aldrich (St. Louis, MO, USA). *n*-Hexane (purity >99%) was obtained from Thermo Fisher Scientific (Waltham, MA, USA).

3.3. Establishment of Odor Profiles

The odor profile was adjusted on the basis of previous experimental methods [35]. All members of the sensory panel are from the Molecular Sensory Science Laboratory, Beijing Technology and Business University (Beijing, China), and have received more than 1 year of sensory skills training. There were 12 members in the panel, including 6 males and 6 females, with an average age of 30. The sensory evaluation room is clean, odorless, noise-free, well-lit, 25 °C, 65% relative humidity, and compartments to ensure an independent assessment process for each team member. The members of the sensory panel discussed and agreed that the following 7 sensory descriptors should be used as odor characteristics of the sample and applied to establish odor profiles. The sensory descriptors and the references agreed upon by the panelists are as follows: "Fermented" refers to fresh cheese, "sweet" reference to diluted 2-heptanone standard, "sour" refers to 0.08% citric acid solution, "milky" refers to fresh whole milk, "creamy" refers to whipped cream, "fishy" refers to the fishy smell of raw milk, "oxidized" refers to the diluted butanoic acid standard. A seven-point scale (0–6, with a difference of 1) was used to quantify the odor characteristics of the sample. The 10 g sample that has been standing at 40 °C for 20 min is presented to the panel members for evaluation, and the evaluation and score are made on the record sheet. The order of sample submission is random, and each sample is guaranteed to be submitted 3 times.

3.4. Volatiles Analysis by Electronic Nose (E-Nose)

A portable electronic nose system PEN3 (Airsense Analytics GmbH., Schwerin, Germany) was used to analyze the volatiles in the samples. Ten receptors with different properties were installed in the electronic nose system and described in Table 1. A 10 g sample of pasteurized yogurt was weighed for the analysis and equilibrated at 40 °C for 20 min before the electronic nose test. The parameters of the electronic nose were set as follows: Cleaning time of 60 s before receptors were tested, preparation time of 5 s, receptor detection time of 180 s, and gas flow rate set to 400 mL/min. Each sample was tested 5 times, and an empty bottle was tested as an empty needle between testing different samples.

3.5. Extraction of Volatile Compounds by Dynamic Headspace Sampling (DHS)

Pasteurized yogurt (50 g) and sodium chloride (7.5 g) were mixed in a dynamic headspace vial, followed by the addition of 5 μL of 2-methyl-3-heptanone (0.816 μg/μL) as an internal standard and then quickly sealed. Place on a magnetic stirrer (Thermo Scientific, USA) at 600 rpm while incubating in a constant circulating water bath at 40 °C for 20 min. Then, the upper space of the vial was purged with nitrogen (99.999% purity) at a flow rate of 150 mL/min for 40 min to allow volatiles to adsorb into the Tenax TA tube inserted into the vial. After purging, the Tenax TA tube was removed and purged with nitrogen to remove water. Finally, the Tenax TA tubes were placed in a thermal desorption unit (TDU) (Gerstel, Germany) for analysis. Each sample was analyzed in triplicate.

3.6. Gas Chromatography–Mass Spectrometry (GC-MS)/Gas Chromatography–Olfactometry (GC-O) Analysis

GC-MS analysis was performed by GC-MS Model 7890A-7000 (Agilent Technologies, Inc., Santa Clara, CA, USA). Two capillary columns of different polarity were installed for the separation of volatile compounds in GC section, respectively, polar capillary column DB-WAX (30 m × 0.25 mm × 0.25 μm; J&W Scientific, Folsom, CA, USA) and medium-polarity DB-5 capillary column (30 m × 0.25 mm × 0.25 μm; J&W Scientific, Folsom, CA, USA). The carrier gas has a constant flow rate of 1 mL/min and uses ultra-high purity helium (99.999%, Beijing AP BAIF Gases Industry Co., Ltd., Beijing, China) as the carrier gas. The initial temperature of the column box was maintained at 40 °C for 5 min and then increased to 230 °C at the rate of 4 °C/min for 5 min. The back GC injector was set to "splitless" mode. The mass spectrometry source temperature was 280 °C. The electron collision mass spectrometry was generated at 70 ev ionization energy, and the scanning range was 33~350 m/z.

The olfactory detection port (ODP4, Gerstel, Germany) was connected to the GC section for GC-O analysis. During the whole process of GC-O analysis, three members of the sensory panel (1 male and 2 females) sniffed at the olfactory detection port and recorded the retention time and odor characteristics of the odor in real time.

3.7. Identification of Volatile Compounds and Odor-Active Compounds

The target compounds were identified by mass spectrometry retrieval (MS), retention index comparison (RI), and sniffing odor description (O). Mass spectrometry retrieval refers to the retrieval of MS results of target compounds in the NIST17 mass spectrometry database, according to the MS matching degree (>800) and MS structure information to identify the compounds. Retention index comparison refers to comparing the actual RI with the standard RI of the target compound. If the difference is less than 50, it is considered to pass the identification. The actual RI value is calculated by the retention time between the target compound and a series of n-alkanes (C_7–C_{30}) [36,37]. Compounds identified only by MS and RI were identified as volatile compounds and were identified as odor-active compounds when they could be detected at the olfactory detection port.

3.8. Quantitation of Compounds

A semi-quantitative method was used to study the changes in volatile compound concentration before and after aging different pasteurized yogurt samples. The concentration of each compound was calculated as follows [38,39]:

$$C_i = \frac{C_{is} \times A_j}{A_{is}}$$

where C_i is the concentration of the compound; C_{is} is the internal standard concentration of 0.816 μg/μL; A_j is the chromatographic peak area of the compound, and A_{is} is the chromatographic peak area of the internal standard.

3.9. Calculation of Relative Odor Activity Value (r-OAV)

Relative odor activity value (r-OAV) was calculated as the ratio of the relative concentration of each compound to its respective odor threshold (OT) [29,39]. The threshold of the compound used in this study is the threshold in water [27].

3.10. Statistical Analysis

The experimental data were analyzed by using Microsoft Excel 2019 (Microsoft Corp., Redmond, WA, USA) in triplicate and expressed as mean ± standard deviation. One-way analysis of variance (ANOVA) and Duncan's multiple range tests were performed using the IBM SPSS Statistics 26 (IBM., Armonk, NY, USA) to analyze the differences between samples. A $p \leq 0.05$ was considered statistically significant. The heat map and radar charts were made by Origin 2019 (Origin Lab Inc, Northampton, MA, USA). The result of E-nose of principal component analysis (PCA) was generated using the software Winmuster (version 1.6.2). The partial least squares discriminant analysis (PLS-DA) was performed by SIMCA 14.1 (MKS Instruments AB, MA, USA).

4. Conclusions

In this study, dynamic headspace sampling (DHS) combined with gas chromatography-olfactometry-mass spectrometry (GC-O-MS) were employed to analyze the flavor changes of pasteurized yogurt with different process combinations before and after aging. A total of 15 odor-active compounds of 38 volatile compounds were detected in six pasteurized yogurt samples. Sensory evaluation and GC-O-MS results showed that under the same fermentation process, the overall odor profile of pasteurized yogurt samples obtained by a smooth pump was better than that obtained by a homogenizer. r-OAV results revealed that hexanal, *(E)*-2-octenal, 2-heptanone, and butanoic acid may be important odor-active compounds responsible for off-odor in aged, pasteurized yogurt samples. PLS-DA and VIP results showed that butanoic acid, hexanal, acetoin, decanoic acid, 1-pentanol, 1-nonanal, and hexanoic acid were differential compounds that distinguish pasteurized yogurt before and after aging. As this study reported the effect of the production process on the overall odor profile of pasteurized yogurt and the variation of odor-active compounds before and after aging, it may provide some useful information for extending shelf life and improving the quality of pasteurized yogurt. Differential odor compounds can be used as odor markers to evaluate the odor quality of pasteurized yogurt and can be used as an evaluation basis to optimize the production process.

Author Contributions: M.Z.: Formal analysis, Writing—original draft. H.L.: Writing—original draft. D.Z.: Methodology. J.L.: Formal analysis. R.W.: Investigation. H.M.: Investigation. T.Z.: Methodology; Writing—Review & Editing. Y.H.: Methodology. H.S.: Conceptualization. All authors have read and agreed to the published version of the manuscript.

Funding: This study was financially supported by Inner Mongolia Mengniu Dairy (Group) Co., Ltd.

Institutional Review Board Statement: Not applicable.

Informed Consent Statement: Not applicable.

Data Availability Statement: The authors will make the raw data supporting the conclusions of this manuscript available to any qualified researcher.

Conflicts of Interest: The authors declare no conflict of interest.

Sample Availability: Not available.

References

1. O'Brien, K.V.; Aryana, K.J.; Prinyawiwatkul, W.; Ordonez, K.C.; Boeneke, C.A. Short communication: The effects of frozen storage on the survival of probiotic microorganisms found in traditionally and commercially manufactured kefir. *J. Dairy Sci.* **2016**, *99*, 7043–7048. [CrossRef]
2. Tian, H.; Shi, Y.; Zhang, Y.; Yu, H.; Mu, H.; Chen, C. Screening of aroma-producing lactic acid bacteria and their application in improving the aromatic profile of yogurt. *J. Food Biochem.* **2019**, *43*. [CrossRef]
3. Miyaji, K.; Maruyama, H.; Kuwano, Y.; Katakura, Y.; Inoue, H.; Azuma, N. Development of a Rapid and Accurate Prediction Model for Whey Separation in Pasteurized Drinking Yogurt Caused by Long-term Ambient Storage. *Food Sci. Technol. Res.* **2020**, *26*, 863–873. [CrossRef]
4. Ding, R.; Li, M.; Zou, Y.; Wang, Y.; Yan, C.; Zhang, H.; Wu, R.; Wu, J. Effect of normal and strict anaerobic fermentation on physicochemical quality and metabolomics of yogurt. *Food Biosci.* **2022**, *46*. [CrossRef]
5. Yang, Y.; Qian, M.C.; Deng, Y.; Yuan, H.; Jiang, Y. Insight into aroma dynamic changes during the whole manufacturing process of chestnut-like aroma green tea by combining GC-E-Nose, GC-IMS, and GC × GC-TOFMS. *Food Chem.* **2022**, *387*, 132813. [CrossRef]
6. Xia, Y.; Yu, J.; Liu, H.; Feng, C.; Shuang, Q. Novel insight into physicochemical and flavor formation in koumiss based on microbial metabolic network. *Food Res. Int.* **2021**, *149*. [CrossRef]
7. Sfakianakis, P.; Tzia, C. Flavour profiling by gas chromatography–mass spectrometry and sensory analysis of yoghurt derived from ultrasonicated and homogenised milk. *Int. Dairy J.* **2017**, *75*, 120–128. [CrossRef]
8. Zhou, T.; Huo, R.; Kwok, L.-Y.; Li, C.; Ma, Y.; Mi, Z.; Chen, Y. Effects of applying Lactobacillus helveticus H9 as adjunct starter culture in yogurt fermentation and storage. *J. Dairy Sci.* **2019**, *102*, 223–235. [CrossRef]
9. Martin, F.; Cachon, R.; Pernin, K.; De Coninck, J.; Gervais, P.; Guichard, E.; Cayot, N. Effect of oxidoreduction potential on aroma biosynthesis by lactic acid bacteria in nonfat yogurt. *J. Dairy Sci.* **2011**, *94*, 614–622. [CrossRef] [PubMed]
10. Moineau-Jean, A.; Raymond, Y.; Sabik, H.; Graveline, N.; Champagne, C.P.; Roy, D.; LaPointe, G. Effect of manufacturing processes and storage on aroma compounds and sensory properties of yoghurt. *Int. Dairy J.* **2020**, *105*. [CrossRef]
11. Rychlik, M.; Sax, M.; Schieberle, P. On the role of short-chain free fatty acids for the development of a cheese-like off-note in pasteurized yoghurt. *Lwt Food Sci. Technol.* **2006**, *39*, 521–527. [CrossRef]
12. Li, Y.H.; Wang, W.J. Short communication: Formation of oxidized flavor compounds in concentrated milk and distillate during milk concentration. *J. Dairy Sci.* **2016**, *99*, 9647–9651. [CrossRef] [PubMed]
13. Li, Y.H.; Zhang, L.W.; Wang, W.J.; Han, X. Differences in particle characteristics and oxidized flavor as affected by heat-related processes of milk powder. *J. Dairy Sci.* **2013**, *96*, 4784–4793. [CrossRef] [PubMed]
14. Zhao, M.; Ma, H.; Hou, Y.; Li, J.; Zou, T.; Zhang, D.; Wen, R.; Li, H.; Song, H. Characterization of Key Odor-Active Off-Flavor Compounds in Aged Pasteurized Yogurt by Sensory-Directed Flavor Analysis. *J. Agric. Food Chem.* **2022**, *70*, 14439–14447. [CrossRef]
15. Cheng, H. Volatile Flavor Compounds in Yogurt: A Review. *Crit. Rev. Food Sci. Nutr.* **2010**, *50*, 938–950. [CrossRef]
16. Liu, C.; Yang, P.; Wang, H.; Song, H. Identification of odor compounds and odor-active compounds of yogurt using DHS, SPME, SAFE, and SBSE/GC-O-MS. *Lwt* **2022**, *154*. [CrossRef]
17. Dong, W.; Hu, R.; Long, Y.; Li, H.; Zhang, Y.; Zhu, K.; Chu, Z. Comparative evaluation of the volatile profiles and taste properties of roasted coffee beans as affected by drying method and detected by electronic nose, electronic tongue, and HS-SPME-GC-MS. *Food Chem.* **2019**, *272*, 723–731. [CrossRef]
18. Vazquez-Landaverde, P.A.; Torres, J.A.; Qian, M.C. Effect of High-Pressure—Moderate-Temperature Processing on the Volatile Profile of Milk. *J. Agric. Food Chem.* **2006**, *54*, 9184–9192. [CrossRef]
19. Gathercole, J.; Reis, M.G.; Agnew, M.; Reis, M.M.; Humphrey, R.; Harris, P.; Clerens, S.; Haigh, B.; Dyer, J.M. Molecular modification associated with the heat treatment of bovine milk. *Int. Dairy J.* **2017**, *73*, 74–83. [CrossRef]
20. Yang, Y.; Zhang, X.; Wang, Y.; Pan, D.; Sun, Y.; Cao, J. Study on the volatile compounds generated from lipid oxidation of Chinese bacon (unsmoked) during processing. *Eur. J. Lipid Sci. Technol.* **2017**, *119*. [CrossRef]
21. Zannou, O.; Kelebek, H.; Selli, S. Elucidation of key odorants in Beninese Roselle (Hibiscus sabdariffa L.) infusions prepared by hot and cold brewing. *Food Res. Int.* **2020**, *133*. [CrossRef]
22. Panseri, S.; Soncin, S.; Chiesa, L.M.; Biondi, P.A. A headspace solid-phase microextraction gas-chromatographic mass-spectrometric method (HS-SPME–GC/MS) to quantify hexanal in butter during storage as marker of lipid oxidation. *Food Chem.* **2011**, *127*, 886–889. [CrossRef]
23. Ott, A.; Hugi, A.; Baumgartner, M.; Chaintreau, A. Sensory Investigation of Yogurt Flavor Perception: Mutual Influence of Volatiles and Acidity. *J. Agric. Food Chem.* **2000**, *48*, 441–450. [CrossRef]
24. Ramos, A.; Jordan, K.N.; Cogan, T.M.; Santos, H. 13 C Nuclear Magnetic Resonance Studies of Citrate and Glucose Cometabolism by Lactococcus lactis. *Appl. Environ. Microbiol.* **1994**, *60*, 1739–1748. [CrossRef]
25. Reis, M.G.; Harris, P.; Berry, C.; Nguyen, H.; Maclean, P.; Weeks, M. Tracking changes in volatile components and lipids after homogenisation and thermal processing of milk. *Int. Dairy J.* **2020**, *103*. [CrossRef]
26. Lozano, P.R.; Miracle, E.R.; Krause, A.J.; Drake, M.; Cadwallader, K.R. Effect of Cold Storage and Packaging Material on the Major Aroma Components of Sweet Cream Butter. *J. Agric. Food Chem.* **2007**, *55*, 7840–7846. [CrossRef] [PubMed]

27. van Gemert, L.J. *Odour Thresholds-Compilations of Odour Threshold Values in Air, Water and Other Media*, 2nd ed.; Oliemans Punter & Partners BV: Utrecht, The Netherlands, 2011.
28. Liu, H.; Hui, T.; Zheng, X.; Li, S.; Wei, X.; Li, P.; Zhang, D.; Wang, Z. Characterization of key lipids for binding and generating aroma compounds in roasted mutton by UPLC-ESI-MS/MS and Orbitrap Exploris GC. *Food Chem.* **2022**, *374*. [CrossRef] [PubMed]
29. Liu, W.; Pu, X.; Sun, J.; Shi, X.; Cheng, W.; Wang, B. Effect of Lactobacillus plantarum on functional characteristics and flavor profile of fermented walnut milk. *Lwt* **2022**, *160*. [CrossRef]
30. Zabbia, A.; Buys, E.M.; De Kock, H.L. Undesirable Sulphur and Carbonyl Flavor Compounds in UHT Milk: A Review. *Crit. Rev. Food Sci. Nutr.* **2012**, *52*, 21–30. [CrossRef]
31. Xu, L.; Mei, X.; Chang, J.; Wu, G.; Zhang, H.; Jin, Q.; Wang, X. Comparative characterization of key odorants of French fries and oils at the break-in, optimum, and degrading frying stages. *Food Chem.* **2022**, *368*. [CrossRef]
32. Yang, S.; Yan, D.; Zou, Y.; Mu, D.; Li, X.; Shi, H.; Luo, X.; Yang, M.; Yue, X.; Wu, R.; et al. Fermentation temperature affects yogurt quality: A metabolomics study. *Food Biosci.* **2021**, *42*. [CrossRef]
33. Wang, X.; Rogers, K.M.; Li, Y.; Yang, S.; Chen, L.; Zhou, J. Untargeted and Targeted Discrimination of Honey Collected by Apis cerana and Apis mellifera Based on Volatiles Using HS-GC-IMS and HS-SPME-GC–MS. *J. Agric. Food Chem.* **2019**, *67*, 12144–12152. [CrossRef] [PubMed]
34. Gurkan, H.; Hayaloglu, A.A. Volatiles and sensory characteristics of yogurt manufactured by incorporating basil (*Ocimum basilicum* L.). *Int. J. Food Prop.* **2017**, *20*, S779–S789. [CrossRef]
35. Gao, X.; Feng, T.; Sheng, M.; Wang, B.; Wang, Z.; Shan, P.; Zhang, Y.; Ma, H. Characterization of the aroma-active compounds in black soybean sauce, a distinctive soy sauce. *Food Chem.* **2021**, *364*. [CrossRef] [PubMed]
36. Gao, X.; Feng, T.; Liu, E.; Shan, P.; Zhang, Z.; Liao, L.; Ma, H. Ougan juice debittering using ultrasound-aided enzymatic hydrolysis: Impacts on aroma and taste. *Food Chem.* **2021**, *345*. [CrossRef] [PubMed]
37. Zhao, M.; Li, T.; Yang, F.; Cui, X.; Zou, T.; Song, H.; Liu, Y. Characterization of key aroma-active compounds in Hanyuan Zanthoxylum bungeanum by GC-O-MS and switchable GC × GC-O-MS. *Food Chem.* **2022**, *385*, 132659. [CrossRef]
38. Wu, W.; Zhan, J.; Tang, X.; Li, T.; Duan, S. Characterization and identification of pork flavor compounds and their precursors in Chinese indigenous pig breeds by volatile profiling and multivariate analysis. *Food Chem.* **2022**, *385*, 132543. [CrossRef]
39. Yang, P.; Song, H.; Lin, Y.; Guo, T.; Wang, L.; Granvogl, M.; Xu, Y. Differences of characteristic aroma compounds in Rougui tea leaves with different roasting temperatures analyzed by switchable GC-O-MS and GC × GC-O-MS and sensory evaluation. *Food Funct.* **2021**, *12*, 4797–4807. [CrossRef]

molecules

Review

Pesticides Identification and Sustainable Viticulture Practices to Reduce Their Use: An Overview

Samuel Tucker [1], Georgiana-Diana Dumitriu (Gabur) [2,*] and Carmen Teodosiu [1,*]

1 Department of Environmental Engineering and Management, "Gheorghe Asachi" Technical University of Iasi, 700050 Iasi, Romania
2 Faculty of Horticulture, Iasi University of Life Sciences, 700490 Iasi, Romania
* Correspondence: diana.gabur@uaiasi.ro (G.-D.D.); cteo@ch.tuiasi.ro (C.T.)

Abstract: The use of pesticides is a necessary practice in the modern era. Therefore, it is impossible to ignore the pesticide market, which has developed into one of the most lucrative in the world. Nowadays, humans are subjected to many potential risks, and significant amounts of toxic compounds enter their bodies through food, drink, and the air itself. Identification and quantification of these hazardous compounds is crucial for the sustainable development of an increasing world population which poses high climatic and political constraints on agricultural production systems. The maximum residue limits for pesticides have been regulated by the Codex Alimentarius Commission and European Union to protect human health. In this review, we have summarized and explained the analytical methods for pesticide extraction and determination. Also, sustainable viticulture practices like organic vineyards, tillage, biopesticides, nanobiopesticides, and precision viticulture are briefly discussed. These new techniques allow wine growers to be more profitable and efficient, while contributing to the reduction of pests and increasing the quality of wines.

Keywords: pesticides; analytical methods; grapes and wines; sustainable viticulture

Citation: Tucker, S.; Dumitriu (Gabur), G.-D.; Teodosiu, C. Pesticides Identification and Sustainable Viticulture Practices to Reduce Their Use: An Overview. *Molecules* **2022**, *27*, 8205. https://doi.org/10.3390/molecules27238205

Academic Editors: Weiying Lu and Yanping Chen

Received: 31 October 2022
Accepted: 21 November 2022
Published: 24 November 2022

1. Introduction

Pesticides are described as "anything that prevents, eliminates, or regulates a hazardous organism ('pest') or illness, or protects plants or plant products throughout production, storage, and transport" by the European Commission [1]. According to the World Health Organization (WHO) pesticides are considered as a special class of chemical compounds used to destroy a broad range of pests that include weeds, insects, and rodents. The Stockholm Convention on Persistent Organic Pollutants has classed these pesticides as persistent organic pollutants (POPs), and their use is rigorously controlled globally.

Pesticides have been used since the Sumerians employed sulphur dust to control insects and mice 4500 years ago, and the Chinese used mercury and arsenic to control lice 3000 years ago [2]. Since then, new discoveries have led to the development of far more effective compounds, and more extensive investigation has resulted in the discontinuation or outright ban of several of these chemicals [3]. Dichloro-diphenyl-trichloroethane (DDT), the first modern synthetic insecticide, is a well-known example of this. During World War II, it was first created to battle malaria, typhus, and other insect-borne infections. After the war, it was widely employed as an insecticide in agriculture, as well as in private homes and gardens, and as a result of its broad use many insects developed resistance to the chemical (Figure 1) [4]. As the use of DDT became more widespread the negative consequences began to emerge, the most serious of which was the significant harm it caused to wildlife populations, particularly birds. DDT-exposed birds lay eggs with abnormally thin shells, raising concerns about the substance's long and short-term impacts on human health, especially given its potential to persist in soil for up to 15 years following application [3]. As a result, several nations throughout the world began banning DDT in the 1970s, and in

2004 the Stockholm Convention categorized DDT as a "restricted" substance that could only be used against mosquitoes in particular countries to prevent malaria [3].

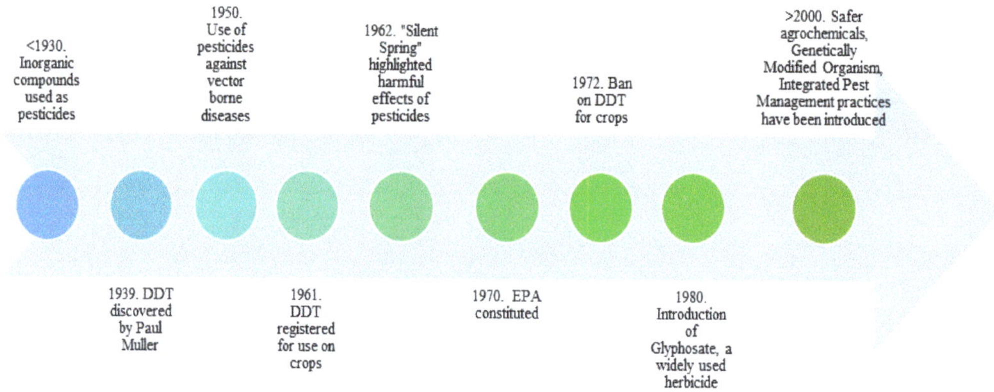

Figure 1. A timeline of pesticide uses since early 1930s.

Pesticides come in a variety of types and classifications, each designed to combat a specific 'pest.' Herbicides are the most frequent type and are used to "lower the density of weeds and stimulate the establishment of beneficial species" in agriculture and wilderness environments [5]. Insecticides are used extensively in agriculture and are designed to control, repel, or kill one or more insect species [6]. Fungicides are another form of pesticide that are used to control and remove mold, mildew, and other fungus [7]. Acaricides, nematicides, molluscicides, rodenticides, growth regulators, repellents, and rodenticides [1] are other common pesticides used in a range of industries for a variety of purposes.

2. Pesticide Trends around the World

According to the U.N., statistical research predicts, with 95 percent accuracy, that by 2030 there will be between 9.4 and 10.1 billion people on the planet, an additional 2 billion to the population estimated for mid-2019 [8]. The use of pesticides and other agrochemicals is therefore a necessary phenomenon in the modern era, due to their chemical interaction with pests and pathogens that lead to safe, high qualitative vegetables and fruits. In these conditions, it is impossible to ignore the pesticides market, which has developed into one of the most lucrative in the world. Pesticide trade volume in 2018 totalled 5.9 million tons, valuing 37.6 billion USD.

With an increase in pesticide exports from 1,992,898 million tons in 2015 to 2,454,480 million tons in 2019, Asia has recently emerged as the world's largest exporter of pesticides. China accounted for two-thirds of all pesticide exports, or 1,468,275 million tons, in 2019. Other significant pesticide exporters in 2019 included Germany with 0.48 million tons, India with 0.44 million tons, France with 0.45 million tons, and Belgium with 0.18 million tons. Contrarily, the top three countries importing pesticides in the same year were Brazil with 0.52 million tons, France with 0.26 million tons, and Canada with 0.21 million tons [9].

The application of pesticides to fields is a sign of the various national or local farming practices. The average amount of pesticides used in agriculture around the world increased from 2.28 kg/ha in 2005 to 2.69 kg/ha in 2019, according to FAO [10].

During this time, pesticide use in Asia and the Americas exceeded the global average, rising from 3.18 and 2.89 kg/ha in 2005 to 3.68 and 3.70 kg/ha in 2019, respectively. Europe used an average of 1.66 kg of pesticides per hectare of agricultural land in 2019, with the highest pesticide application rates in the Netherlands (8.88 kg/ha), Belgium (6.96 kg/ha), Montenegro (6.07 kg/ha), Ireland (5.97 kg/ha), Italy (5.21 kg/ha), and France (4.46 kg/ha).

While pesticides have a number of extremely useful attributes, they also have a *"deleterious effect on humans and the environment and their presence in food is particularly dangerous"* [11]. Many pesticides have been found to have severe effects with regards to endocrine disorders, reproductive problems, cancers, diabetes, obesity, and cardiovascular diseases [12].

According to a World Bank report, pesticide poisoning is thought to be the cause of 355,000 annual deaths [13]. According to their mode of action and level of exposure, different pesticides obviously have varying degrees of toxicity in humans and other biota. They consequently have a variety of consequences on living things. According to Muhammad et al. [14] and Sidhu et al. [15] pesticides affect the nervous systems of aquatic and terrestrial fauna and humans, resulting in endocrine, metabolic, and neurological disorders as well as various cancers like leukemia and bladder cancer.

To safeguard human health, while still facilitating world trade, the WHO and the FAO have set up a joint Codex Alimentarius Commission in order to coordinate food standards, as well as establishing universal Maximum Residue Levels (MRLs) of pesticides legally permitted in or on food or animal feed. Unfortunately, the MRLs for a particular pesticide used on a particular commodity can vary from country to country, causing a lack of harmonization in international trade. Farmers will produce food according to their own countries standards but are then penalized when they try to sell their product internationally and it does not meet the required MRLs. In addition, this poses a potential health risk to consumers in countries with stricter MRL standards when importing from a country with lower MRLs [16].

Although there are broad rules for pesticide residues in fruits, vegetables, or drinking water, scarce attention is still dedicated to derivate products, as grape must and grape-based products, which may contain these commodities as an ingredients. The maximum residue limits for grapes along with the acceptable daily intake (ADI), Acute Reference Doses (ARfT) and Acceptable Operator Exposure Level (AOEL) are presented in this review (Table 1) [17].

Table 1. Maximum residue limit (MRL) in grapes [table adapted from ref. [17]].

Pesticide	**MRL mg/kg**	**Year of Adoption**	**EU Pesticides Data**			**Regulation European Union, European Food Safety Authority Joint Meeting on Pesticide Residues**
			ADI (mg/kg bw/day)	**ARfT (mg/kg bw)**	**AOEL (mg/kg bw/day)**	
Acetamiprid	0.5	2012	0.025	0.025	0.025	Reg. (EU) 2018/113
Aldicarb	0.2	1997	Not approved Reg. (EC) No. 1107/2009			
Ametoctradin	6	2013	10.0		2.0	EFSA 2012
Amitrole	0.05	2004	0.001	0.015	0.001	EFSA 2015
Azocyclotin	0.3	2006	0.003	0.02		JMPR 2005
Azoxystrobin	2	2009	0.2		0.2	EFSA 2010
Benalaxyl	0.3	2010	0.04	0.5	0.05	Reg. (EU) 2020/1280
Benzovindiflupyr	1	2017	0.05	0.1	0.04	Reg. (EU) 2016/177
Bifenazate	0.7	2007	0.01		0.0028	05/58/EC
Bifenthrin	0.3	2016	0.015	0.03	0.0075	Reg. (EU) 2018/291
Boscalid	5	2010	0.04		0.1	08/44/EC
Bromopropylate	2	1997	Not approved Reg. (EC) No. 1107/2009			
Buprofezin	1	2010	0.01	0.5	0.04	Reg. (EU) 2017/360
Captan	25	2008	0.1	0.3	0.1	Dir 07/5, SCoFCAH July 08
Carbendazim	3	2008	0.02	0.02	0.02	Dir 06/135
Chlormequat	0.04	2018	0.04	0.09	0.04	EFSA 08
Chlorothalonil	3	2011	0.015	0.05	0.003	Reg. (EU) 2019/677

Table 1. *Cont.*

Pesticide	MRL mg/kg	Year of Adoption	EU Pesticides Data			Regulation European Union, European Food Safety Authority Joint Meeting on Pesticide Residues
			ADI (mg/kg bw/day)	ARfT (mg/kg bw)	AOEL (mg/kg bw/day)	
Chlorpyrifos	0.5	2003	-	-	-	
Chlorpyrifos-Methyl	1	2010	-	-	-	
Clofentezine	2	2008	0.02		0.01	Dir 08/69
Clothianidin	0.7	2012	0.097	0.1	0.1	06/41/EC
Cyazofamid	1.5	2016	0.17		0.3	03/23/EC
Cycloxydim	0.3	2013	0.07	2.0	0.1	EFSA 10
Cyflumetofen	0.6	2015	0.17		0.11	Reg. (EU) No. 2019/716
Cyhexatin	0.3	2006	0.003	0.02		JMPR 2005
Cypermethrins (including alpha- and zeta-)	0.2	2009	0.05	0.2	0.06	Dir 05/53
Cyprodinil	3	2005	0.03		0.03	Dir 06/64
Deltamethrin	0.2	2004	0.01	0.01	0.0075	Dir 03/5
Dichlobenil	0.05	2015	Not approved Reg. (EC) No. 1107/2009			
Dichloran	7	2004	0.005	0.025	0.005	EFSA 10
Difenoconazole	3	2014	0.01	0.16	0.16	Dir 08/69
Dimethomorph	3	2015	0.05	0.6	0.15	Dir 07/25
Dinotefuran	0.9	2013	-	-	-	
Dithiocarbamates	5	2005	-	-	-	
Emamectin benzoate	0.03	2012	0.0005	0.01	0.0003	EFSA 2012
Ethephon	0.8	2016	0.03	0.05	0.03	Dir 06/85, SCoFCAH December 08
Etofenprox	4	2013	0.03	1.0	0.06	EFSA 08
Etoxazole	0.5	2011	0.04		0.03	Reg. (EU) 2020/2105
Famoxadone	2	2005	0.006	0.1	0.0024	Reg. (EU) 2021/1379
Fenamidone	0.6	2015	-	-	-	
Fenarimol	0.3	1999	0.01	0.02		Dir 06/134
Fenbuconazole	1	1999	0.006	0.3	0.02	EFSA 10
Fenbutatin Oxide	5	1995	0.05	0.1		EFSA 10
Fenhexamid	15	2006	0.2		0.2	Reg. (EU) 2015/1201
Fenpyrazamine	3	2018	0.13	0.3	0.2	Reg. (EU) No. 595/2012
Fenpyroximate	0.1	2018	0.01	0.02	0.005	EFSA 2013
Fluazifop-p-butyl	0.01	2017	-	-	-	
Flubendiamide	2	2011	0.017	0.1	0.006	EFSA 2013
Fludioxonil	2	2006	0.37		0.59	Dir 07/76
Flumioxazin	0.02	2016	0.009	0.05	0.018	02/81/EC
Fluopicolide	2	2010	0.08	0.18	0.05	2010/15/EU
Fluopyram	2	2011	0.012	0.5	0.05	Reg. (EU) No. 802/2013
Flupyradifurone	3	2017	0.064	0.15	0.064	Reg. (EU) 2015/2084
Flusilazole	0.2	2008	0.002	0.005	0.005	Dir 06/133
Flutriafol	0.8	2013	0.01	0.05	0.05	11/42/EU
Fluxapyroxad	3	2016	0.02	0.25	0.04	EFSA 12
Folpet	10	2006	0.1	0.2	0.1	Dir 07/5, SCoFCAH July 08

Table 1. *Cont.*

Pesticide	MRL mg/kg	Year of Adoption	EU Pesticides Data			Regulation European Union, European Food Safety Authority Joint Meeting on Pesticide Residues
			ADI (mg/kg bw/day)	ARfT (mg/kg bw)	AOEL (mg/kg bw/day)	
Fosetyl Al	60	2018	-	-	-	
Glufosinate-Ammonium	0.15	2013	0.021	0.021	0.0021	Dir 07/25
Haloxyfop	0.02	2011	0.00065	0.075		EFSA 06
Hexythiazox	1	2010	0.03		0.009	11/46/EU
Imidacloprid	1	2004	0.06	0.08	0.08	Dir 08/116
Indoxacarb	2	2006	0.006	0.125	0.004	06/10/EC
Iprodione	10	-	0.02	0.06	0.04	Reg. (EU) 2017/2091
Kresoxim-Methyl	1.5	2019	0.4		0.9	99/1/EC
Malathion	5	2006	0.03	0.3	0.03	Reg. (EU) 2018/1495
Mandipropamid	2	2009	0.15		0.17	EFSA 2018
Meptyldinocap	0.2	2011	0.016	0.12	0.008	Reg. (EU) No. 1330/2014
Metalaxyl	1	-	0.08	0.5	0.08	2010/28/EU
Methidathion	1	1999	0.001	0.01		JMPR 1997
Methomyl	0.3	2009	0.0025	0.0025	0.0025	EFSA 06
Methoxyfenozide	1	2005	0.1	0.1	0.06	Reg. (EU) 2019/158
Metrafenone	5	2015	0.25		0.43	07/6/EC
Myclobutanil	0.9	2015	0.025	0.31	0.03	EFSA 10
Oxathiapiprolin	0.9	2017	0.14		0.04	Reg. (EU) 2017/239
Parathion-Methyl	0.5	2004	Not approved Reg. (EC) No. 1107/2009			
Penconazole	0.4	2017	0.03	0.5	0.03	Dir 09/77
Permethrin	2	-	Not approved Reg. (EC) No. 1107/2009			
Phosmet	10	1999	0.01	0.045	0.02	Dir 07/25
Propargite	7	2004	0.03	0.06		EFSA 2018
Pyraclostrobin	2	2006	0.03	0.03	0.015	04/30/EC
Pyrimethanil	4	2008	0.17		0.12	Dir 06/74
Quinoxyfen	2	2007	0.2		0.14	Directive 2004/60/EC
Saflufenacil	0.01	2012	Not approved Reg. (EC) No. 1107/2009			
Spinetoram	0.3	2013	0.025	0.1	0.0065	EFSA 2013
Spirodiclofen	0.2	2010	0.015		0.009	EFSA '09
Spirotetramat	2	2009	0.05	1.0	0.05	EFSA 2013
Sulfoxaflor	2	2013	0.04	0.25	0.06	Reg. (EU) 2015/1295
Tebuconazole	6	2012	0.03	0.03	0.03	Dir 08/125, EFSA 08
Tebufenozide	2	2004	0.02		0.008	ESFA 10
Teflubenzuron	0.7	2017	0.01		0.016	EFSA 08
Triadimefon	0.3	2015	0.03	0.08		JMPR 2004
Triadimenol	0.3	2015	0.05	0.05	0.05	EFSA 08, Dir 08/125
Trifloxystrobin	3	2006	0.1	0.5	0.06	Reg. (EU) 2018/1060
Triflumizole	3	2014	0.05	0.1	0.05	2010/27/EU
Zoxamide	5	2008	0.5		0.3	Reg. (EU) 2018/692

3. Pesticides Extraction and Detection

3.1. Pesticide Extraction

In general, pesticide determination using analytical methods involves a number of main steps such as: sample preparation, homogenization, extraction, and clean-up procedures including, separation, detection and data analysis. Identification of pesticides residues includes 2 steps: extraction of analytes from the bulk of the samples, and clean-up of the analytes from any co-extractives present in the samples [18]. Figure 2 shows a visual example of this process.

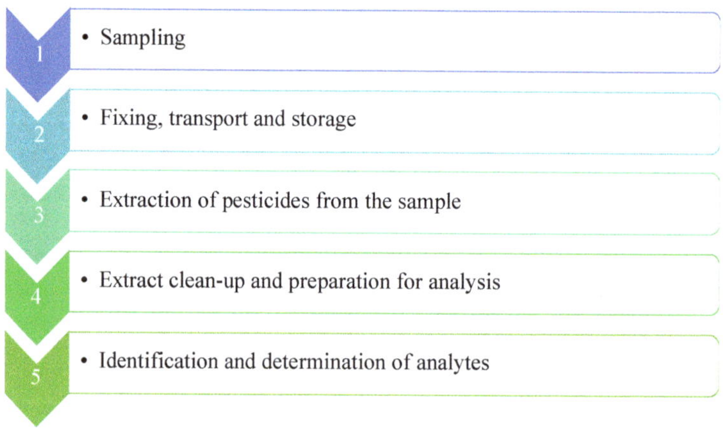

Figure 2. The main stages of analytical procedures for determining pesticides [11].

There are many different kinds of extraction techniques, ranging from traditional extraction techniques like the soxhlet extractor and quick, easy, cheap, effective, rugged, and safe (QuEChERS) to microextraction techniques, which are divided by different sorbents, such as the liquid-phase microextraction (LPME), solid-phase microextraction (SPME), and stir bar sorptive extraction (SBSE) [19].

The advancement in analytical procedures brought about by the evolution of extraction techniques has decreased the complexity of sample treatment while simultaneously increasing the accuracy and precision of the analysis [16]. The choice relies on the analytical problem at hand because each technique has advantages and disadvantages of its own.

The Mills method was created in 1963 and is based on the use of acetonitrile to remove organochlorine insecticides and other nonpolar pesticides from low-fat meals. Following extraction, partitioning takes place with the addition of sodium chloride into a nonpolar solvent, such as petroleum. The extract is cleaned using a Florisil column. It should be noted that when analyzed with a nonpolar solvent, moderately polar pesticides like organophosphorous insecticides largely evaporate [16].

The Stoherr method is a small variation of the Mills method that seeks to broaden the procedure's analytical applicability to compounds with various polarities [16]. Once more, acetonitrile is used for extraction, but this time, Florisil is replaced by acid-treated charcoal and nonpolar petroleum is swapped out for dichloromethane, which has a higher polarity. The vast majority of organophosphorous pesticides found in fruits and vegetables, including grapes, can be removed using this technique.

The Luke method, developed in 1975, centres around acetone as an extractant, little cleaning, and a variety of GC systems with element-specific and element-selective detectors. With this technique, Florisil is used for both the combined cleanup of organochlorine and organophsphorous pesticides. In order to fully saturate the aqueous phase, sodium chloride is also added. This increases the amount of acetone and raises its polarity, which results in excellent polar analyte recoveries [16].

Following on from the Luke Method is liquid-liquid extraction (LLE), also known as solvent extraction and partitioning, considered as a green analytical chemistry method. Even though this method is compatible with the sustainable development concepts, LLE is not employed in multiresidue procedures [16]. This approach, which separates chemicals based on their relative solubility in water and an organic solvent, is typically employed for sample cleanup. Dichloromethane is added to a solvent combination after acetone extraction. This procedure works with grapes and their by-products, but it is time-consuming, labor-intensive, and requires a lot of hazardous solvents that are dangerous to use around people [16]. The limits of detection (LODs) range obtained in one study [20] which combined the LLE method with HPLC-MS/MS analytical systems to determine four pesticides (malathion, diazinon, imidacloprid, and triamedimefon) in fruit juice samples was from 3×10^{-4} to 3×10^{-2} mg/L, with a correlation coefficient of 0.995. Another study by Farajzadeh et al. [21] devised a straightforward, quick, and affordable approach for pyrethroid pesticide determination using the LLE and dispersive liquid-liquid microextraction (DLLME), with detection limits ranging from 0.02 to 0.17 mg/kg.

An alternative method to LLE, known as solid-phase extraction (SPE), was created in the middle of the 1970s for the separation, purification, preconcentration, and solvent exchange of solutes for solutions. A sample is isolated, concentrated, purified, and cleaned up using this technique. Given that it involves less time, less solvent, fewer stages, and is more cost-effective than LLE, it offers a number of major advantages. Additionally, SPE can be used on materials in conjunction with other analytical techniques to enhance the process; however, the main drawback of this technique is a significant degree of variability in the results [16].

Solid-Phase Microextraction (SPME) was developed in 1990 to further optimize SPE/LLE and redress their limitations. This method requires negligible solvent for sample preparation and uses a fused silica fiber coated with a stationary phase attached to a micro syringe. It is important to note that the extraction temperature, time and ample agitation must be optimized, and operating conditions need to be consistent. The main advantages are the reduction of solvent use, the combination of sampling and extraction into one step, the ability to examine smaller sample sizes, the possibility to use fibers many times without the loss of the adsorbate, the possibility to rerun the analysis of any given sample, and major design changes with respect to chromatographs are not necessary. However, the disadvantages include the fact that there is no way of ensuring a sufficiently broad analytical range in a single analysis, there remain some problems with reproducibility, method optimization problems, low recoveries of analytes, as well as limited volume of stationary phase to be applied to the fiber which can lead to incomplete extraction [16].

Matrix Solid-Phase Dispersion (MSPD) was developed in 1989, and was a modification of SPE based on the use of a sorbent which acts as an abrasive in order to produce a modified opening of the solid matrix, allowing for extraction. It is based on a mixer of fine dispersion of the matrix with a sorbent material such as alumina or silica, and many MSPD procedures use co-columns to obtain further fractionation. Advantages of MSPD include the small amount of sample and solvent, fewer experimental steps, direct handling of samples, ease of implementation, reduced solvent consumption, and the low overall cost. The main disadvantages are the insufficiently wide analytical range of a single sample, its unsuitability for dry samples or samples with a high lipid content, and the fact that it requires an additional cleanup step. A rapid and sensitive multiresidue method for the analysis of pesticides in fruits (acetamipride, carbendazim, carbaryl, carbofuran, imidacloprid, malathion, propazine, dimethoate and tebufenozide was developed by Radisic et al. [22]. The method involves an extraction procedure based on MSPD using diatomaceous earth as a dispersant and dichloromethane as the eluent. In addition, according to European and Brazilian monitoring programs, most of the selected pesticides are frequently detected pesticides in fruits and vegetables, and that MSPD requires approximately 95% less solvent and can be performed in 90% less time when compared to such classical methods.

Stir Bar Sorptive Extraction (SBSE) was developed in 1999 and attempted to overcome the limited extraction capacity of SPME fibers. According to Urkude et al. [16], this method is a solvent-less sample preparation method for extraction and enrichment of organic compounds from aqueous matrices. A large surface area is created by a thick bonded absorbent layer on a glass stirrer bar, leading to a higher phase ration and thus a better recovery and sample capacity. It has a high effectiveness for nonpolar and medium polarity compounds from liquid samples, it is easy to apply and automate, and it is highly flexible, sensitive, reputable, and reproduceable. However its main disadvantage is the fact that it can only be applied to medium-high volatility and medium-high thermo-stability analytes, and sampling times and cost of instrumentation can be high.

In order to separate desired analytes from the sample matrix and introduce them into the solvent, microwave radiation is used in microwave-assisted extraction (MAE). As a result, the solvent may be heated quickly, and extraction typically lasts 15 to 30 min. It has a high sample throughput, uses less solvent, operates at low temperatures, has great automation and extraction rates, and allows for the uninterrupted extraction of multiple samples at once. But only thermally stable chemicals can be used with this method, and they need to be dissolved in a polar solvent like water [16].

QuEChERS is the name of a method in which many pesticides can be analysed simultaneously in different food matrices, and similar to many of the techniques previously outlined, it involves an extraction, separation and cleanup phase. Two predominant methods of QuEChERS arose, the European Committee for Standardization proposed a citrate-buffered method, while Association of Official Analytical Collaboration (AOAC) International proposed an acetate-buffered method. While both were especially effective in terms of lipid coextractives and therefore well suited towards extraction of the high-sugar grape, the acetate-buffered method was concluded to be more appropriate for use with grapes. The main benefits of this QuEChERS include high recoveries with a wide range of polarity and volatility, high sample throughput, the need for simple equipment for sample preparation, the need for a smaller amount of organic solvent, lower reagent costs, ruggedness, and the removal of organic acids and other potential contaminants during cleanup [16]. It also offers a significant advantage over traditional methods which require the use of multistage procedures, large samples, and one or more extract cleanup steps. The number one disadvantage of QuEChERS is that the final extract has to be concentrated to a greater extent in order to achieve the necessary sensitivity and thus to achieve the limits of quantification desired. Many modifications were proposed to the QuEChERS model, including the use of different solvents such as graphitized carbon black, or the use of low-temperature precipitation, which allowed for the extraction of large numbers of pesticides from different classes and matrices, as well as for advanced cleanup stage processes [23–26].

There are other forms of microextraction methods used for the pesticide detection which include dispersive liquid liquid microextraction (DLLME), single drop microextraction (SDME), continuous flow microextraction (CFME), hollow fiber-liquid phase microextraction (HF-LPME), as well as a combination of SPE, DLLME, and solid liquid extraction (SLE) [18]. While there are a number of different extraction techniques used, each have their own distinct advantages and disadvantages depending on a number of criteria. These advantages and disadvantages are summarized for each technique below in Table 2 according to research carried out by Samsidar et al. [18], and Wilkowska and Biziuk [27].

Table 2. Pesticide identification techniques and associated advantages and drawbacks [table adapted from ref. [18,27]].

Technique	Advantages	Drawbacks
Microwave-assisted extraction (MAE)	• realized easily • extraction of numerous samples can occur simultaneously • short extraction time required for small quantities of solvents	• extraction selectivity is not sufficient • need to separate extract from post-extraction residue • requires clean-up step • requires time for the vessels to cool down
Accelerated solvent extraction (ASE)	• ability to automate extraction • all process steps can be carried out identically • short extraction time • moderate solvent consumption • preparation of sample prior to analysis is straightforward	• purchasing and maintaining apparatus is costly • poor extraction selectivity • requires clean-up of extracts and equipment after each use which can be time-consuming
Matrix solid-phase dispersion (MSPD)	• cost per analysis is relatively low • straightforward equipment • several analyses can be performed simultaneously • can be carried out under in-situ conditions • requires only small quantities of solvents	• it is not possible to have a sufficiently wide analytical range in a single procedure • dry samples or samples with high lipids content are not suited • adsorbent consumption is relatively high • requires an additional clean-up step • low recoveries of analytes can occur
Solid-phase microextraction (SPME)	• use of solvents can be eliminated • suspended matter sensitivity is not present • adsorbent capacity is limited • one fibre can be utilised many times without loss of adsorbate • chromatographs with ordinary injectors can be used	• ensuring a sufficiently broad analytical range in a single procedure is not possible • reproducibility can yield difficulties • recoveries of analytes are relatively low
Supercritical fluid extraction (SFE)	• substantial reduction of solvent consumption • extraction of thermolabile compounds is possible • analysed compounds are not degraded • short extraction time • relatively low labour intensity • extraction permitted in semi-automatic mode with a special device	• purchasing and maintaining apparatus is costly • poor extraction selectivity • requires clean-up of extracts and equipment after each use which can be time-consuming • relatively complicated compared to other extraction techniques

Table 2. *Cont.*

Technique	Advantages	Drawbacks
Membrane extraction techniques	• untreated samples introduced directly • little to no use of solvent • samples with very complex matrixes can be analysed • high selectivity • elimination of interferences • analyte enrichment is high • easily automated	• high time consumption • low efficiency • membrane pores can be easily clogged by solid contaminants which leads to extended time of analysis
Liquid liquid extraction (LLE)	• reliable and simple • can be adapted to numerous sample types and analytes	• large volume of hazardous solvent required • time consuming technique
Solid phase extraction (SPE)	• less time consuming than LLE • purification and pre-concentration procedures are effective	• requires pre-treatment • requires toxic organic solvent
Traditional column-based SPE	• ensures more effective sample clean-up	• requires plastic cartridges containing a sorbent material and vacuum manifolds • requires a larger sample • requires multiple solvents, manual operation, column preconditioning, and solvent evaporation steps • generates solvent waste components
Dispersive SPE (d-SPE)	• recoveries of analytes with acidic or basic properties are greater and have more reproducibility • uses less sorbent, sample, and equipment and thus is a cheaper and easier technique	• can only be used when the SPE sorbent removes matrix components and not the analytes
Quick, Easy, Cheap, Rugged, Effective and Safe (QuEChERS)	• broad scope of analytes • low volume of solvents and glassware required • straightforward instrumentation • effective and flexible	• enrichment factors are low • to achieve the necessary sensitivity and thus to achieve the limits of quantification, the final extract must be concentrated to a greater extent
Dispersive liquid-liquid microextraction (DLLME)	• simplicity • low volume of toxic solvents • rapid extraction • inexpensive	• low efficiency of extraction
Single Drop Microextraction (SDME)	• rapid and inexpensive • straightforward to operate • requires little organic solvents and therefore is environmentally friendly • renewability of extraction phase	• less stability of the suspending drop • extraction times can be lengthy

Table 2. *Cont.*

Technique	Advantages	Drawbacks
Hollow fiber-liquid phase microextraction (HF-LPME)	• inexpensive • volumetric ratio of the acceptor and the sample phases greatly reduced	
Continuous Flow Microextraction (CFME)	• reduced solvent consumption • requires inexpensive instrument • rapid and straightforward • efficient pesticide extraction in complex matrices	• limited volume of micro-drop • inserting the drop into extraction glass chamber can lead to difficulty

3.2. Pesticide Detection

Chromatography, which has been employed in the detection and analysis of a range of pesticides, consists of a mobile phase (gas or solvent) and a stationary phase, such as a column or capillary tube. Gas chromatography (GC), liquid chromatography (LC), high-performance liquid chromatography (HPLC), and supercritical fluid chromatography (SFC), as well as mass spectrometry (MS), are different types of chromatography based on the mobile phases used [28]. In the future, chromatography and its combination with mass spectrometry will be widely used.

With its beginnings in the 1950s and current widespread use, GC is a significant detection method. In order to assess the sample's composition after gasification, the inert gas is transferred into the separation apparatus. For non-polar, highly volatile, and quickly vaporized chemicals, the GC technique is appropriate. To estimate pesticides using the GC method, experts from all around the world have recently used a variety of extraction techniques.

Many compounds in plant-derived food that are rarely studied or difficult to identify, like highly polarized and non-volatile and/or thermally labile pesticides, can be quickly and effectively identified using the liquid chromatography-mass spectrometry (LC-MS) method, even those that are not GC-amenable. It is now possible to detect pesticide traces in complex systems like fruits and vegetables, grains, and animal-derived foods thanks to better LC-MS/MS.

In HPLC, the solvent moves under high pressure that is generated by a pump to get around the pressure drop challenge and shorten the separation time. In terms of pesticide detection, HPLC-MS technology has produced a wealth of qualitative and quantitative data. According to Bletsou, Jeon, Hollender, Archontaki, and Thomaidis [29], HPLCQqQ-MS in multiple reaction monitoring (MRM) has demonstrated great sensitivity, selectivity, and low detection limits for studies.

With advantages in speed, sensitivity, and low cost, the supercritical fluid chromatography-tandem mass spectrometry (SFC-MS/MS) method is frequently used for separations involving non-volatile or thermally labile pesticides as well as to quantify chiral or achiral chemical compounds in biological samples [30].

The most used method for detecting pesticide residues in food samples generated from plants is LC or GC combined with MS. However, these methods necessitate specialized equipment used by skilled individuals, which is quite expensive and hostile to the environment as many chemical agents may be consumed during the detection [31]. Spectrum analysis is more effective than chromatography methodology at detecting pesticide residues due to its high sensitivity and quick process.

Spectrum analysis is a complementary chromatography method that is effective for detection of pesticide residues due to its high sensitivity and quick process. The most popular spectroscopic techniques are based on the Raman spectrum, near-infrared spectroscopy, and fluorescence spectrum. Resonance Raman spectroscopy (RRS), coherent anti-Stokes

Raman spectroscopy (CARS), stimulated Raman spectroscopy (SRS), surface-enhanced Raman spectroscopy (SERS), and tip-enhanced Raman spectroscopy (TERS) are some of the most advanced Raman spectroscopy techniques available today [32].

The pesticide detection process calls for significant human resources and intricate pre-treatment techniques. Numerous quick detection techniques for pesticide residues have been created in such situations, making it simple, quick, and accurate to check pesticide residues. Some sensors, such as electrochemical and optical methods, can measure pesticides with adequate accuracy and over the proper time. The efficacy of electrochemical sensors, which use working electrodes as a transducer, depends on the analyte's potential redox state and the working potential. Nanosensors have been proposed for the detection of pesticides as a result of the development of nanotechnology. With carbon nanotubes as an example, the development of nanosensors has increased their sensitivity, stabilizing the effect on suppressing acetylcholine esterase (AChE) activity. Gas chromatography (GC), gas chromatography mass spectrometry (GC/MS), and gas chromatography tandem mass spectrometry (GC/MS/MS) are often used because of their high separation power, selectivity, and identification capabilities of MS.

Different studies have outlined the effectiveness of LC-MS/MS and GC/MS respectively [33]. In addition, the various sensitive detectors coupled with GC such as a nitrogen phosphorus detector (NPD), a flame ionization detector (FID), a flame photometric detector (FPD), and an electron capture detector (ECD), have improved the detection and quantification of pesticides, with ECD being especially useful for the organochlorine pesticides, NPD for organophosphorus and nitrogenated pesticides, and FPD for sulpher and phosphorous pesticides [16]. Enzyme-linked immunosorbent assay (ELISA) is another conventional method and due to its reliable high-throughput immunoassay it is currently known as the most prevalent form of immunoassay for pesticide monitoring tools. Capillary electrophoresis (CE) is another valuable analysis technique and relatively applicable for various practices due to the fact that it requires small volumes of reagents and samples, and has great separation efficiency [18].

4. Vineyard and Wine and Specific Research on Pesticides

In research undertaken by Nieto-Garcia et al. [34], GC-QqQ-MS/MS was used to optimize a new approach for determining pesticide residues at trace levels in dietary supplements from grape seed extracts. Because of the matrix's complexity, numerous cleansing stages must be included in the extraction operation to eliminate interferences and coextractive compounds, hence improving sensitivity and reducing GC maintenance. In this regard, it was discovered that using a single sorbent is insufficient to provide satisfactory results, and that a combination of sorbents should be utilized instead. The validation criteria (intra-day and inter-day precision, recovery, linearity, limits of quantification (LOQs), and LODs) were assessed and appropriate results were obtained. Given that there is still no European legislation on pesticide residues in nutraceuticals, the LODs and LOQs were deemed adequate.

He et al. [35] proposed an analytical method for wine samples which provides a broad pesticide screen and quantification methodology. Through multiple reaction monitoring (MRM) and isotope dilution analysis mass spectrometry (IDA-MS/MS) collection, liquid chromatography-tandem quadrupole-linear ion trap (LC-QqLIT MS) was used to analyze and screen target pesticides. Pesticides and other organic contaminants were screened for both target and non-target pesticides and other organic contaminants using LC-QTOF MS in an automated IDA-MS/MS. For target and non-target screening and quantification of pollutants, the combination of LC-QTOF MS and LC-QqLIT MS proved to be superior. The combination of both methodologies yielded excellent results in terms of precise quantification and unambiguous confirmation. Quantification was done using an LC-QqLIT MS in MRM mode, which was used as a supplement to LC-QTOF MS quantification at low concentrations. LC-QqLIT MS working in EPI mode and LC-QTOF MS operating in IDA-MS/MS mode, respectively, provided unequivocal detection of target and non-target pollutants. This technology for wine quality control was made feasible and efficient by direct injection of wine samples and wide-ranging contaminants screening combined with MRM measurement. In environmental science and food chemistry, this technology provides a new perspective on pesticide and other contamination screening and quantification.

Vaquero-Fernandez et al. [36] presented a simple, rapid method for the determination of pyrimethanil during the winemaking process from grape to bottled wine. Gas chromatography with nitrogen–phosphorus detection (GC-NPD) was used to make the determination, which was later verified by gas chromatography/mass spectrometry (GC/MS). For different portions of the fruit, the overall process included three methods: surface, skin, and pulp. After a short sample extraction, the proposed SPE-GC-NPD approach permitted rapid determination of pyrimethanil, suitable for monitoring the fungicide in must and wine from red grapes. The procedures for grapes, must, fermenting must, and wine were highly sensitive and offered good recoveries, linearity, precision, and accuracy. The quantification was done with a matrix-matched calibration to avoid matrix effects.

In a paper presented by Pérez-Ortega et al. [37], a generic sample treatment approach based on solid-phase extraction (SPE) using polymeric-type SPE cartridges was developed for large-scale simultaneous assessment of multiclass pesticides and mycotoxins in wines. The sample treatment procedure was evaluated using a liquid chromatography electrospray time-of-flight mass spectrometry method with 60 representative multiclass pesticides and 9 mycotoxins. The results in terms of sensitivity, extract cleanliness, and matrix effects were comparable to earlier studies, with good recovery rates obtained for several pesticide and mycotoxins classes, demonstrating the adaptability and broad applicability of the suggested technique. The method was successfully applied to the analysis of 24 red wine samples acquired on the open market in Spain. Aflatoxin B2 and metalaxyl were the most detected compounds, in 75% and 50% of the studied samples, respectively.

Navarro et al. [38] outlined a rapid multiresidue gas chromatographic method using both electron capture detector (ECD) and nitrogen phosphorus detector (NPD). After a simple extraction of the sample, the suggested approach allows for the rapid assessment of 17 fungicides often employed in vineyards, and may be utilized for their determination in grapes, must, and wine, according to quality control and Good Laboratory Practice (GLP) requirements. Because chromatograms of untreated grape, must, and wine samples are free of interfering peaks, no clean-up is required. The linearity regression coefficients were all at least 0.994. The percentage of spiked grapes, must, and wine samples recovered varied from 78 to 107 percent, with relative standard deviations of less than 14 percent. Individual detection limits ranged from 0.02 to 0.1 ng. Quantification levels ranged from 0.01 to 0.05 mg/kg, all of which were lower than the maximum residue limits imposed by the European Union's principal wine-producing countries, Spain, France, and Italy. The quantification limits only coincide with the maximum residue limits (0.05 mg/kg) specified by Spanish legislation for fludioxonil and hexaconazole.

A study carried out by Pelajic et al. [39] used GC/MS to establish a new multiresidue approach for determining 25 pesticide residues in red wine. Solid phase extraction was employed to extract samples from wine, with a washing solution of methanol and water, and elution solvents of acetonitrile and n-hexane. For most pesticides, the LOQs were much below 10 lg/L, and recoveries ranged from 70 to 120 percent. Pesticides were detected in 30 of the 32 red wine samples from Croatia, with a total of 15 pesticides discovered, seven of which were at high concentrations.

González-Rodríguez et al. [40] outlined a specific and sensitive method based on ethyl acetate/hexane extraction followed by SPE clean-up with GCB/PSA followed by GC-ITMS and LC-DAD identification for the analysis of fungicides tested in samples (grapes, musts, pomaces, lees, distilled spirits, and wines). Fungicide concentrations in grapes harvested at the legal preharvest time were lower than the EU MRL values; however, new fungicide concentrations in grapes, present in phytosanitary treatments to control downy mildew applied under critical agricultural practices, were higher or close to the EU MRL values. The dissipation of fungicide residues observed during all steps of the white wine-making process was possible, with the pressing and settling stages being the most important in their removal. Except for valifenalate, each fungicide had a very high decrease rate (ranging from 90% to 99%). Estimated MRLs for white wines were proposed for future EU legislation to restrict the level of fungicides in wines based on data acquired during the vinification process.

Recentlly, Yang et al. [41] showcased a novel, simple, and successful method of using Ultra Performance Liquid Chromatography Mass Spectrometry (UPLC-MS/MS) to simultaneously assess the presence of pyraclostrobin, dimethomorph, cymoxanil and cyazofamid in grapes. The kinetics of fungicide degradation and terminal residue levels in grapes using field tests in Zhejiang Province and Tianjin in 2017 was investigated, finding half-lives in grapes ranging from 0.9 to 13.3 days. Cymoxanil degraded the fastest of all fungicides, and the maximum grape terminal residue levels for pyraclostrobin, dimethomorph, cymoxanila and cyazofamid observed during three monitoring intervals were all below the respective MRLs in China.

González-Rodríguez et al. [42] utilized gas chromatography with an ion trap mass spectrometry detector (GC–IT MS) to determine the presence of tebuconazole residues in grapes, musts, and wines. Tebuconazole remained on the solid matter (cakes and lees) as well as the clarifying agent. Tebuconazole was removed from 86 percent of the finished wine. According to these findings, the MRL for tebuconazole in red wines might be set at eight times lower (0.25 mg/L) than the MRL for wine grapes (2 mg/kg). Tebuconazole did not alter alcoholic or malolactic fermentations in vitro, according to the results of in vitro tests. At the same time, neither the degradation nor the adsorption of tebuconazole was affected by these two fermentative processes.

In a study carried out by Heshmati et al. [43], the QuEChERS extraction method was developed and validated in conjunction with GC-MS/MS to assess penconazole, hexaconazole, diazinon, ethion, and phosalone in grapes. The half-life of triazole fungicides was shown to be longer than that of phosphorus compounds in dissipation experiments, which could be a contributing cause to the pesticides' high preharvest interval (PHI). Meeting spraying criteria in vineyards, such as setting a controlled dose for these pesticides and paying attention to their PHI, can have a major impact on residual pesticides in grapes. The PHI of penconazole, hexaconazole, diazinon, ethion, and phosalone concentration in grape was 15, 23, 12, 13 and 15 days after spraying, according to the current study results. In addition to taking pesticide PHI into account, immersing grapes in a sodium bicarbonate solution could considerably limit pesticide exposure for consumers.

In research carried out during the winemaking process, the fate of zoxamide and its enantiomers was studied in depth by Pan et al. [44]. After each processing method, including washing, peeling, fermentation, and clearing, the enantiomers of zoxamide were separated and identified using ultra-high-performance liquid chromatography coupled with tandem mass spectrometry (UHPLC–MS/MS). All three treatments showed significant enantioselectivity, and the results showed that R-zoxamide deteriorated faster than S-zoxamide during the fermentation process. Each procedure's processing factors (PFs) were frequently less than 1, and the total process's PF ranged from 0.019 to 0.051, indicating that the entire process may significantly reduce zoxamide residue in red and white wine. The findings could aid in more precise zoxamide risk assessments during the winemaking process.

Paya et al. [45] set out to determine the in vitro bioavailability of pesticides that control and inhibit insect growth–flufenoxuron, lufenuron, pyriproxyfen, and fenoxycarb–in grapes grown under good agricultural practice (GAP) while adhering to pre-harvest intervals (PHI) for critical conditions (CAP) in the most unfavonurable conditions. In order to determine matrix-related variations, the bioavailability of wines made from grapes was investigated in each test and in standard solutions. The human gastric, intestinal, and absorption processes were replicated. The researchers employed porcine pepsin, pancreatin, bile salts, and semipermeable cellulose dialysis tubing. The residues of the pesticides investigated were extracted using the QuEChERS technique, and the determination was done using HPLC-

MS. Fenoxycarb (3.27 percent) and pyriproxifen (2.04 percent) in wine had the highest percentages of dialyzation for grape and wine matrices.

In research carried out by Cus et al. [46], pesticide residues in the vinification process of two white and two red grapevine varietals were monitored. During the vinification process, crushed grapes, cake, must, lees, and wine were all sampled. During the ripening period, grapes were also sampled. All of the samples were taken in triplicate and tested for 117 pesticides. Three internal analytical methods were used to determine pesticide residues: the multi-residual GC-MS method (71 pesticides), the multiresidual LC-MS-MS method (45 pesticides), and the GC-MS method for dithiocarbamate determination. During ripening, the insecticides boscalid and phosalone were the most persistent. Separations during the solid and liquid phases of the vinification process, particularly the pressing of crushed grapes and wine racking following alcoholic fermentation, considerably reduced pesticide residual quantities in must and wine. Boscalid, cyprodinil, dimethomorph, fenhexamid, metalaxyl, and procymidone were the most persistent pesticides in grapes during ripening.

Golge and Kabak [47] examined the levels of 172 pesticide residues in table grapes in Turkey from August to October 2016. 280 table grape samples were collected from supermarkets, bazaars, and greengrocer shops throughout four Turkish provinces. Liquid chromatography with tandem mass spectrometry was used to examine the samples. Quantification limits varied from 0.002 to 0.010 mg/kg. The validation data demonstrated good recoveries, repeatability, and reproducibility, as well as meeting the rest of the European SANTE/11945/2015 Guideline's standards. In 59.6 percent of the table grapes, pesticide traces were discovered. 20.4 percent of the samples had residues over the EU limit residue values. Azoxystrobin, chlorpyrifos, boscalid, and cyprodinil were the most common pesticide residues. Lower bound, middle bound, and upper bound values were replaced for left-censored results (40.4 percent of the results). The hazard index (HI) for adults was 3.37 percent and 9.42 percent for children in the worst-case scenario. Chlorpyrifos was the leading cause of HI (65 percent).

In another study undertaken by Castro et al. [48], liquid chromatography with tandem mass spectrometry using triple quadrupole (QqQ) and quadrupole time-of-flight (QTOF) MS instruments was used to investigate the coexistence of the anilinopyrimidine fungicides pyrimethanil (PYR), cyprodinil (CYP), and suspected metabolites in wine samples. For the first time, quantitative data acquired from wine samples after solid-phase extraction (SPE) revealed the systematic existence of 4-hydroxyanilino derivatives of PYR and CYP in wines bearing parent fungicide residues at concentrations ranging from 0.2 to 58 ng/mL. Red wines had higher concentration ratios (hydroxylated derivative/active fungicide) than white wines, especially in the case of PYR. PYR-4OH concentrations were twice as high as PYR concentrations in red wines on average. In the structure of both anti-botrytis fungicides, a targeted search of hydroxyl derivatives in wine extracts using LC-QTOF-MS revealed the presence of additional hydroxylation locations in the pyrimidine ring and/or in the alkyl substituents bonding to this cycle. Furthermore, free and glycosylated forms of both fungicides' hydroxylated metabolites coexist in wine samples. This research established that hydroxylated and glycosylated metabolites are present in grapes prior to vinification in the case of CYP.

In a study published in 2006, Pose-Juan, Cancho-Grande, Rial-Otero, and Simal-Gándara [49] examined the rates of degradation of four drugs in grape juice: cyprodinil, fludioxonil, procymidone, and vinclozolin. These pesticides were removed using a dichloromethane/acetone solution (4:1, v/v, 75 mL), and their identities were ascertained using gas chromatography mass spectrometry (GC/MS).

5. Sustainable Viticulture Practices to Reduce Pesticides Use

- No-tillage

In perennial agroecosystems such as vineyards, tillage has been shown to decrease plant [50,51] and animal diversity [52]. Tillage and non-chemical weed control (harrowing, mulching), nutrient application, and other interventions affect soil functioning to varying extents Capowiez et al. [53] (Figure 3). In response to over soil erosion and export of agrochemicals which become more acute, more farmers have adopted conservation practices including no-till. In the USA, the rate of no-till adoption has grown from 26% in 1990 to 41% in 2008, while conventional tillage has decreased from 49 to 37% during that same period.

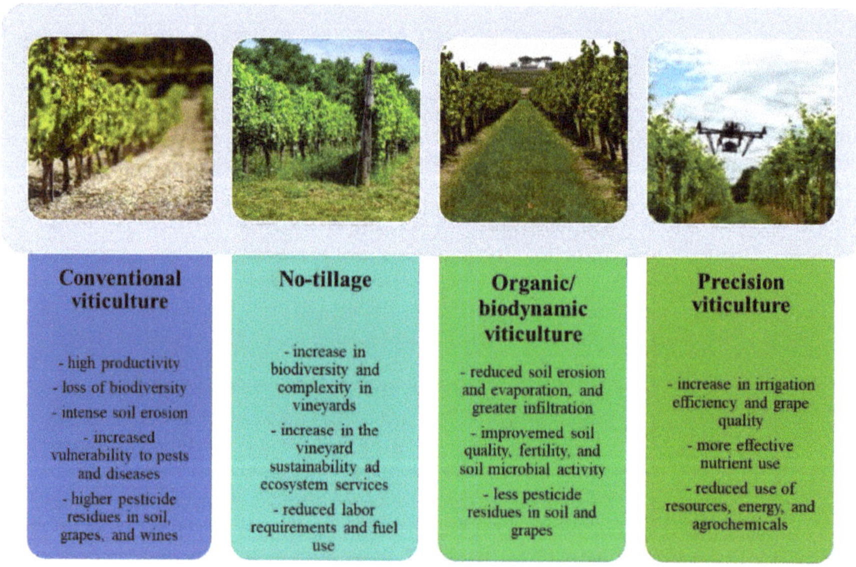

Figure 3. Sustainability of viticulture practices.

- Organic vineyards

Organic viticulture is a production method that underwent significant expansion at the end of the 20th century and has continued to grow ever since. Organic pest management primarily focuses on enhancing the presence of beneficial arthropods to the detriment of pests, using economical and low-impact practices that consider the ecosystem [54]. Organic farming consists of a low-input agro-ecosystem in which crop productivity is based on the natural availability of plant nutrients, the use of green manure and biological pathogen control. Biological control may be realized in various ways such as classical biological control, augmentation, and conservation. Biological control in organic viticulture obtains good results in controlling pests through the use of periodic discharges of biological control agents (augmentation), and by using ecosystem management techniques (conservation). Physical control methods refers to the elimination of insect pests through the application of physical barriers such as nets [55] and kaolin clay [56,57]. Semiochemicals such as pheromones and kairomones attract insects and have a high insect specificity.

In 2019, 63 countries engaged in organic viticulture and certified organic vineyards summed a surface area estimated at approximately 454 kha, or 6.2% of global vine cultivated area [58]. Among all the countries containing organic vineyards, 10 countries exploit 91% of organic vine surfaces, and while only 3 of these 10 countries are European, namely Spain, Italy, and France, EU countries account for 75% of the world's certified organic vineyard surface area.

- Biodynamic vineyards

Biodynamic agriculture was developed in the 1920s based on a set of conferences performed by the philosopher Rudolf Steiner [59]. This type of agriculture considers a holistic approach concerning the exploitation of natural resources, taking into consideration the sustainability of different elements, such as the crops themselves, animal life preservation, or the maintenance of a high-quality soil, in order to recover, preserve, or improve ecological harmony [60]. By significantly reducing the number of external inputs into the production system, utilizing set preparations to apply to crops which aid fertilization, and the additional application of other homeopathic treatments derived from infusions or plant extracts, this perspective can be achieved. Villanueva-Rey et al. [61] analysed biodynamic viticulture from a life-cycle perspective, and compared it with two other types of viticulture techniques: conventional viticulture and biodynamic-conventional viticulture. The obtained results do not only confirm prior findings that the environmental impact linked to a specific viticulture surface can have relevant variations on an inter annual basis, but also demonstrate strong variability between viticulture practices. In fact, biodynamic viticulture, and to a lesser extent, intermediate biodynamic-conventional vineyards, showed substantially lower environmental profiles for all the environmental impacts assessed.

- Biopesticides and nanobiopesticide

Pesticides that are naturally created by living things like bacteria, herbs, plants, etc. are referred to as biopesticides [62]. Since they are less hazardous to living systems, they are generally safer to employ than synthetic pesticides. The application of pesticides is essential for good crop production since pest infestations in agricultural fields significantly harm crops. Because of the high cost and ongoing usage of synthetic pesticides which has resulted in insect resistance, these chemical substances are no longer effective. *Bacillus thuringiensis* is one of the microbes that has been used to combat several insect pests. The plant *Azadirachta indica* has been discovered to be a powerful pesticide with anti-carcinogenic qualities [63].

Since they are not intentionally generated anywhere, biopesticides are biodegradable. These organic compounds successfully eradicate the intended pest and offer a variety of additional advantages [64,65]. They can increase the nutrients that are available to plants in the soil, and can support plant drought tolerance. Consequently, they are a crucial component of integrated pest control (IPM) techniques. For example, fungi like *Beauveria bassiana* are used in place of insecticides [66]. The amount of pollution produced by using these natural pesticides is minimal.

- Precision Viticulture

Crop monitoring and pesticide spraying are very important aspects within precision agriculture. New autonomous aerial and land vehicles in the near future could result in significant benefits to Agriculture 4.0. The drone was initially created as a military device, including Unmanned Aerial Vehicles (UAVs), Flying Mini Robots and Miniature Pilotless Aircraft. However, the utilization of UAVs in recent years is expanding quickly in agribusiness [67]. These devices incorporate the use of cameras and sensors, and can be grouped into three types: Fixed-wing, Helicopter, and Multi-copter. Semi-controlled drones have been combined with artificial intelligence (Al) in order to monitor farms, which is a remarkably useful device for real-time data analysis. Drones can carry out soil and crop health monitoring scans, as well as assist in irrigation, fertilizer application, and estimate farming yield [68] (Figure 3).

In traditional pesticide application, a manual mechanical sprayer is used which comes with many disadvantages, such as: environmental pollution, less area coverage, increased chemical use, farm labour shortages, lower spray uniformity, and higher costs. Moreover, manual spraying can significantly affect human health through hypersensitivity, asthma, cancer and other diseases [69]. Therefore, it is necessary to improve these deficiencies through the use of the modern drone-mounted sprayer. The advantages of drone use numerous, including: enhanced coverage ability, faster and more straighforward spraying application, increased chemical effectiveness, and the ability to access areas which mechanical sprayers cannot access. One of the most researched and widespread precision technologies is Variable-rate application (VRA), which, through its combined use with Global Positioning Systems (GPS), Geographic Information Systems (GIS), soil sampling, and integrated pest management (IPM), can greatly increase fertilizer input efficiency. It can be applied to seeding, weed and pest control, lime distribution, and fertilizer application [70].

6. Conclusions

The monitoring of pesticides has received significant attention over the years due to their toxic effects to both horticultural crops and human safety. Therefore, several techniques have been developed for the extraction and determination of pesticide residues. As this review demonstrates, a wide range of pesticide extraction methods are available, including: liquid-liquid extraction (LLE), solid phase extraction (SPE), solid phase micro extraction (SPME), matrix solid phase dispersion (MSPD), stir bar sorptive extraction (SBSE), microwave-assisted extraction (MAE), Quick, Easy, Cheap, Rugged, Effective and Safe (QuEChERS), etc. Analytical techniques such as gas chromatography or liquid chromatography in conjunction with mass spectrometry (GC-MS, or LC-MS), high-performance liquid chromatography (HPLC), and supercritical fluid chromatography (SFC), are the most frequently used in order to quantify pesticide residues.

Nowadays, the identification and quantification of pesticide residues in grapes and wines is generally carried out by the QuEChERS method and validated in conjunction with gas chromatography/mass spectrometry (GC-MS/MS) and ultra performance liquid chromatography/mass spectrometry (UPLC-MS/MS).

In addition, sustainable management practices to improve vineyard performance within a more sustainable farming system were considered. However, this field of research remains largely unexplored, despite the potential positive effects on vine growth and productivity. Among sustainable viticulture practices, dedicated fertilizers, precision agriculture, and ad-hoc policies will invariably shape the future of this economical area.

Author Contributions: Conceptualisation, S.T., G.-D.D. and C.T.; Methodology, S.T., G.-D.D. and C.T.; Writing—original draft preparation, S.T.; Writing—review and editing, C.T. and G.-D.D.; Supervision, C.T. and G.-D.D., Funding acquisition, C.T. and G.-D.D. All authors have read and agreed to the published version of the manuscript.

Funding: This study was supported by a grant from the Romanian Ministry of Education and Research, CNCS—UEFISCDI, project number PN-III-P1-1.1-PD-2019-065 and a grant of the Romanian Ministry of Education and Research, CCCDI—UEFISCDI, project number PN-III-P2-2.1-PED-2019-0175, within PNCDI III.

Institutional Review Board Statement: Not applicable.

Informed Consent Statement: Not applicable.

Data Availability Statement: The data presented in this study are available on request from the first author.

Acknowledgments: We are also grateful to "Gheorghe Asachi" Technical University of Iasi, Romania.

Conflicts of Interest: The authors declare no conflict of interest.

References

1. European Commission. Pesticides. 2021. Available online: https://ec.europa.eu/food/plant/pesticides_en (accessed on 22 September 2022).
2. International Union of Pure and Applied Chemistry. History of Pesticide Use. 2010. Available online: http://agrochemicals.iupac.org/index.php?option=com_sobi2&sobi2Task=sobi2Details&catid=3&sobi2Id=31 (accessed on 22 September 2022).
3. Stockholm Convention. All POPs listed in the Stockholm Convention. 2019. Available online: http://www.pops.int/TheConvention/ThePOPs/AllPOPs/tabid/2509/Default.aspx (accessed on 22 September 2022).
4. Environmental Protection Agency. DDT-A Brief History and Status. 2017. Available online: https://www.epa.gov/ingredients-used-pesticide-products/ddt-brief-history-and-status (accessed on 22 September 2022).
5. Holt, J.S. *Herbicides, Encyclopedia of Biodiversity*, 2nd ed.; Academic Press: Waltham, MA, USA, 2013; pp. 87–95.
6. National Pesticide Information Center. Insecticides. 2019. Available online: http://npic.orst.edu/ingred/ptype/insecticide.html (accessed on 3 October 2022).

7. Canadian Centre for Occupational Health and Safety. Pesticides–General. 2019. Available online: https://www.ccohs.ca/oshanswers/chemicals/pesticides/general.html (accessed on 3 October 2022).
8. United Nations. Department of Economic and Social Affairs. 2019. Available online: https://www.un.org/en/desa/world-population-projected-reach-98-billion-2050-and-112-billion-2100 (accessed on 3 October 2022).
9. FAO. FAOSTAT Pesticides Trade Dataset. Annual. 2020. Available online: http://www.fao.org/faostat/en/#data/RT (accessed on 3 October 2022).
10. FAO. FAOSTAT Agri-Environmental Indicators/Pesticides [www Document]. Annual. 2020. Available online: http://www.fao.org/faostat/en/#data/EP. (accessed on 3 October 2022).
11. Fenik, J.; Tankiewicz, M.; Biziuk, M. Properties and determination of pesticides in fruits and vegetables. *Trends Anal. Chem.* **2011**, *30*, 814–826. [CrossRef]
12. Thakur, M.; Deepak, P. Environmental fate of organic pollutants and effect on human health. *Abat. Environ. Pollut.-Trends Strateg.* **2020**, *12*, 245–262. [CrossRef]
13. World Bank. *Agriculture for Development-World Development Report*; The World Bank: Whashington, DC, USA, 2008.
14. Muhammad, G.; Rashid, I.; Firyal, S. Practical aspects of treatment of organophosphate and carbamate insecticide poisoning in animals. *Matrix Sci. Pharma.* **2017**, *1*, 10–11. [CrossRef]
15. Sidhu, G.K.; Singh, S.; Kumar, V.; Dhanjal, D.S.; Datta, S.; Singh, J. Toxicity, monitoring and biodegradation of organophosphate pesticides: A review. *Crit. Rev. Environ. Sci. Technol.* **2019**, *49*, 1135–1187. [CrossRef]
16. Urkude, R.; Dhurvey, V.; Kochhar, S. Pesticide Residues in Beverages. In *Quality Control in the Beverage Industry: The Science of Beverages*; Academic Press: Cambridge, MA, USA, 2019; Volume 17, pp. 529–560.
17. FAO. Pesticide Database Codex., Codex Alimentarius. 2022. Available online: https://www.fao.org/fao-who-codexalimentarius/codex-texts/dbs/pestres/pesticides/en/ (accessed on 19 July 2022).
18. Samsidar, A.; Siddiquee, S.; Shaarani, S.M. A review of extraction, analytical and advanced methods for determination of pesticides in environment and foodstuffs. *Trends Food Sci. Technol.* **2018**, *71*, 188–201. [CrossRef]
19. Nasiri, M.; Ahmadzadeh, H.; Amiri, A. Sample preparation and extraction methods for pesticides in aquatic environments: A review. *TrAC Trends Anal. Chem.* **2020**, *123*, 115772. [CrossRef]
20. Timofeeva, I.; Shishov, A.; Kanashina, D.; Dzema, D.; Bulatov, A. On-line insyringe sugaring-out liquid-liquid extraction coupled with HPLC-MS/MS for the determination of pesticides in fruit and berry juices. *Talanta* **2017**, *167*, 761–767. [CrossRef] [PubMed]
21. Farajzadeh, M.A.; Khoshmaram, L.; Nabil, A.A.A. Determination of pyrethroid pesticides residues in vegetable oils using liquid-liquid extraction and dispersive liquid-liquid microextraction followed by gas chromatography-flame ionization detection. *J. Food Compos. Anal.* **2014**, *34*, 128–135. [CrossRef]
22. Radišić, M.M.; Vasiljević, T.M.; Dujaković, N.N.; Laušević, M.D. Application of matrix solid-phase dispersion and liquid chromatography–ion trap mass spectrometry for the analysis of pesticide residues in fruits. *Food Anal. Methods* **2013**, *6*, 648–657. [CrossRef]
23. Anastassiades, M.; Lehotay, S.J.; Stajnbaher, D.; Schenck, F.J. Fast and easy multiresidue method employing acetonitrile extraction/partitioning and "dispersive solid-phase extraction" for the determination of pesticide residues in produce. *J. AOAC Int.* **2003**, *86*, 412–431. [CrossRef]
24. Lehotay, S.J.; de Kok, A.; Hiemstra, M.; Van Bodegraven, P. Validation of a fast and easy method for the determination of residues from 229 pesticides in fruits and vegetables using gas and liquid chromatography and mass spectrometric detection. *J. AOAC Int.* **2005**, *88*, 595–614. [CrossRef] [PubMed]
25. Perestrelo, R.; Silva, P.; Porto-Figueira, P.; Pereira, J.A.M.; Silva, C.; Medina, S.; Câmara, J.S. QuEChERS-fundamentals, relevant improvements, applications and future trends. *Anal. Chim. Acta* **1070**, *2019*, 1–28. [CrossRef] [PubMed]
26. Dumitriu Gabur, G.-D.; Gabur, I.; Cucolea, E.I.; Costache, T.; Rambu, D.; Cotea, V.V.; Teodosiu, C. Investigating Six Common Pesticides Residues and Dietary Risk Assessment of Romanian Wine Varieties. *Foods* **2022**, *11*, 2225. [CrossRef] [PubMed]
27. Wilkowska, A.; Biziuk, M. Determination of pesticide residues in food matrices using the QuEChERS methodology. *Food Chem.* **2011**, *125*, 803–812. [CrossRef]
28. D'Orazio, G.; Fanali, C.; Asensio-Ramos, M.; Fanali, S. Chiral separations in food analysis. *TrAC Trends Anal. Chem.* **2017**, *96*, 151–171. [CrossRef]
29. Bletsou, A.A.; Jeon, J.; Hollender, J.; Archontaki, E.; Thomaidis, N. Targeted and non-targeted liquid chromatography-mass spectrometric workflows for identification of transformation products of emerging pollutants in the aquatic environment. *Trends Anal. Chem.* **2015**, *66*, 32–44. [CrossRef]
30. Chen, L.X.; Dean, B.; Liang, X.R. Evaluation of polysaccharide-based chiral stationary phases in modern SFC-MS/MS for enantioselective bioanalysis. *Bioanalysis* **2019**, *11*, 251–266. [CrossRef]
31. Koukouvinos, G.; Tsialla, Z.; Petrou, P.S.; Misiakos, K.; Goustouridis, D.; Moreno, A.U.; Fernandez-Alba, A.R.; Raptis, I.; Kakabakos, S.E. Fast simultaneous detection of three pesticides by a White Light Reflectance Spectroscopy sensing platform. *Sens. Actuators B-Chem.* **2017**, *238*, 1214–1223. [CrossRef]
32. Weng, S.; Zhu, W.; Zhang, X.; Yuan, H.; Zheng, L.; Zhao, J.; Huang, L.; Han, P. Recent advances in Raman technology with applications in agriculture, food and biosystems: A review. *Artif. Intell. Agric.* **2019**, *3*, 1–10. [CrossRef]
33. Kailani, M.H.; Al-Antary, T.M.; Alawi, M.A. Monitoring of pesticides residues in soil samples from the southern districts of Jordan in 2016/2017. *Toxin Rev.* **2019**, *40*, 198–214. [CrossRef]

34. Nieto-Garcia, A.; Romero-Gonzalez, R.; Frenich, A.G. Multi-pesticide residue analysis in nutraceuticals from grape seed extracts by gas chromatography coupled to triple quadrupole mass spectrometry. *Food Control* **2014**, *47*, 369–380. [CrossRef]
35. He, Z.; Xu, Y.; Wang, L.; Peng, Y.; Luo, M.; Cheng, H.; Liu, X. Wide-scope screening and quantification of 50 pesticides in wine by liquid chromatography/quadrupole time-of-flight mass spectrometry combined with liquid chromatography/quadrupole linear ion trap mass spectrometry. *Food Chem.* **2016**, *196*, 1248–1255. [CrossRef]
36. Vaquero-Fernandez, L.; Sanz-Asensio, J.; Fernandez-Zurbano, P.; Lopez-Alonso, M.; Martinez-Soria, M. Determination of fungicide pyrimethanil in grapes, must, fermenting must and wine. *J. Sci. Food Agric.* **2012**, *93*, 1960–1966. [CrossRef] [PubMed]
37. Perez-Ortega, P.; Gilbert-Lopez, B.; Garcia-Reyes, J.F.; Ramos-Martos, N.; Molina-Diaz, A. Generic sample treatment method for simultaneous determination of multiclass pesticides and mycotoxins in wines by liquid chromatography–mass spectrometry. *J. Chromatogr. A* **2012**, *1249*, 32–40. [CrossRef]
38. Navarro, S.; Barba, A.; Navarro, G.; Vela, N.; Olivia, J. Multiresidue method for the rapid determination–in grape, must and wine–of fungicides frequently used on vineyards. *J. Chromatogr. A* **2000**, *882*, 221–229. [CrossRef] [PubMed]
39. Pelajic, M.; Pecek, G.; Pavlovic, D.M.; Cepo, D.V. Novel multiresidue method for determination of pesticides in red wine using gas chromatography–mass spectrometry and solid phase extraction. *Food Chem.* **2016**, *200*, 98–106. [CrossRef]
40. Gonzalez-Rodriguez, R.M.; Rial-Otero, R.; Cancho-Grande, B.; Gonzalez-Barreiro, C.; Simal-Gandara, J. A review on the fate of pesticides during the processes within the food-production Chain. *Crit. Rev. Food Sci. Nutr.* **2011**, *51*, 99–114. [CrossRef]
41. Yang, M.; Luo, F.; Zhang, X.; Zhoa, X.; Lou, Z.; Zhao, M.; Chen, Z. Dissipation and Risk Assessment of Multiresidual Fungicides in Grapes under Field Conditions. *J. Agric. Food Chem.* **2020**, *68*, 1071–1078. [CrossRef]
42. González-Rodríguez, R.M.; Cancho-Grande, B.; Torrado-Agrasar, A.; Simal-Gándara, J.; Mazaira-Pérez, J. Evolution of tebuconazole residues through the winemaking process of Mencía grapes. *Food Chem.* **2009**, *117*, 529–537. [CrossRef]
43. Heshmati, A.; Nili-Ahmadabadi, A.; Rahimi, A.; Vahidinia, A.; Taheri, M. Dissipation behavior and risk assessment of fungicide and insecticide residues in grape under open-field, storage and washing conditions. *J. Clean. Prod.* **2020**, *270*, 122287. [CrossRef]
44. Pan, X.; Dong, F.; Lie, N.; Cheng, Y.; Xu, J.; Liu, X.; Wu, X.; Chen, Z.; Zheng, Y. The fate and enantioselective behavior of zoxamide during wine-making process. *Food Chem.* **2018**, *248*, 14–20. [CrossRef]
45. Paya, P.; Mulero, J.; Oliva, J.; Camara, M.A.; Barba, A. Influence of the matrix in bioavailability of flufenoxuron, lufenuron, pyriproxyfen and fenoxycarb residues in grapes and wine. *Food Chem. Toxicol.* **2013**, *60*, 419–423. [CrossRef] [PubMed]
46. Cus, F.; Cesnik, H.B.; Bolta, S.V.; Gregoric, A. Pesticide residues in grapes and during vinification process. *Food Control* **2010**, *21*, 1512–1518. [CrossRef]
47. Golge, O.; Kabak, B. Pesticide Residues in Table Grapes and Exposure Assessment. *J. Agric. Food Chem.* **2018**, *66*, 1701–1713. [CrossRef]
48. Castro, G.; Perez-Mayan, L.; Carpinteiro, I.; Ramil, M.; Cela, R.; Rodriguez, I. Residues of anilinopyrimidine fungicides and suspected metabolites in wine samples. *J. Chromatogr. A* **2020**, *1622*, 461104. [CrossRef] [PubMed]
49. Pose-Juan, E.; Cancho-Grande, B.; Rial-Otero, R.; Simal-Gándara, J. The dissipation rates of cyprodinil, fludioxonil, procymidone and vinclozoline during storage of grape juice. *Food Control* **2006**, *17*, 1012–1017. [CrossRef]
50. Hall, R.M.; Penke, N.; Kriechbaum, M.; Kratschmer, S.; Jung, V.; Chollet, S.; Guernion, M.; Nicolai, A.; Burel, F.; Fertil, A.; et al. Vegetation man- agement intensity and landscape diversity alter plant species richness, functional traits and community composition across European vineyards. *Agric. Syst.* **2020**, *177*, 102706. [CrossRef]
51. Kazakou, E.; Fried, G.; Richarte, J.; Gimenez, O.; Violle, C.; Metay, A. A plant trait-based response-and-effect framework to assess vineyard inter-row soil management. *Bot. Lett.* **2016**, *163*, 373–388. [CrossRef]
52. Sánchez-Moreno, S.; Castro, J.; Alonso-Prados, E.; Alonso-Prados, J.L.; García-Baudín, J.M.; Talavera, M.; Durán-Zuazo, V.H. Tillage and herbicide decrease soil biodiversity in olive orchards. *Agron. Sustain. Dev.* **2015**, *35*, 691–700. [CrossRef]
53. Capowiez, Y.; Cadoux, S.; Bouchant, P.; Ruy, S.; Roger-Estrade, J.; Richard, G.; Boizard, H. The effect of tillage type and cropping system on earthworm communities, macroporosity and water infiltration. *Soil Till. Res.* **2009**, *105*, 209–216. [CrossRef]
54. Bruggisser, O.T.; Schmidt-Entling, M.H.; Bacher, S. Effects of vineyard management on biodiversity at three trophic levels. *Biol. Conserv.* **2010**, *143*, 1521–1528. [CrossRef]
55. Mannini, F. Hot water treatment and field coverage of mother plant vineyards to prevent propagation material from phytoplasma infections. *Bull. Insect.* **2007**, *60*, 311–312.
56. Valizadeh, H.; Abbasipour, H.; Farazmand, H.; Askarianzadeh, A. Evaluation of kaolin application on oviposition control of the Vine Cicada, Psalmocharias alhageos in Vineyards (Homoptera: Cicadidae). *Entomol. Ger.* **2013**, *34*, 279–286. [CrossRef]
57. Dumitriu (Gabur), G.-D.; Teodosiu, C.; Cotea, V.V. Management of Pesticides from Vineyard to Wines: Focus on Wine Safety and Pesticides Removal by Emerging Technologies. In *Grapes and Wine*; Morata, A., Loira, I., González, C., Eds.; IntechOpen: London, UK, 2021.
58. OIV. Focus OIV the World Organic Vineyard. 2021. Available online: https://www.oiv.int/public/medias/8514/en-focus-the-world-organic-vineyard.pdf (accessed on 3 October 2022).
59. Steiner, R. *Spiritual Foundations for the Renewal of Agriculture: A course of lectures held at Koberwitz, Silesia, June 7 to 16, 1924*; Creeger, C., Gardner, M., Translators; Bio-Dynamic Farming and Gardening Association Inc.: Kimberton, PA, USA, 1993.
60. Lotter, D.W. Organic agriculture. *J. Sustain. Agric.* **2003**, *21*, 59–128. [CrossRef]
61. Villanueva-Rey, P.; Vázquez-Rowe, I.; Moreira, M.T.; Feijoo, G. Comparative life cycle assessment in the wine sector: Biodynamic vs. conventional viticulture activities in NW Spain. *J. Clean. Prod.* **2014**, *65*, 330–341. [CrossRef]

62. Chandler, D.; Bailey, A.S.; Tatchell, G.M.; Davidson, G.; Greaves, J.; Grant, W.P. The development, regulation and use of biopesticides for integrated pest management. *Philos. Trans. R. Soc. B-Biol. Sci.* **2012**, *366*, 1987–1998. [CrossRef]
63. Chaudhary, S. Progress on *Azadirachta indica* Based Biopesticides in Replacing Synthetic Toxic Pesticides. *Front. Plant Sci.* **2017**, *8*, 610. [CrossRef]
64. Tamhane, V.A.; Chougule, N.P.; Giri, A.P.; Dixit, A.R.; Sainani, M.N.; Gupta, V.S. In vivo and in vitro effects of Capsicum annum proteinase inhibitors on Helicoverpa armigera gut proteinases. *Biochim. Biophys. Acta.* **2005**, *1722*, 156–167. [CrossRef]
65. Telang, M.A.; Giri, A.P.; Sainani, M.N.; Gupta, V.S. Characterization of two midgut proteinases of Helicoverpa armigera and their interaction with proteinase inhibitors. *J. Insect Physiol.* **2005**, *51*, 513–522. [CrossRef]
66. Kanzok, S.M.; Jacobs-Lorena, M. Entomopathogenic fungi as biological insecticides to control malaria. *Trends Parasitol.* **2006**, *22*, 49–51. [CrossRef]
67. Gayathri Devi, K.; Sowmiya, N.; Yasoda, K.; Muthulakshmi, K.; Kishore, B. Review on application of drones for crop health monitoring and spraying pesticides and fertilizer. *J. Crit. Rev.* **2020**, *7*, 667–672. [CrossRef]
68. Geetharamani, G.; Arun, P.J. Identification of plant leaf diseases using a nine-layer deep convolutional neural network. *Comput. Electr. Eng.* **2019**, *76*, 323–338. [CrossRef]
69. Koc, C. Design and Development of a Low-cost UAV for Pesticide Applications. *J. Agric Fac. Gaziosmanpasa Univ.* **2017**, *34*, 94–103. [CrossRef]
70. Pallottino, F.; Biocca, M.; Nardi, P.; Figorilli, S.; Menesatti, P.; Costa, C. Science mapping approach to analyze the research evolution on precision agriculture: World, EU and Italian situation. *Precis. Agric.* **2018**, *19*, 1011–1026. [CrossRef]

molecules

Article

The Impact of Different Pretreatment Processes (Freezing, Ultrasound and High Pressure) on the Sensory and Functional Properties of Black Garlic (*Allium sativum* L.)

Kai-Hui Chan [1,†], Chao-Kai Chang [1,†], Mohsen Gavahian [2], Bara Yudhistira [1,3], Shella Permatasari Santoso [4,5], Kuan-Chen Cheng [6,7,8,9,*] and Chang-Wei Hsieh [1,9,*]

1 Department of Food Science and Biotechnology, National Chung Hsing University, 145 Xingda Rd., South Dist., Taichung City 40227, Taiwan
2 Department of Food Science, National Pingtung University of Science and Technology, Pingtung 91201, Taiwan
3 Department of Food Science and Technology, Sebelas Maret University, Surakarta City 57126, Indonesia
4 Department of Chemical Engineering, Widya Mandala Surabaya Catholic University, Surabaya 60114, Indonesia
5 Department of Chemical Engineering, National Taiwan University of Science and Techology, Daan Dist., Taipei 10607, Taiwan
6 Institute of Biotechnology, National Taiwan University, Taipei 10617, Taiwan
7 Graduate Institute of Food Science Technology, National Taiwan University, Taipei 10617, Taiwan
8 Department of Optometry, Asia University, Taichung City 413305, Taiwan
9 Department of Medical Research, China Medical University Hospital, Taichung City 404333, Taiwan
* Correspondence: kccheng@ntu.edu.tw (K.-C.C.); welson@nchu.edu.tw (C.-W.H.); Tel.: +886-4-22840385 (ext. 5010) (C.-W.H.)
† These authors contributed equally to this work.

Citation: Chan, K.-H.; Chang, C.-K.; Gavahian, M.; Yudhistira, B.; Santoso, S.P.; Cheng, K.-C.; Hsieh, C.-W. The Impact of Different Pretreatment Processes (Freezing, Ultrasound and High Pressure) on the Sensory and Functional Properties of Black Garlic (*Allium sativum* L.). *Molecules* **2022**, *27*, 6992. https://doi.org/10.3390/molecules27206992

Academic Editors: Weiying Lu and Yanping Chen

Received: 27 September 2022
Accepted: 16 October 2022
Published: 18 October 2022

Abstract: Black garlic (BG) is an emerging derivative of fresh garlic with enhanced nutritional properties. This study aimed to develop functional BG products with good consumer acceptance. To this end, BG was treated with freezing (F-BG), ultrasound (U-BG), and HHP (H-BG) to assess its sensory and functional properties. The results showed that F-BG and H-BG had higher S-allyl-cysteine (SAC), polyphenol, and flavonoid contents than BG. H-BG and F-BG displayed the best sensory quality after 18 days of aging, while 5-hydroxymethylfurfural (5-HMF), SAC, and polyphenols were identified as the most influential sensory parameters. Moreover, the F-BG and H-BG groups achieved optimal taste after 18 days, as opposed to untreated BG, which needed more than 24 days. Therefore, the proposed approaches significantly reduced the processing time while enhancing the physical, sensory, and functional properties of BG. In conclusion, freezing and HHP techniques may be considered promising pretreatments to develop BG products with good functional and sensory properties.

Keywords: black garlic; S-allyl-cysteine; sensory quality; polyphenol; freezing; high hydrostatic pressure

1. Introduction

Black garlic (BG) is an emerging functional food that has recently become popular in China and some other regions of the world [1], mainly because of its functional properties. BG can decrease the risk of developing diseases and contains various health-promoting phytoconstituents (e.g., phenolics, flavonoids, S-allyl cysteine, hydroxycinnamic acid derivatives), which may make it superior to fresh garlic [2]. Dietary supplementation with SAC has health benefits, as shown in some previous studies, including anti-inflammatory and anti-apoptosis effects [3]. Another study indicated that SAC has functional properties such as anti-neuroinflammatory, insulinotropic, antidiabetic, hepatoprotective, and cytoprotective effects [3].

For the traditional processing of BG, fresh garlic is cooked and incubated under high-temperature and high-humidity conditions [1], which not only changes the original pungent flavor but also provides greater functionality. Bae et al. [4] also indicated that after garlic was made into BG at 70 °C for 45 days, the main active substance in BG increased by 5.5-fold. The polysaccharides in garlic account for 75% of its dry weight. During the aging process, polysaccharides are decomposed into monosaccharides and disaccharides and increase to levels 10 times more than those of raw garlic, thereby leading to BG's sweet flavor [5]. Phan et al. [6] noted that bound polyphenols can degrade into free polyphenols because of heat treatment during processing, which further improves the antioxidant properties and the levels of S-allyl-cysteine (SAC)—the main active substance in garlic. Heat treatment for 45 days at 40 °C resulted in an SAC concentration of 124.67 mg/g [4]. SAC can prevent indomethacin-induced gastric damage and increase the total antioxidant concentration, although the dose of SAC affects this preventive effect [3]. Moreover, researchers are exploring approaches that can further enhance the functional properties of BG. 5-Hydroxymethylfurfural (5-HMF), a furan compound with both aldehyde and alcohol functional groups, is present in many heat-treated foods, such as coffee and BG [7]. While some studies highlight the hazards of 5-HMF (e.g., carcinogenicity, liver and kidney toxicity, reducing glutathione content in cells, and DNA damage) [7], recent studies have revealed that 5-HMF has beneficial physiological activities (e.g., anti-hypoxic injury and improving blood circulation) and can act as an antioxidant and anti-allergen [2]. This compound is also an important intermediate product of the Maillard reaction [1], and its accumulation is closely related to the speed of garlic's blackening during BG processing [8,9]. Therefore, the 5-HMF content can be used as an important monitoring indicator for the quality of BG.

Using different techniques to increase food safety and functionality has become a recent trend. Ultrasound and high-pressure processing have been reported, showing that tyrosinase activity be enhanced by ultrasound, further promoting the production of unique organoleptic substances, as in cocoa and fermented tea leaves [10,11]. High-pressure pretreatment can destroy intracellular structures, enhancing the Maillard reaction [8]. In our previous research, freezing, ultrasound, or high-pressure pretreatments were combined with low-temperature aging (40 °C for 6–9 days), which was found to be effective in modifying the biological structure of garlic, promoting the enzymatic activity, and increasing the functional ingredients by 4–10-fold [12,13]. The effects of processing methods on SAC compounds are generally only focused on the content of SAC [3]. However, further information on the effects of such pretreatments on the functional components and sensory attributes of the final product is needed. To date, many studies related to the combination of emerging technologies have been carried out to improve the characteristics of materials, to assist in processing, and to extend the shelf life of materials or products [14–16].

Therefore, the present study aimed to produce BG from fresh garlic samples pretreated by freezing, ultrasound, and high-pressure processing, as well as to assess the variations in SAC, total polyphenols, flavonoids, reducing sugars, and 5-HMF, along with sensory attributes. The pretreatment methods used were freezing, ultrasonic, and high hydrostatic pressure, because previous studies suggested that these methods can increase the SAC content in garlic. According to a previous study, garlic samples that were frozen in liquid nitrogen and stored at −80 °C had a greater concentration of SAC (1.03 mg/g) than untreated samples (0.85 mg/g) [13]. In comparison to raw garlic (0.85 mg/g), ultrasonic pretreatment at 28 kHz for 1 h can enhance the SAC concentration in garlic (1.07 mg/g) [13]. A recent study on high hydrostatic pressure pretreatment revealed that HHP increased SAC concentrations by about 7–10-fold, and the processing at 300 MPa for 15 min was reported to represent the best processing conditions for increasing SAC formation (from 0.51 to 5.60 mg/g) [12].

2. Results and Discussion

2.1. Effects of Different Pretreatments on Black Garlic's Appearance, Reducing Sugar Contents, and pH Value

The results for color change (ΔE) and appearance are shown in Figure 1a,b. On day 0 of the heat treatment, the degree of ΔE in the freezing and high-pressure pretreatment groups was higher than that in the ultrasonic and control groups. According to Che, Chen et al. [12], freezing and high-pressure pretreatment cause structural deformation and porosity in garlic, and enzymatic browning may be the cause of garlic browning caused by freezing and HHP pretreatments. Moreover, polyphenols and related oxidative enzymes such as polyphenol oxidase are bound to the cell wall and cell membrane. Once the plant structure is damaged, the substrate and enzymes are released, causing an enzymatic browning reaction. The ΔE of all groups increased rapidly on day 3 of heat treatment. In terms of the appearance, the color of the garlic was brown or black on the day 3. The reason for this is that under the environment of high temperature and high humidity, fructan is thermally cleaved to fructose and glucose, which react with amino acids to produce menerioids, which darken the garlic and lead to increased ΔE [8]. On days 6 and 9 of heat treatment, although the appearance of the garlic in all groups was black, the degree of ΔE varied with the severity of cell structure damage in the different pretreatments, ΔE in the freezing and high-pressure pretreatment groups was higher than that in the ultrasonic pretreatment group and the control group. The ΔE of all groups gradually leveled off after the 9th day of heat treatment, and the color was black.

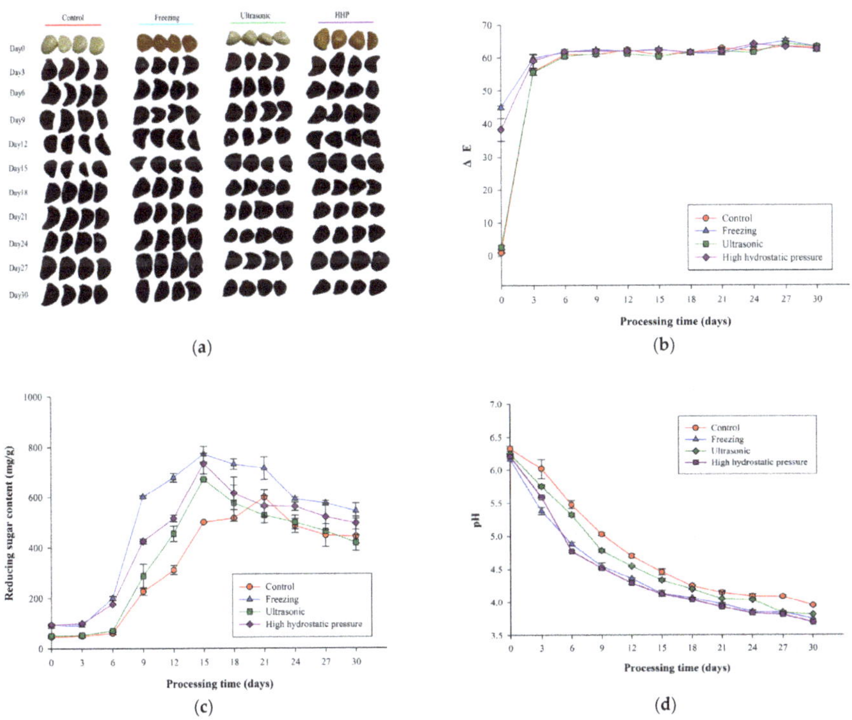

Figure 1. (**a**) Appearance, (**b**) ΔE, (**c**) reducing sugar content (mg/g dry weight), and (**d**) pH of garlic subjected to different pretreatment methods during thermal processing (70 °C, 80% RH for 30 days). All values are means ± standard deviations ($n = 3$). Control: only heat treatment at 40 °C, 80% RH for 6 days. Freezing: frozen at −20 °C for 30 h and then heat-treated at 40 °C with 80% RH for 6 days. Ultrasonic: 28 kHz for 2 h and then heat-treated at 40 °C with 80% RH for 6 days. High hydrostatic pressure: 300 MPa for 15 min and then heat-treated at 40 °C with 80% RH for 9 days.

According to previous studies, through microstructural analysis, it was found that freezing and high-pressure pretreatments have obvious destructive effects on garlic, causing structural deformation and porosity of the garlic [12,13]. Therefore, before the 12th day of heat treatment, the ΔE was higher than that of the ultrasonic pretreatment group and the control group, due to the severe degree of damage to the cell structure caused by the freezing and high-pressure pretreatments; after day 12, the ΔE of all groups plateaued.

The changes in reducing sugar contents in F-BG, U-BG, and H-BG aged for 30 days are shown in Figure 1c. It was found that after 15 d, the reducing sugar contents were highest, and those of the pretreated groups were higher than that of the control group (F-BG, 772.18 $\pm$ 30.85 mg/g dry weight; U-BG, 671.47 $\pm$ 1.89 mg/g dry weight; H-BG, 733.83 $\pm$ 42.31 mg/g dry weight; control, 501.46 $\pm$ 2.69 mg/g dry weight). Because garlic was aged under high temperatures after pretreatment, such conditions destroyed the cell structure, causing intracellular substances to flow, and decomposed more polysaccharides into monosaccharides and disaccharides, resulting in a higher reducing sugar content than that of the untreated group [9]. After 15 days, the reducing sugar contents of F-BG, U-BG, and H-BG began to decrease, indicating that as the aging time increased, the reducing sugar content decreased, which may have been because of the Maillard reaction [5]. Finally, the reducing sugar content was highest in the F-BG group (544.79 $\pm$ 29.92 mg/g dry weight), followed by the H-BG group (496.3 $\pm$ 26.26 mg/g dry weight). However, the control group reached the highest reducing sugar content after 21 days. This inflection point is called the reducing sugar balance point (RSBP), which may be positively associated with the improved processing efficiency of BG. The earlier equilibrium point indicated that the maturation time of BG was shorter, which could improve the processing efficiency. Before reaching the RSBP, the reducing sugar generation rate is faster than the consumption rate. After reaching the RSBP, the consumption rate of reducing sugars gradually becomes greater than the production rate [8].

As the processing time of BG increases, the pH value gradually decreases, and the increase in organic acid contents makes the taste of BG slightly sour [5,17]. The results of pH changes are shown in Figure 1d. During the heat treatment, the pH value of garlic gradually decreased. In addition to the release of organic acids due to the destruction of the garlic cell structure, organic acids were also generated during the heat treatment process, resulting in a decrease in the pH value of BG [5]. Liang et al. [18] noted that during the heating of garlic, the contents of acetic acid and formic acid increased significantly, mainly as a result of the cleavage of α-dicarbonyl and β-dicarbonyl of five-carbon or six-carbon sugars. During heat treatment at 70 °C, the pH values of the control, freezing, HPP, and ultrasonic pretreatment groups decreased from 6.33 to 3.95, 6.16 to 3.75, 6.21 to 3.69, and 6.25 to 3.81, respectively. Overall, the pH values of the pretreated samples were lower than that of the control group, indicating that the cell structure of these samples was damaged more severely, resulting in the release of more organic acids and lower pH values. Similarly, the results of Choi et al. [19] showed that the pH value of garlic after heat treatment at 70 °C for 28 days decreased from 6.33 to 4.07, similar to the trend of the control group in this experiment.

2.2. *Effects of Different Pretreatments on Black Garlic SAC and 5-HMF Content*

Chen indicated that the SAC content in garlic could be increased 6-fold, 4-fold, and 10-fold through freezing pretreatment (−20 °C for 30 h), ultrasonic pretreatment (28 kHz for 2 h), and high hydrostatic pressure (HHP) pretreatment (300 MPa for 15 min), respectively [12,13]. Furthermore, we increased the SAC content in garlic via pretreatment and aged the BG under high-temperature and high-humidity conditions, observing the changes in SAC content under different pretreatment conditions. F-BG denotes BG subjected to freezing pretreatment, U-BG denotes BG subjected to ultrasound pretreatment, and H-BG denotes BG subjected to HHP pretreatment. The changes in SAC content in BG, F-BG, U-BG, and H-BG aged for 30 d are shown in Figure 2a. On day 0, the SAC content in all groups was similar to Chen's findings [12,13]. During the aging period of F-BG, U-BG, and H-BG, there were no significant changes in the SAC contents of F-BG and H-BG because of the high-temperature conditions. It was speculated that during pretreatment γ-GTP had completely reacted with most of the GSAC, so that during the subsequent high-temperature and high-humidity aging stage the SAC formation tended to be flat. On day 3, the decreased SAC content may have been incurred by oxidation itself [20]. The literature indicates that γ-GTP will inactivate within 30 min at temperatures over 75 °C [4]; hence, SAC formation is less significant during the high-temperature aging stage of BG. Nevertheless, SAC formation in U-BG still gradually increased during aging—presumably because of the degree of cell disruption caused by ultrasound pretreatment—and the promotion of γ-GTP activity was lower than that of the freezing and HHP pretreatment methods, causing the GSAC and γ-GTP in the ultrasound pretreatment group to not fully react during the pretreatment period. In the subsequent aging stage, the cell walls were gradually destroyed because of the persistently high temperature, so the intracellular γ-GTP outflow contacting the GSAC gradually increased the SAC content. As a result, the SAC content in the U-BG group was higher than that in the F-BG and H-BG groups.

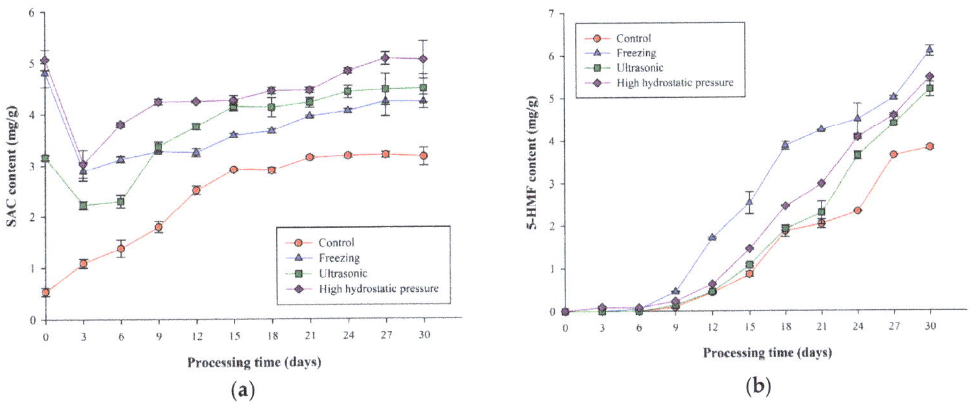

Figure 2. (**a**) SAC content (mg/g dry weight) and (**b**) 5-HMF content (mg/g) in garlic subjected to different pretreatment methods during thermal processing (70 °C, 80% RH for 30 days). All values are means ± standard deviations (n = 3). Control: only heat treatment at 40 °C, 80% RH for 6 days. Freezing: frozen at −20 °C for 30 h and then heat-treated at 40 °C with 80% RH for 6 days. Ultrasonic: 28 kHz for 2 h and then heat-treated at 40 °C with 80% RH for 6 days. High hydrostatic pressure: 300 MPa for 15 min and then heat-treated at 40 °C with 80% RH for 9 days.

The final SAC contents of BG subjected to the three different pretreatment methods during the aging process were not significantly different (F-BG, 4.23 ± 0.13 mg/g; H-BG, 5.05 ± 0.37 mg/g; and U-BG, 4.49 ± 0.27 mg/g), but they were significantly higher than the SAC content of the control group (3.16 ± 0.17 mg/g). The main reason for this was that through the different pretreatment methods, the cell structure of garlic was destroyed, and SAC formation in the garlic was promoted [12,13]. The untreated group had an aging temperature (70 °C) that was higher than the optimal temperature of γ-GTP (40 °C). After prolonged exposure to high temperatures, the enzymes were gradually inactivated; thus, the SAC content of the control group did not change significantly after 12 days.

The results of changes in 5-HMF content are shown in Figure 2b. With the increase in the aging treatment time, the contents of 5-HMF in all groups showed an upward trend. On the ninth day of aging, the content of 5-HMF in the freezing pretreatment group began to increase most significantly, from 0.5 mg/g to 6.3 mg/g, followed by the HHP pretreatment group, from 0.3 mg/g to 5.8 mg/g. The 5-HMF content in the ultrasonic pretreatment group increased from 0.2 mg/g to 5 mg/g, while that in the control group increased from 0.2 mg/g to 3.8 mg/g.

Since the rate of 5-HMF formation is proportional to the rate of the Maillard reaction which, in turn, is proportional to the reducing sugar content. Li, Cao et al. and Li, Lu et al. [8,9] showed garlic with high-pressure pretreatment, the cell structure of black garlic is damaged, which accelerates the Maillard reaction rate, makes the rate of 5-HMF formation faster, and promotes the earlier aging of black garlic. It also can be seen from Figure 1 that compared with the other treatments, the freezing pretreatment accelerated the aging process of black garlic and was also the pretreatment group that produced the most reducing sugars. Therefore, this part of the results also showed that black garlic in the freezing pretreatment group produced more 5-HMF than the other pretreatment groups, probably because the damage to the garlic's structure in the freezing pretreatment group was greater than that caused by the high-pressure and ultrasonic pretreatments.

2.3. Effects of Different Pretreatments on Black Garlic's Polyphenol and Flavonoid Contents

The changes in total polyphenol contents in F-BG, U-BG, and H-BG aged for 30 days are shown in Figure 3a. The total polyphenol contents of F-BG, U-BG, and H-BG increased with time until reaching their highest values at 18 days (F-BG, 15.66 ± 0.15 mg GAE/g dry weight; U-BG, 14.84 ± 0.36 mg GAE/g dry weight; H-BG, 15.55 ± 0.49 mg GAE/g dry weight). There were no significant differences between these three groups. Afterwards, the polyphenol contents began to decline, but were still higher than that of the control group. This was related to the obvious destruction of the cell structures of garlic following pretreatment [12,13]; hence, more free polyphenols were formed, making the total polyphenol contents higher than that in the control group. Several reports have also confirmed that heat treatment can release polyphenols, increasing the free components of phenolic acids while reducing esterification, glycoside formation, ester binding, and formation of other phenol-containing macromolecular components, resulting in increased free phenols [19,21]. Moreover, the increase in polyphenol contents may have been due to the precursors of phenol molecules passing through phenolic non-enzymatic mutual conversion between molecules or to the release of polyphenols that are bound to the cell wall [22,23], whereas the decreased polyphenol contents and flattened rate after 18 days may have been due to the thermally sensitive polyphenols being easily degraded in high-temperature environments [24].

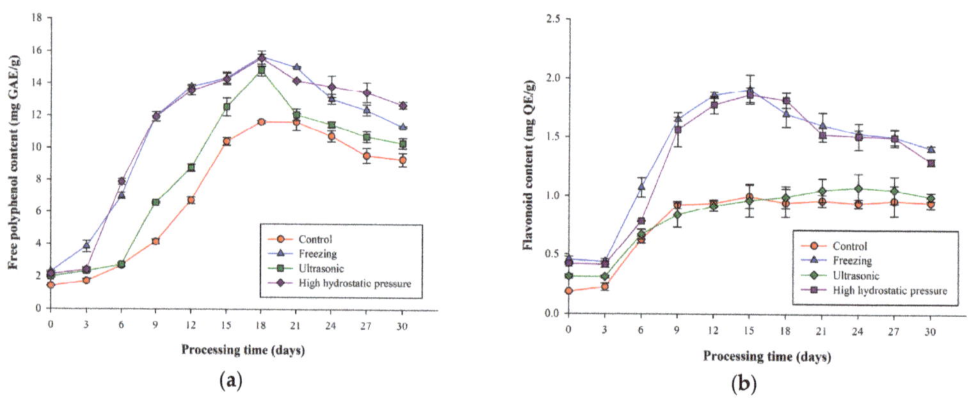

Figure 3. (**a**) Free polyphenol contents (mg GAE/g) and (**b**) flavonoid contents (mg QE/g dry weight) in garlic subjected to different pretreatment methods during thermal processing. All values are means ± standard deviations (n = 3).

The changes in total flavonoid contents in F-BG, U-BG, and H-BG aged for 30 days are shown in Figure 3b. The flavonoid contents of F-BG and H-BG were significantly higher than those of the control and U-BG groups, and they increased with time until they reached their highest values after 15 days (F-BG, 1.91 ± 0.12 mg QE/g dry weight; H-BG,

1.86 ± 0.06 mg QE/g dry weight). The data revealed that the flavonoid contents of the F-BG and H-BG groups were two times higher than those of the control (1.00 ± 0.10 mg QE/g dry weight) and U-BG groups (0.96 ± 0.14 mg QE/g dry weight). The main reason for this phenomenon was that the freezing and high-pressure pretreatment methods caused a high degree of cell wall damage to garlic [12,13], and the contents of flavonoids in the F-BG and H-BG groups were significantly higher than those in the control and U-BG groups. Sukrasno et al. [25] noted that the endogenous bioconversion of flavonoid precursors or intermediates may have caused increased flavonoid contents—e.g., conversion of phenylpropane into flavones, or the polyphenol oxidase that was gradually inactivated at high temperatures—so the polyphenols and flavones were not oxidized but were accumulated in the samples. Among them, the total flavonoid contents in the F-BG, U-BG, and H-BG were higher than that in the control group. The reason for this phenomenon was that, in addition to the tissue damage caused by pretreatment, more flavones were released. Another reason may be that during the pretreatment period, the activity of chalcone synthase was promoted, which is a key enzyme for the production of flavonoids in plants [26], leading to the increased flavonoid content. The flavonoid content began to decrease after 15 days, presumably because of the thermal destruction of heat-sensitive components. Chaaban et al. [27] indicated that the reactivity of flavones to heat differs depending on their structures, and glycosylated flavones have better heat tolerance. Therefore, the rate of formation is less than the rate of degradation, resulting in reduced flavonoid content.

2.4. *Principal Component Analysis of Functional Components of Black Garlic by Different Processes*

A previous report used PCA to study the correlations between the variety, aging process, and functional properties or flavor of BG [28]. This study further explored the effects of different processing pretreatments on the functional components of BG during the ripening process. As shown in Figure 4, PC1 (F1) has 71.98% of the variance, PC2 (F2) has 14.17%, and the total is 82.16%.

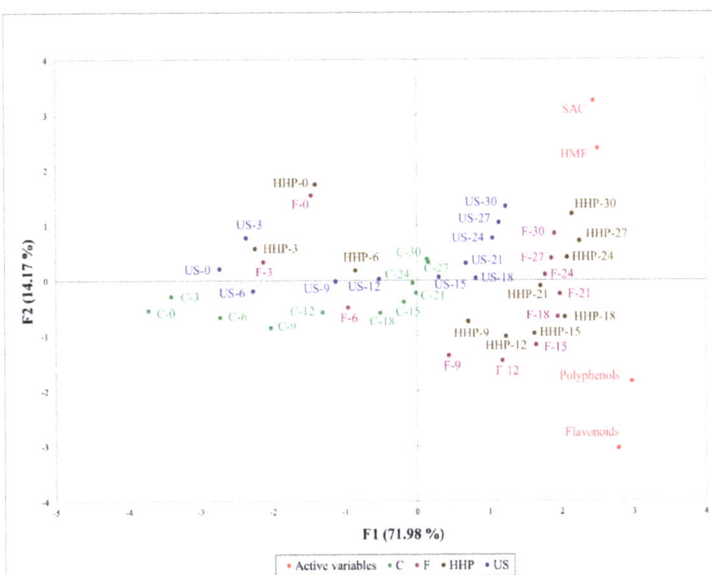

Figure 4. Principal component analysis of functional components of black garlic subjected to different processes.

The ultrasonic, freezing, and high hydrostatic pressure pretreatment groups all showed considerable influence on the generation of SAC and 5-HMF, but the effect of ultrasonic treatment on the generation of polyphenols and flavonoids was not as strong as that of the high hydrostatic pressure and freezing pretreatments. In our previous study, it was noted that ultrasound, HPP, and freezing can cause irreversible physical changes in the intercellular structure of garlic, further promoting the production of SAC, 5-HMF, polyphenols, and flavonoids [12,13]. This study also found that HPP and freezing had a more severe impact on the integrity of garlic cell nodules, so the use of these two pretreatment methods can affect the production of polyphenols and flavonoids more strongly than the ultrasonic pretreatment.

2.5. Correlation Analysis of Sensory Evaluation and Functional Components of Black Garlic by Different Processes

According to Figure 5a, the HHP, ultrasonic, and freezing pretreatments can significantly improve the acceptability of BG. Although these pretreatments did not significantly improve the appearance of BG (Figure 5c), they did significantly change the taste (Figure 5b) and odor (Figure 5d) of BG.

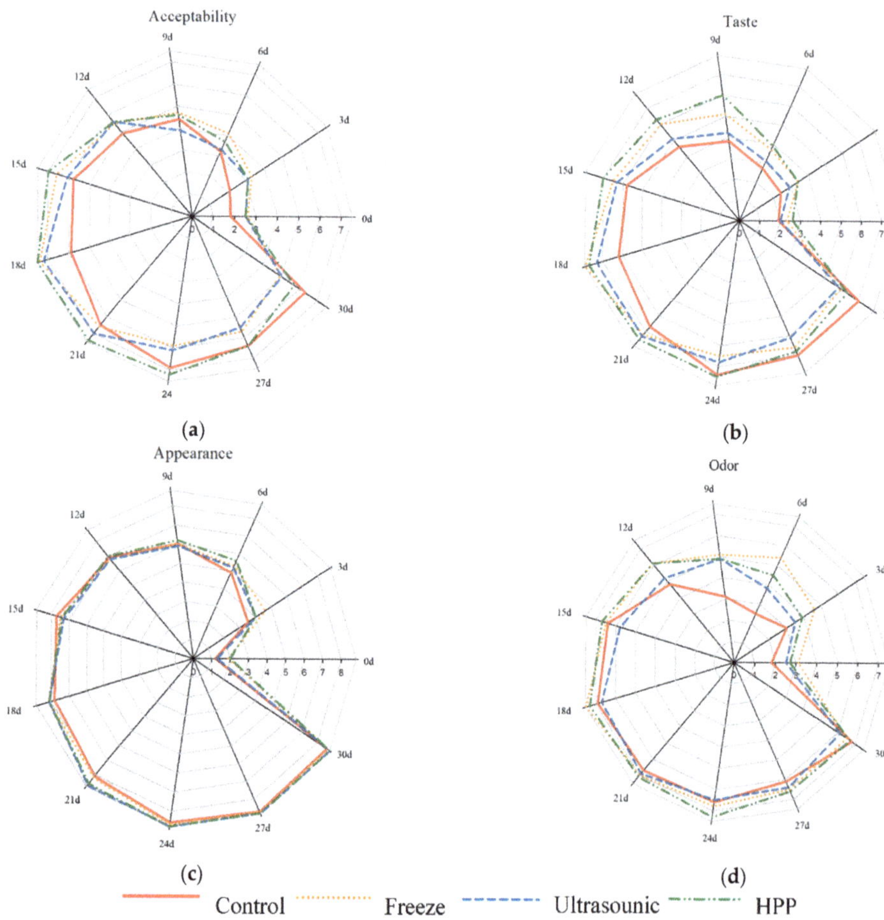

Figure 5. Effects of different processes on the sensory evaluation of black garlic: (**a**) acceptability; (**b**) taste; (**c**) appearance; (**d**) odor. $n = 50$.

Figure 5 also shows that the highest total scores for BG samples were recorded after 18 days for F-BG and H-BG, and there were no significant differences between 18 and 21 days for U-BG and 24 days for the control group. The literature indicates that BG has the best taste after 21–24 days [9], which is consistent with the observations from control group in the present study. Moreover, it was shown that BG under these conditions is acceptable to consumers, F-BG and H-BG having the best taste after 18 days of aging. Sensory evaluation and functional ingredients, including polyphenol and reducing sugar contents or the end of blackening of BG, have been the usual quality indicators [1,8].

To clarify Figure 5, the significance of the treatment in terms of sensory parameters is shown in Table 1. Based on Table 1, there was an increase in sensory acceptability scores over time (days), although each treatment had a different trend. In the control sample, on day 21, the panelists found that there were no significant differences in terms of sensory values up to day 30. However, for the freezing, ultrasound, and HPP samples, this trend started on days 24, 27, and 27, respectively. According to the results, all sensory parameters tended to increase with increasing aging time.

Table 1. Results of the sensory evaluation of black garlic based on a nine-point scoring method.

Treatment	Day	Acceptability	Taste	Odor	Appearance
	0	1.80 ± 0.63^{e}	1.90 ± 0.57^{g}	1.80 ± 1.23^{d}	1.20 ± 0.63^{h}
	3	2.10 ± 0.74^{e}	2.40 ± 0.52^{fg}	3.00 ± 1.15^{c}	3.50 ± 1.08^{g}
	6	3.20 ± 0.63^{d}	2.70 ± 0.67^{f}	2.50 ± 0.85^{cd}	4.90 ± 1.20^{f}
	9	4.40 ± 0.97^{c}	3.80 ± 0.63^{e}	3.10 ± 1.20^{c}	6.00 ± 1.05^{e}
	12	4.90 ± 0.57^{c}	4.60 ± 1.07^{d}	4.80 ± 1.03^{b}	6.90 ± 1.29^{d}
Control	15	5.80 ± 0.92^{b}	5.80 ± 1.03^{c}	6.40 ± 1.35^{a}	7.70 ± 0.48^{c}
	18	5.90 ± 1.10^{b}	6.20 ± 0.92^{bc}	6.90 ± 1.37^{a}	7.80 ± 0.42^{bc}
	21	6.50 ± 0.71^{ab}	6.70 ± 0.82^{ab}	6.70 ± 1.25^{a}	8.10 ± 0.57^{bc}
	24	6.90 ± 0.74^{a}	7.40 ± 0.70^{a}	6.60 ± 1.58^{a}	8.60 ± 0.70^{ab}
	27	6.40 ± 1.26^{ab}	7.00 ± 1.15^{a}	6.10 ± 1.52^{a}	8.70 ± 0.48^{a}
	30	6.30 ± 1.06^{ab}	7.00 ± 0.47^{a}	6.80 ± 0.92^{a}	8.60 ± 0.70^{ab}
	0	2.60 ± 1.17^{g}	2.30 ± 0.95^{f}	3.10 ± 1.10^{e}	1.90 ± 0.74^{g}
	3	3.30 ± 0.82^{fg}	3.40 ± 1.26^{e}	4.60 ± 0.52^{d}	4.60 ± 1.96^{f}
	6	4.10 ± 1.60^{ef}	3.70 ± 1.25^{e}	5.40 ± 1.07^{cd}	5.30 ± 1.25^{ef}
	9	4.70 ± 1.25^{de}	5.10 ± 1.37^{d}	5.10 ± 1.45^{d}	6.00 ± 1.41^{de}
	12	5.50 ± 1.08^{bcd}	6.00 ± 1.33^{cd}	6.10 ± 1.20^{bc}	6.80 ± 0.79^{cd}
Freezing	15	6.60 ± 0.70^{ab}	6.50 ± 0.71^{bc}	6.50 ± 0.97^{ab}	7.50 ± 0.53^{bc}
	18	7.40 ± 0.70^{a}	7.90 ± 0.57^{a}	7.50 ± 0.71^{a}	8.10 ± 0.57^{ab}
	21	6.60 ± 0.70^{ab}	7.20 ± 0.63^{ab}	7.00 ± 0.94^{ab}	8.20 ± 0.63^{ab}
	24	5.90 ± 0.88^{bc}	6.50 ± 0.71^{bc}	6.80 ± 1.23^{ab}	8.70 ± 0.67^{a}
	27	5.70 ± 1.06^{bcd}	6.60 ± 0.84^{bc}	6.50 ± 0.97^{ab}	8.70 ± 0.67^{a}
	30	4.90 ± 1.79^{cde}	5.90 ± 1.52^{cd}	6.50 ± 0.97^{ab}	8.70 ± 0.67^{a}
	0	2.50 ± 0.85^{f}	2.00 ± 0.82^{f}	2.50 ± 1.18^{f}	1.30 ± 0.67^{f}
	3	3.10 ± 0.88^{ef}	2.90 ± 1.20^{ef}	3.50 ± 0.53^{e}	4.00 ± 1.70^{e}
	6	3.30 ± 0.67^{ef}	3.10 ± 0.57^{e}	3.80 ± 0.79^{e}	5.20 ± 1.23^{d}
	9	3.90 ± 0.88^{e}	4.20 ± 1.48^{d}	4.90 ± 0.57^{d}	5.90 ± 1.52^{d}
	12	5.60 ± 1.07^{cd}	5.10 ± 0.99^{cd}	5.20 ± 0.92^{cd}	6.80 ± 1.03^{c}
Ultrasound	15	6.10 ± 0.88^{bc}	6.30 ± 0.95^{ab}	5.80 ± 0.79^{bcd}	7.30 ± 0.67^{bc}
	18	7.20 ± 0.63^{a}	7.30 ± 0.95^{a}	6.70 ± 0.82^{ab}	8.10 ± 0.57^{ab}
	21	7.00 ± 1.05^{ab}	7.30 ± 0.95^{a}	6.90 ± 1.10^{a}	8.70 ± 0.48^{a}
	24	6.10 ± 1.10^{bc}	6.80 ± 1.03^{ab}	6.50 ± 1.27^{ab}	8.80 ± 0.42^{a}
	27	5.50 ± 1.35^{cd}	6.10 ± 1.29^{bc}	6.40 ± 1.26^{ab}	8.80 ± 0.42^{a}
	30	5.00 ± 1.56^{d}	5.80 ± 1.48^{bc}	6.10 ± 1.20^{ab}	8.80 ± 0.42^{a}
	0	2.50 ± 0.85^{f}	2.60 ± 0.84^{f}	2.70 ± 1.88^{e}	2.00 ± 0.94^{g}
	3	3.10 ± 0.74^{ef}	3.40 ± 0.70^{e}	3.90 ± 2.08^{d}	4.00 ± 1.63^{f}
	6	3.70 ± 1.06^{e}	3.90 ± 1.10^{e}	4.50 ± 2.07^{cd}	5.60 ± 1.65^{e}
	9	4.60 ± 1.07^{d}	6.00 ± 1.05^{d}	4.90 ± 0.53^{c}	6.20 ± 1.69^{de}
	12	5.60 ± 1.07^{c}	6.30 ± 0.67^{cd}	6.10 ± 1.45^{b}	7.00 ± 1.05^{cd}
HPP	15	7.00 ± 0.94^{ab}	7.00 ± 0.47^{abc}	6.70 ± 1.15^{ab}	7.40 ± 0.70^{bc}
	18	7.50 ± 0.85^{a}	7.70 ± 0.48^{ab}	7.30 ± 1.32^{a}	8.10 ± 0.57^{ab}
	21	7.40 ± 0.70^{a}	7.50 ± 0.71^{ab}	7.10 ± 1.25^{a}	8.60 ± 0.52^{a}
	24	7.20 ± 0.92^{ab}	7.50 ± 0.53^{ab}	7.30 ± 1.10^{a}	8.80 ± 0.42^{a}
	27	6.40 ± 1.07^{bc}	6.80 ± 0.42^{bc}	6.60 ± 1.05^{ab}	8.80 ± 0.42^{a}
	30	5.70 ± 1.42^{c}	6.30 ± 0.95^{cd}	6.70 ± 1.03^{ab}	8.80 ± 0.42^{a}

Values are means ± SD ($n = 3$). In the same pretreatments, values with different superscripts within the same column are significantly different ($p < 0.05$).

The relationships between functional ingredients and sensory evaluation were analyzed using Pearson's correlation coefficient, as presented in Table 2.

Table 2. Correlation of quality indicators of black garlic.

Variables *	SAC	5-HMF	Polyphenols	Flavonoids	Acceptability	Taste	Appearance	Odor
SAC	**1**	0.539	0.692	0.618	0.637	0.567	−0.053	0.262
5-HMF	0.539	**1**	0.649	0.525	0.627	**0.709**	**0.775**	**0.732**
Polyphenols	0.692	0.649	**1**	**0.790**	0.550	0.585	−0.343	0.647
Flavonoids	0.618	0.525	**0.790**	**1**	0.412	0.491	−0.461	0.598
Acceptability	0.637	0.627	0.550	0.412	**1**	**0.888**	0.199	0.300
Taste	0.567	**0.709**	0.585	0.491	**0.888**	**1**	0.381	0.550
Appearance	−0.053	**0.775**	−0.343	−0.461	0.199	0.381	**1**	−0.005
Odor	0.262	**0.732**	0.647	0.598	0.300	0.550	−0.005	**1**

* Values in bold are different from 0 with a significance level alpha = 0.05.

According to the results, 5-HMF was highly correlated with acceptability, taste, appearance, and odor, with values of 0.627, 0.709, 0.770, and 0.732, respectively. In addition, SAC and polyphenols were also highly correlated with acceptability (0.637 and 0.550, respectively); other correlations included SAC and taste (0.567), polyphenols and taste (0.585), polyphenols and odor (0.647), and flavonoids and odor (0.598). These observations suggest that the functional components of BG have a relevant effect on consumer preferences, while 5-HMF and polyphenols have a more significant correlation with consumer preferences.

3. Materials and Methods

3.1. Materials

Garlic (*Allium sativum*) was harvested from Cihtong Township, Yunlin County, Taiwan and provided by Xiluo Agricultural Products Market Co., Ltd. Garlic weighing 30 ± 5 g was selected for the experimental samples, and 3 bulbs of garlic were used in each group of experiments, each with three replicates.

3.2. Sample Pretreatment

The garlic samples were divided into four groups: control, freezing, ultrasound, and high hydrostatic pressure (HHP) pretreatments. Samples were put in vacuum packages before the pretreatments. In the freezing pretreatment, the samples were frozen at −20 °C (Frt-U6009mfzw, Frigidaire Appliance Company, Charlotte, NC, USA) for 30 h. In the ultrasound pretreatment, the samples were processed at an ultrasonic frequency of 28 kHz (TST-TP, Taiwan Supercritical Technology Co., Ltd., Tainan, Taiwan) for 2 h. In the HHP pretreatment, the garlic samples were treated at 300 MPa (HPP600MPa/6.2 L, Kuentai International Co., Ltd., Yunlin, Taiwan) for 15 min. After processing, the samples were removed from their packages. The freezing and HHP pretreatment groups were heat-treated at 40 °C with 80% relative humidity (RH) for 6 days, the ultrasonic group was heat-treated at 40 °C with 80% RH for 9 days, and the control samples were heat-treated at 40 °C with 80% RH for 6 days without any pretreatment for the generation of SAC. Then, the different processing groups and the control group were placed in a chamber at a constant temperature of 70 °C with 80%RH for 30 days to age the garlic into black garlic [12,13]. All of the abovementioned heat treatment procedures used a programmable constant temperature and humidity testing machine (CH-TH-5BP-A, E. Chung Machinery Company, Taoyuan, Taiwan).

Sampling was performed every 3 days, and the BG samples were peeled and stored in a refrigerator at −20 °C to be analyzed later. For chemical and functional tests, samples were ground using a mortar and then homogeneously mixed. Then, 1 ± 0.5 g of BG mash was mixed with 10 mL of distilled water, processed with ultrasound for 30 min, and the extract was collected by gravity filtration.

3.3. Measurement of the Color Changes, Reducing Sugar Contents, and pH Value of Garlic

The color values of samples were measured using a colorimeter (NE-4000, Denshoku, Tokyo, Japan). The color of the samples was divided into *L*, *a*, and *b*, which represent white/black, red/green, and yellow/blue, respectively. Each full black garlic sample underwent five replications. Each individual black garlic bulb's surface color was measured. Before taking the sample measurements, the equipment was calibrated against a reference white tile. Results were presented in the coordinate system of the CIELab tristimuli where a defines the red–green color range ($a > 0$ indicates redness; $a < 0$ suggests greenness), b describes the yellow–blue color range ($b > 0$ indicates yellowness; $b < 0$ indicates blueness), and L is a measure of brightness, from black (0) to white (100) [29]. The color difference (ΔE) was calculated with Equation (1), where $L0$, $a0$, and $b0$ are the L, a, and b values, respectively, of fresh garlic and BG [5]:

$$\Delta E = \sqrt{(L1 - L0)^2 + (a1 - a0)^2 + (b1 - b0)^2} \quad (1)$$

The analysis of reducing sugars was performed using the phenol–sulfuric acid method [9]. Briefly, 1 mL extraction dilutions (diluted 1000 times) of BG samples from different treatment groups were poured into test tubes, and 0.6 mL of 3,5-dinitrosalicylic acid reagent (98%, Merck & Co., Inc., Kenilworth, NJ, USA) and 0.8 mL of 1 M NaOH (96%, Katayama Chemical Co., Ltd., Osaka, Japan) were added sequentially. The test tube was heated in boiling water for 15 min. After cooling, 5 mL of double-distilled water was detected by a microplate spectrophotometer (51119200, Thermo Fisher, Waltham, MA, USA) at 500 nm. Glucose (99%, Sigma-Aldrich, Burlington, MA, USA) solutions at different concentrations (0, 31.25, 62.5, 125, 250, 500, and 1000 μg/g) were used as standards. In the pH value analysis, 6 mL of the extraction solution was put into a test tube, the glass electrode of the pH meter (EL-20, Greifensee, Switzerland) was inserted into the test tube and dipped below the extraction solution, and the pH value was read [5].

3.4. Quantification of 5-HMF in Black Garlic by HPLC

We added BG puree (1 ± 0.5 g) to 10 mL of 80% methanol. After mixing, the mixture was sonicated for 30 min. The post-extraction solution was filtered through Whatman No. 4 filter paper and a 0.22 μm syringe filter. The filtered samples were analyzed by high-performance liquid chromatography (HPLC)-UV (L-7400, Hitachi, Hitachi shi, Japan). The column was a C18 column (250 mm × 4.6 mm ID, 5 μm, Nacalai Tesque Inc., Kyoto, Japan) with a flow rate of 1.0 mL/min. The mobile phase consisted of deionized water and acetonitrile (Merck & Co., Inc., Darmstadt, Germany) (88:12, *v*/*v*). The injected sample volume was 20 μL, and the detection wavelength was 284 nm [1].

3.5. Quantification of SAC in Garlic by HPLC

The BG samples were ground to a mash using a mortar and homogeneously mixed. Then, a fixed weight of BG mash (1 ± 0.5 g) was added to 10 mL of distilled water. After mixing, the mixture was processed with ultrasound for 30 min. The extraction solution was filtered through Whatman No. 4 filter paper and a 0.22 μm syringe filter. The filtered samples were subjected to HPLC with an ultraviolet detector (L-7400, Hitachi, Hitachi shi, Japan). The samples were separated on a C18 column (250 mm × 4.6 mm ID, 5 μm, Waters, Milfor, MA, USA) at a flow rate of 0.6 mL/min. Distilled water and acetonitrile (88:12, *v*/*v*) were used as the mobile phase. The sample injected volume was 20 μL, and the detection wavelength was 210 nm [30].

3.6. Measurement of Total Polyphenol and Flavonoid Contents

To assess the total phenolic content, the extract solution was diluted 100 times. Then, 40 μL of the test solution was taken and transferred into a microcentrifuge tube, and 400 μL of distilled water was added to it. Next, 20 μL of Folin–Ciocâlteu reagent (PanReac AppliChem, Hesse, Germany) was added to the test tube, and the tube was protected from light before adding 500 μL of 20% Na_2CO_3 (99%, Katayama Chemical Co., Ltd.) solution. After vortexing, the solutions were incubated in the dark for 20 min. The absorbance of the solution was measured at 750 nm. A standard curve was developed using various concentrations of gallic acid (99%, Scharlab, Barcelona, Spain) (0, 6.25, 12.5, 25, 50, and 100 μg/g) [31].

For analysis of flavonoid content, 0.1 mL of the diluted sample of the extraction solution was poured into a microcentrifuge tube, and then 0.3 mL of 99% ethanol (99%, Merck & Co., Inc., Darmstadt, Germany), 0.02 mL of 10% $AlCl_3$ (97%, Katayama Chemical Co., Ltd.), 0.02 mL of 1 M CH_3COOK (99%, PanReac AppliChem, Chicago, IL, USA), and 0.56 mL of deionized water were sequentially added. Thereafter, the solution was left to react at room temperature for 40 min. Absorbance was then detected at 415 nm. A standard curve was developed with quercetin, and the results were expressed as mg QE/g dw. The analyses were carried out in triplicate [6].

3.7. Principal Component Analysis (PCA)

The relationships between the different pretreatments (i.e., control, freezing, ultrasound and HHP) in term of the changes in the main functional components—i.e., SAC, 5-HMF, polyphenol, and flavonoid contents—during 30-day heat treatment of BG were analyzed by PCA. Sampling of BG was performed every three days, named C-Sampling Day, F-Sampling Day, US-Sampling Day, and HHP-Sampling Day. The PCA analysis was performed using a method described by Vidal, Manful [30] for Pearson's correlation. XLSTAT software (Microsoft, Washington, DC, USA) was used to analyze these variables, and the significance level of the test was 5%.

3.8. Sensory Evaluation

Sensory evaluation was measured using the method previously described by Li, Lu et al. [9], with slight modifications. The BG samples pretreated using various methods were peeled. Sensory evaluation was conducted by 50 students and researchers with a food science background trained in the sensory evaluation course. The ratio of male to female respondents was 1:1, and the age distribution was 25–40 years. The appearance, taste, odor, and acceptability were evaluated and scored according to their preference. We used a nine-point scoring method to evaluate each feature, with a midpoint of 5, where 9 is the highest score and 1 is the lowest score.

3.9. Statistical Analysis

Data were expressed as means ± standard deviations. Statistical data processing was implemented through dispersion analysis using the SPSS 20 software. Statistical analysis was performed using one-way ANOVA and Duncan's multiple range test, and statistical significance was set at $p < 0.05$. The XLSTAT 20 software was used to perform Pearson's correlation coefficient analysis [19].

4. Conclusions

In this study, suitable processing technologies were established to enhance the quality of BG products. F-BG and H-BG had higher S-allyl-cysteine (SAC), polyphenol, and flavonoid contents than BG. Moreover, H-BG and F-BG displayed the best sensory quality after 18 days of aging, while 5-hydroxymethylfurfural (5-HMF), SAC, and polyphenols were identified as the most influential sensory parameters. Furthermore, the F-BG and H-BG groups optimal taste after 18 days, as opposed to untreated BG, which needed more than 24 days; therefore, the proposed approaches can significantly reduce the processing time while enhancing the physical, sensory, and functional properties of BG. In the future, researchers could produce BG products with high functionality and good sensory properties based on the findings of this study. Moreover, this technology could also be applied to the production of related fermented products to increase the biologically active components of functional foods. The emerging food processing technologies showed good potential in enhancing the product quality, while freezing and HHP techniques may be considered the most promising pretreatments to develop BG products with good functional and sensory properties.

Author Contributions: Conceptualization, K.-H.C. and C.-W.H.; methodology, K.-H.C., C.-K.C., M.G., C.-W.H. and K.-C.C.; software, K.-H.C., S.P.S. and B.Y.; validation, K.-H.C., C.-K.C., C.-W.H. and K.-C.C.; formal analysis, K.-H.C. and C.-K.C.; investigation, K.-H.C. and C.-K.C.; resources, C.-W.H.; data curation, K.-H.C. and B.Y.; writing—original draft preparation, K.-H.C., B.Y., M.G. and C.-K.C.; writing—review and editing, B.Y., S.P.S. and C.-W.H.; visualization, M.G.; supervision, C.-W.H.; project administration, C.-K.C.; funding acquisition, C.-W.H. All authors have read and agreed to the published version of the manuscript.

Funding: This work was supported by the National Science and Technology Council, Republic of China (Grant No. NEC 109-2221-E-005-031-MY3).

Institutional Review Board Statement: Not applicable.

Informed Consent Statement: Not applicable.

Data Availability Statement: Not applicable.

Acknowledgments: This research was partially supported by the Ajins Biomedical Co., Ltd. (AS-2019002-a2).

Conflicts of Interest: The authors declare no conflict of interest.

Sample Availability: Samples of the compounds are available from the authors.

References

1. Zhang, X.; Li, N.; Lu, X.; Liu, P.; Qiao, X. Effects of temperature on the quality of black garlic. *J. Sci. Food Agric.* **2016**, *9*, 2366–2372. [CrossRef] [PubMed]
2. Vinayagam, R.; Lee, K.E.; Ambati, R.R.; Gundamaraju, R.; Ramadan, M.F.; Kang, S.G. Recent development in black garlic: Nutraceutical applications and health-promoting phytoconstituents. *Food Rev. Int.* **2021**, 1–21. [CrossRef]
3. Yudhistira, B.; Punthi, F.; Lin, J.A.; Sulaimana, A.S.; Chang, C.K.; Hsieh, C.W. S-Allyl cysteine in garlic (*Allium sativum*): Formation, biofunction, and resistance to food processing for value-added product development. *Compr. Rev. Food Sci. Food Saf.* **2022**, *21*, 2665–2687. [CrossRef] [PubMed]
4. Bae, S.E.; Cho, S.Y.; Won, Y.D.; Lee, S.H.; Park, H.J. Changes in S-allyl cysteine contents and physicochemical properties of black garlic during heat treatment. *LWT* **2014**, *55*, 397–402. [CrossRef]
5. Yuan, H.; Sun, L.; Chen, M.; Wang, J. An analysis of the changes on intermediate products during the thermal processing of black garlic. *Food Chem.* **2018**, *239*, 56–61. [CrossRef]
6. Phan, A.D.T.; Netzel, G.; Wang, D.; Flanagan, B.M.; D'Arcy, B.R.; Gidley, M.J. Binding of dietary polyphenols to cellulose: Structural and nutritional aspects. *Food Chem.* **2015**, *171*, 388–396. [CrossRef]
7. Capuano, E.; Fogliano, V. Acrylamide and 5-hydroxymethylfurfural (HMF): A review on metabolism, toxicity, occurrence in food and mitigation strategies. *LWT* **2011**, *44*, 793–810. [CrossRef]
8. Li, F.; Cao, J.; Liu, Q.; Hu, X.; Liao, X.; Zhang, Y. Acceleration of the Maillard reaction and achievement of product quality by high pressure pretreatment during black garlic processing. *Food Chem.* **2020**, *318*, 126517. [CrossRef]
9. Li, N.; Lu, X.; Pei, H.; Qiao, X. Effect of freezing pretreatment on the processing time and quality of black garlic. *J. Food Process Eng.* **2015**, *38*, 329–335. [CrossRef]
10. Seo, S.-Y.; Sharm, V.K.; Sharma, N. Mushroom tyrosinase: Recent prospects. *J. Agric. Food Chem.* **2003**, *51*, 2837–2853. [CrossRef]
11. Yu, Z.-L.; Zeng, W.C.; Lu, X.L. Influence of ultrasound to the activity of tyrosinase. *Ultrason. Sonochem.* **2013**, *20*, 805–809. [CrossRef] [PubMed]
12. Chen, Y.-T.; Chen, Y.A.; Lee, C.H.; Wu, J.T.; Cheng, K.C.; Hsieh, C.W. A strategy for promoting γ-glutamyltransferase activity and enzymatic synthesis of S-allyl-(L)-cysteine in aged garlic via high hydrostatic pressure pretreatments. *Food Chem.* **2020**, *316*, 126347. [CrossRef] [PubMed]
13. Chen, Y.-T.; Lee, C.H.; Chen, Y.A.; Wu, J.T.; Tsai, M.S.; Cheng, K.C.; Hsieh, C.W. Preparation of S-allyl cysteine-enriched garlic by two-step processing. *LWT* **2020**, *124*, 109130. [CrossRef]
14. Lung, C.T.; Chang, C.K.; Cheng, F.C.; Hou, C.Y.; Chen, M.H.; Santoso, S.P.; Yudhistira, B.; Hsieh, C.W. Effects of pulsed electric field-assisted thawing on the characteristics and quality of Pekin duck meat. *Food Chem.* **2022**, *390*, 133137. [CrossRef] [PubMed]
15. Sulaimana, A.S.; Chang, C.K.; Hou, C.Y.; Yudhistira, B.; Punthi, F.; Lung, C.T.; Cheng, K.C.; Santoso, S.P.; Hsieh, C.W. Effect of Oxidative Stress on Physicochemical Quality of Taiwanese Seagrape (*Caulerpa lentillifera*) with the Application of Alternating Current Electric Field (ACEF) during Post-Harvest Storage. *Processes* **2021**, *9*, 1011. [CrossRef]
16. Punthi, F.; Yudhistira, B.; Gavahian, M.; Chang, C.K.; Cheng, K.C.; Hou, C.Y.; Hsieh, C.W. Pulsed electric field-assisted drying: A review of its underlying mechanisms, applications, and role in fresh produce plant-based food preservation. *Compr. Rev. Food Sci. Food Saf.* **2022**, *21*, 1–22. [CrossRef]
17. Lee, S.; Yoo, M.; Kim, S.; Shin, D. Identification and quantification of S-allyl-L-cysteine in heated garlic juice by HPLC with ultraviolet and mass spectrometry detection. *LWT* **2014**, *57*, 516–521. [CrossRef]

18. Liang, T.; Wei, F.; Lu, Y.; Kodani, Y.; Nakada, M.; Miyakawa, T.; Tanokura, M. Comprehensive NMR analysis of compositional changes of black garlic during thermal processing. *J. Agric. Food Chem.* **2015**, *63*, 683–691. [CrossRef]
19. Choi, I.S.; Cha, H.S.; Lee, Y.S. Physicochemical and antioxidant properties of black garlic. *Molecules* **2014**, *19*, 16811–16823. [CrossRef]
20. Yamaguchi, Y.; Kumagai, H. Characteristics, biosynthesis, decomposition, metabolism and functions of the garlic odour precursor, S-allyl-l-cysteine sulfoxide (Review). *Exp. Ther. Med.* **2019**, *19*, 1528–1535. [CrossRef]
21. Ríos-Ríos, K.L.; Montilla, A.; Olano, A.; Villamiel, M. Physicochemical changes and sensorial properties during black garlic elaboration: A review. *Trends Food Sci. Technol.* **2019**, *88*, 459–467. [CrossRef]
22. Vega-Gálvez, A.; Scala, K.D.; Rodríguez, K.; Lemus-Mondaca, R.; Miranda, M.; López, J.; Perez-Won, M. Effect of air-drying temperature on physico-chemical properties, antioxidant capacity, colour and total phenolic content of red pepper (*Capsicum annuum*, L. var. Hungarian). *Food Chem.* **2009**, *117*, 647–653. [CrossRef]
23. Gan, R.Y.; Lui, W.Y.; Chan, C.K.; Corke, H. Hot air drying induces browning and enhances phenolic content and antioxidant capacity in mung bean (*Vigna radiata* L.) sprouts. *J. Food Process. Preserv.* **2017**, *41*, 12846. [CrossRef]
24. Moreno, J.; Gonzales, M.; Zuniga, P.; Petzold, G.; Mella, K.; Munoz, O. Ohmic heating and pulsed vacuum effect on dehydration processes and polyphenol component retention of osmodehydrated blueberries (cv. Tifblue). *Innov. Food Sci. Emerg. Technol.* **2016**, *36*, 112–119. [CrossRef]
25. Sukrasno, S.; Fidriany, I.; Anggadiredja, K.; Handayani, W.A.; Anam, K. Influence of drying method on flavonoid content of *Cosmos caudatus* (Kunth) leaves. *Res. J. Med. Plant* **2011**, *5*, 189–195. [CrossRef]
26. Ghasemzadeh, A.; Nasiri, A.; Jaafar, H.J.E.; Baghdadi, A.; Ahmad, I. Changes in phytochemical synthesis, chalcone synthase activity and pharmaceutical qualities of Sabah snake grass (*Clinacanthus nutans* L.) in relation to plant age. *Molecules* **2014**, *19*, 17632–17648. [CrossRef]
27. Chaaban, H.; Ioannou, I.; Chebil, L.; Slimane, M.; Gérardin, C.; Paris, C.; Charbonnel, C.; Chekir, L.; Ghoul, M. Effect of heat processing on thermal stability and antioxidant activity of six flavonoids. *J. Food Process. Preserv.* **2017**, *41*, 13203. [CrossRef]
28. Qiu, Z.; Zheng, Z.; Zhang, B.; Lu, X.; Qiao, X. Characterization of the growth properties of garlic endophytes and their roles in the formation of black garlic. *LWT* **2021**, *147*, 111537. [CrossRef]
29. Hajas, L.; Sipos, L.; Csobod, C.; Bálint, M.V.; Juhász, R.; Benedek, C. Lentil (*Lens culinaris* Medik.) Flour Varieties as Promising New Ingredients for Gluten-Free Cookies. *Foods* **2022**, *11*, 2028. [CrossRef]
30. Vidal, N.P.; Manful, C.F.; Pham, T.H.; Stewart, P.; Keoug, D.; Thomas, R.H. The use of XLSTAT in conducting principal component analysis (PCA) when evaluating the relationships between sensory and quality attributes in grilled foods. *MethodsX* **2020**, *7*, 100835. [CrossRef]
31. Granato, D.; de Araújo Calado, V.M.; Jarvis, B. Observations on the use of statistical methods in Food Science and Technology. *Int. Food Res. J.* **2014**, *55*, 137–149. [CrossRef]

molecules

Article

Assessment of the Hypoglycemic and Hypolipidemic Activity of Flavonoid-Rich Extract from *Angelica keiskei*

Lanlan Tu [1], Rui Wang [2], Zheng Fang [3], Mengge Sun [1], Xiaohui Sun [1], Jinhong Wu [1,*], Yali Dang [2,*] and Jianhua Liu [4]

1 Department of Food Science and Engineering, School of Agriculture and Biology, Shanghai Jiao Tong University, Shanghai 200240, China
2 State Key Laboratory for Managing Biotic and Chemical Threats to the Quality and Safety of Agro-Products, College of Food and Pharmaceutical Sciences, Ningbo University, Ningbo 315211, China
3 Department of Laboratory Medicine, School of Medicine, Institute of Molecular Medicine, Shanghai Jiao Tong University, Shanghai 200240, China
4 International Faculty of Applied Technology, Yibin University, Yibin 644000, China
* Correspondence: wujinhong@sjtu.edu.cn (J.W.); dangyali@nbu.edu.cn (Y.D.); Tel.: +86-021-3420-6614 (J.W.); +86-136-5671-0410 (Y.D.)

Citation: Tu, L.; Wang, R.; Fang, Z.; Sun, M.; Sun, X.; Wu, J.; Dang, Y.; Liu, J. Assessment of the Hypoglycemic and Hypolipidemic Activity of Flavonoid-Rich Extract from *Angelica keiskei*. *Molecules* **2022**, *27*, 6625. https://doi.org/10.3390/molecules27196625

Academic Editors: Jolanta Sereikaitė, Weiying Lu and Yanping Chen

Received: 12 September 2022
Accepted: 3 October 2022
Published: 6 October 2022

Abstract: *Angelica keiskei* contains a variety of bioactive compounds including chalcone, coumarin, and phytochemicals, endowing it with pharmacological effects such as lipid-lowering activity, antitumor activity, liver protection, and nerve protection. This study aims to study the hypoglycemic and hypolipidemic effects of the flavonoid-rich extract from *Angelica keiskei* (FEAK) in an effort to exploit new applications of FEAK and increase its commercial value. In this paper, flavonoid compounds in *Angelica keiskei* were extracted using 50% ethanol, and the contents of the flavonoid compounds were analyzed by UPLC-MS/MS. Then, the hypoglycemic and hypolipidemic activities of the FEAK were investigated through in vitro enzyme activity and cell experiments as well as establishing in vivo zebrafish and Caenorhabditis elegans (*C. elegans*) models. The UPLC-MS/MS results show that the major flavonoid compounds in the FEAK were aureusidin, xanthoangelol, kaempferol, luteolin, and quercetin. The inhibitory rates of the FEAK on the activity of α-amylase and cholesterol esterase were 57.13% and 72.11%, respectively. In cell lipid-lowering experiments, the FEAK significantly reduced the total cholesterol (TC) and total triglyceride (TG) levels in a dose-dependent manner, with 150 μg/mL of FEAK decreasing the intracellular levels of TC and TG by 33.86% and 27.89%, respectively. The fluorescence intensity of the FEAK group was 68.12% higher than that of the control group, indicating that the FEAK exhibited hypoglycemic effects. When the concentration of the FEAK reached 500 μg/mL, the hypoglycemic effect on zebrafish reached up to 57.7%, and the average fluorescence intensity of *C. elegans* in the FEAK group was 17% lower than that of the control group. The results indicate that the FEAK had hypoglycemic and hypolipidemic activities. The findings of this study provide theoretical references for the high-value utilization of *Angelica keiskei* and the development of natural functional food with hypoglycemic and hypolipidemic activities.

Keywords: hypoglycemic activity; hypolipidemic activity; *Angelica keiskei*; flavonoid; function evaluation

1. Introduction

Diabetes is a metabolic disease characterized by abnormal glucose metabolism, and type 2 diabetes accounts for about 90% of all diabetes patients [1]. Insulin resistance, one of the most prominent features of type 2 diabetes, mainly results from disorders of lipid metabolism, which is, in turn, one of the frequent complications of diabetes. According to a China Health and Nutrition Survey, a total of 20~90% of diabetes patients suffer from hyperlipidemia [2]. Therefore, the prevention and treatment of type 2 diabetes as well as lipid metabolism disorders are of utmost importance. At present, metformin and acarbose tablets are proven to be relatively reliably safe long-term in prediabetic patients, while there is insufficient evidence for the long-term safety of other drugs [3]. The drugs used to regulate blood lipids, including statins and bast, may cause adverse reactions such as liver injury and rhabdomyolysis, despite their high efficacy [4]. It has been reported that traditional Chinese medicine treatment is a useful therapeutic option for regulating and treating type 2 diabetes patients with hyperlipidemia because of its safety and efficacy [5]. In addition to conventional drug therapy, functional foods have become increasingly attractive options for helping lower blood sugar and fat.

Angelica keiskei is a perennial herb plant from the Umbelliferae family that can be used as both medicine and food [6]. Praised in "Foods for Health in the 21st century" [7], *Angelica keiskei* has pharmacological functions such as antitumor and anti-inflammatory activities, and it can lower hypertension, hyperlipidemia, and hyperglycemia, prevent cardiovascular and cerebrovascular diseases, regulate the intestinal tract, and improve human immunity and sleep quality.

Angelica keiskei is a rich source of flavonoid compounds. Studies have shown that flavonoid compounds possess a wide range of pharmacological activities, such as vasodilation, lipid-lowering activity, anticoagulation, anti-inflammation, antitumor activity, analgesia, and nonenzymatic glycosylation, and these properties have become one of the focuses of phytochemical studies. Riezki Amalia et al. showed that extract from *Angelica keiskei* stems protects HEK293 cells from N-acetyl-*p*-benzoquinone imine (NAPQI) damage and has nephroprotective properties [8]. Zhang studied the effect of chalcone from *Angelica keiskei* on the lipid metabolism in rats with type 2 diabetes. Compared with the diabetic model group, the serum levels of the TG, TC and FFA contents were significantly lower in the intervention groups [9]. Zhang found that a flavonoid-rich ethanol extract from *Angelica keiskei* leaves in a dose of 800 mg/kg could exhibit the same effect on TC with metformin [10]. Liu studied the effect of chalcones extracted from *Angelica keiskei* (AC) on the hepatocytes of rats with type 2 diabetes. Compared with rats in the diabetic control group, the levels of blood glucose and serum insulin in the 10 mg/kg AC group were decreased [11].

Most of the reports were about the single functional study of *Angelica keiskei*, and the simultaneous glucose-lowering and lipid-lowering effects of the *Angelica keiskei* extract were not investigated. However, further development of *Angelica keiskei* as a natural functional food requires systematic and comprehensive studies of its hypoglycemic and hypolipidemic activities, including its detailed flavonoid composition and the dose–effect relationship of the hypoglycemic and hypolipidemic activities of the flavonoid-rich extract. Therefore, this study aimed to assess the hypoglycemic and hypolipidemic effects of the flavonoid extract obtained from *Angelica keiskei* by in vitro and in vivo experiments so as to lay the foundation for further development of the functional food or drugs from *Angelica keiskei*.

2. Results and Discussions

2.1. The Flavonoid Compositions of FEAK

UPLC-MS/MS was used to analyze the components of flavonoids in the FEAK. As can be seen from Table 1, the most abundant flavonoids were aureusidin, xanthoangelol, kaempferol, luteolin, and quercetin. Studies have shown that aureusidin, xanthoangelol, and quercetin play a crucial role in regulating the cardiovascular system through the inhibition of blood lipids, prevention of the elevation of low-density lipoprotein in serum, and reduction in blood sugar and serum cholesterol. Wang Junbo [10] showed that 10 μg/mL luteolin and quercetin cholesterol significantly reduce the TG content in HepG2 cells. Busu et al. [12] showed that quercetin and kaempferol significantly inhibit the expression of C/EBPα, PPARγ, and SPEBP-1c at high concentrations, thus improving lipid metabolism and preventing lipid overaccumulation. The flavonoid composition of FEAK, therefore, supports its potential application in the development of functional material with hypoglycemic and lipid-lowering activities.

Table 1. Component analysis of flavonoids in the ethanol extract of *Angelica keiskei*.

Compounds	Class	Relative Quantification (%) (Mean ± SE)
Aureusidin-4-*O*-glucoside	Aurones	5.395 ± 0.082
Xanthoangelol	Chalcones	3.995 ± 0.041
4-Hydroxyderricin	Chalcones	3.767 ± 0.015
Kaempferol-3-*O*-(6″-malonyl)glucoside	Flavonols	3.622 ± 0.116
Kaempferol-3-*O*-(6″-malonyl)galactoside	Flavonols	3.482 ± 0.065
Luteolin-7-*O*-rutinoside	Flavones	3.311 ± 0.212
Kaempferol-3-*O*-glucoside-7-*O*-rhamnoside	Flavonols	3.251 ± 0.007
Luteolin-7-*O*-neohesperidoside (Lonicerin)	Flavones	3.172 ± 0.003
Kaempferol-3-*O*-neohesperidoside	Flavonols	3.170 ± 0.072
Kaempferol-3-*O*-glucorhamnoside	Flavonols	3.125 ± 0.058
Diosmetin-7-*O*-galactoside	Flavones	2.888 ± 0.034
Quercetin-5-*O*-β-D-glucoside	Flavonols	2.805 ± 0.095
6-*C*-MethylKaempferol-3-glucoside	Flavones	2.755 ± 0.009
Diosmetin-7-*O*-glucoside	Flavones	2.750 ± 0.028
Luteolin-7-*O*-glucoside (Cynaroside)	Flavones	2.739 ± 0.406
Hispidulin-7-*O*-Glucoside	Flavones	2.678 ± 0.027
Chrysoeriol-7-*O*-(6″-malonyl)glucoside	Flavones	1.997 ± 0.326
Diosmetin-7-*O*-rutinoside (Diosmin)	Flavones	1.995 ± 0.015
Luteolin-4′-*O*-glucoside	Flavones	1.851 ± 0.171
Quercetin-3-*O*-galactoside (Hyperin)	Flavonols	1.785 ± 0.062
Cyanidin-3-*O*-glucoside (Kuromanin)	Anthocyanidins	1.688 ± 0.191
Quercetin-3-*O*-glucoside (Isoquercitrin)	Flavonols	1.683 ± 0.038
Hispidulin-7-*O*-(6″-*O*-*p*-Coumaroyl)Glucoside	Flavones	1.650 ± 0.055
Kaempferol-3-*O*-glucoside (Astragalin)	Flavonols	1.644 ± 0.036
Dihydrokaempferide	Flavanonols	1.527 ± 0.080
Luteolin-3′-*O*-glucoside	Flavones	1.506 ± 0.018
Luteolin-7,3′-di-*O*-glucoside	Flavones	1.394 ± 0.053
Isobavachalcone	Chalcones	1.36 ± 0.028
Kaempferol-3-*O*-galactoside-4′-*O*-glucoside	Flavonols	1.297 ± 0.041
Xanthoangelol F	Chalcones	1.277 ± 0.044
Yuanhuanin	Flavones	1.216 ± 0.147
Quercetin-7-*O*-glucoside	Flavonols	1.172 ± 0.014
Quercetin-4′-*O*-glucoside (Spiraeoside)	Flavonols	1.165 ± 0.027
Hesperetin-5-*O*-glucoside	Flavanones	1.145 ± 0.028

2.2. The Inhibition Activities on the α-Amylase and Cholesterol Esterase

α-amylase inhibitors destroy the α-1,4-glycosidic bond of starch in the intestinal tract and are considered as therapeutic targets for type 2 diabetes [13]. Therefore, the inhibition of α-amylase activity is of great significance for the study of hypoglycemia. Figure 1 shows that, with increasing concentrations of the FEAK, the inhibitory rate of the FEAK on α-amylase significantly increased overall. When the concentration was 40 mg/mL, the inhibition rate of α-amylase reached 57.13 ± 1.88%.

Moreover, pancreatic cholesterol esterase contributes to the bile salt-dependent hydrolysis of dietary cholesterol esters as well as the hydrolysis of triglyceride-phospholipids which are potential targets for the prevention of dietary cholesterol absorption [14]. Figure 1 shows that the inhibitory rate of the FEAK on the activity of cholesterol esterase first increased and then decreased with the increasing concentration. When the concentration of the FEAK was increased to 15 mg/mL, the inhibition rate reached a maximum value of 72.11 ± 1.69%. These results demonstrate that FEAK acts as both an α-amylase inhibitor and a cholesterol esterase inhibitor.

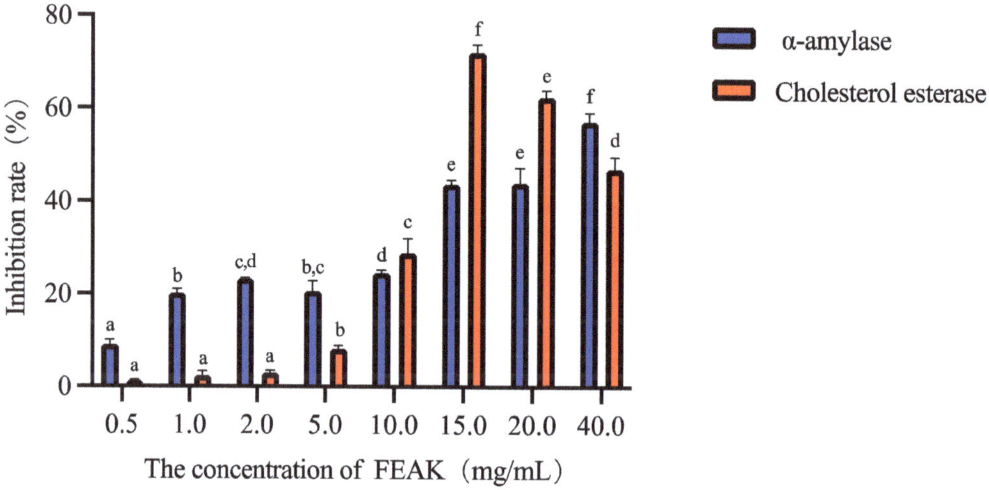

Figure 1. The inhibitory rate of the FEAK on the activity of α-amylase and cholesterol esterase. Note: different lowercase letters in the same enzyme activity test group represent the existence of significant differences at $p < 0.05$ in the different concentration group of the FEAK.

2.3. Effect of FEAK on the Intracellular Levels of TC and TG in HepG2 Cells

Blood fat is the general name for various lipid substances in blood, among which TC and TG are the most important and abundant, so they are important routine clinical test indices of blood lipid levels [15]. There have been many studies on the hypolipidemic activity of flavonoids. The TG and FFA levels in cells and FFA levels in the cell supernatant in cells treated with low, medium, and high doses of total chamomile flavonoids are significantly reduced [16]. Kobayashi et al. found that persimmon flavone regulates the expression of cholesterol 7α strengthening the enzyme gene and the cholesterol regulatory primary binding protein (SREBP) gene to affect the lipoprotein receptor [17]. Liu Chang showed that 200 μg/mL of mangrove berry extract reduces lipid and TG levels in HepG2 cells by 25.93% and 37.23%, respectively [18]. Lv Yichun et al. showed that 100 μg/mL of blueberry polyphenols decrease TG levels in liver cells by 40%, demonstrating that these compounds can clear fat accumulation in liver cells [19].

Figure 2 shows the effect of the FEAK on the TC and TG levels in the HepG2 cells. Compared with the normal control groups, the TC and TG levels in the sodium oleate and sodium palmitate model groups were significantly increased ($p < 0.05$), while the levels in the groups treated with the FEAK were significantly decreased. Compared with the model control groups, the change in TC and TG levels showed a dose-dependent relationship. When the concentration of the FEAK increased to 150 μg/mL, the intracellular levels of TC in the HepG2 cells decreased by 33.86% ($p < 0.001$). When the concentration of the FEAK increased to 150 μg/mL, the intracellular levels of TG in the HepG2 cells decreased by 27.89% ($p < 0.001$).

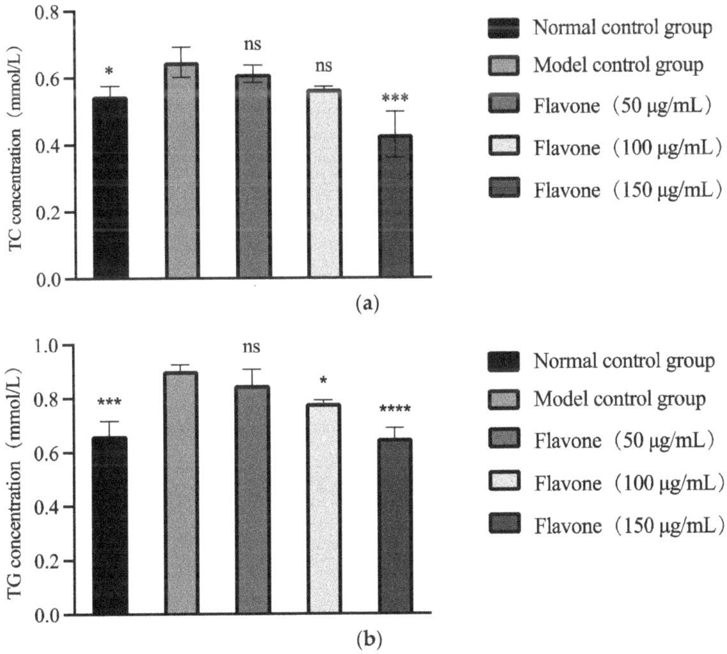

Figure 2. Changes of TC (**a**) and TG (**b**) values in HepG2 cells after treatment of FEAK. Compared with the model control groups, * $p < 0.05$, *** $p < 0.001$, **** $p < 0.0001$.

2.4. Effect of FEAK on Glucose Uptake in HepG2 Cells

In a glucose uptake model using HepG2 cells, hypoglycemic ingredients were added, the cells in each group were treated with fluorescent-labeled glucose, and the fluorescence intensity of the cells was measured using flow cytometry. The fluorescence intensity of the treated cells was higher than that of the control group, indicating that the cells treated with hypoglycemic ingredients absorbed more glucose. These observations reflect the increase in the glucose utilization rate of cells in the human body and the subsequent decrease in blood glucose, demonstrating that the samples were exhibiting hypoglycemic effects [20].

According to Table 2, increasing the FEAK leads to stronger fluorescence intensity, and the change due to the FEAK treatment was higher than that of the contrast group. When the concentration of the FEAK reached 100 mg/mL, the contrast value reached the highest value of 68.12% ($p < 0.01$). This result indicates that the FEAK absorbs the glucose in cells and exerts a hypoglycemic effect.

Table 2. Results of cell fluorescence intensity.

Sample	Fluorescence Intensity	δ (Contrast Value)
0 mg/mL	0.1208 ± 0.0211 [a]	
10 mg/mL	0.1237 ± 0.0272 [a]	2.40%
25 mg/mL	0.1573 ± 0.0108 [b]	26.66%
50 mg/mL	0.1711 ± 0.0082 [b]	41.64%
100 mg/mL	0.2031 ± 0.0008 [c]	68.12%

The sample without FEAK was used as the control group used to calculate the δ value according to the Equation (2) shown in the Section of Materials and Methods. Note: the different lowercase letters in the fluorescence intensity test group represent the existence of significant differences at $p < 0.05$ in the different concentration group of the FEAK.

2.5. Evaluation of Hypoglycemic Efficacy of FEAK In Vivo in a Zebrafish Model

Zebrafish are often used as a hypoglycemic model for the research of human diabetes drugs because they employ similar mechanisms of blood glucose regulation to mammals [21]. Zebrafish are an ideal model with which to study diabetes because of their high carbohydrate diet. Sugar in food is metabolized under catalytic enzymes, such as hexokinase and glucose 6 phosphatases, via glycometabolism. Therefore, the lack of any intermediate steps in the carbohydrate metabolism will have irreversible effects on the nervous system and the development of zebrafish.

Several zebrafish models of diabetes have been established. Gleeson et al. [22] induced type 2 diabetes in zebrafish by soaking the fish in a 2% glucose solution, but the induction period of this method is long (28 days). Capiotti et al. [23] established a hyperglycemia model of zebrafish by soaking the fish in a 0.111 mol/L glucose solution for 14 days. This induction cycle is relatively short, but the model is unstable, and the glucose concentration is high, which can lead to the death of the zebrafish. In this study, a zebrafish hyperglycemia model was established by combining glucose solution immersion and egg yolk powder feeding. The hypoglycemic effect of 125~500 μg/mL of FEAK in vivo was evaluated with this model.

To determine the maximum tolerance concentration (MTC) of the FEAK in the zebrafish model experiment, the mortality rate of adult zebrafish was calculated 2 days after the FEAK administration (Table 3). No death was found in the concentration range from 31.2 μg/mL to 500 μg/mL, and the phenotypic appearance, feeding and swimming abilities of the treated fish were the same as those of the control group (Figure 3), indicating that the FEAK is not toxic to adult zebrafish. As 500 μg/mL is a large dose, the MTC of the FEAK was considered to be 500 μg/mL and was used to determine the experimental doses.

Table 3. Evaluation of MTC (n = 30).

Group	Concentration (μg/mL)	Mortality (%)	Phenotype
Normal control group	-	0	No obvious abnormality
Model control group	-	0	No obvious abnormality
FEAK group	31.2	0	The state is similar to that of the model control group
	62.5	0	
	125	0	
	250	0	
	500	0	

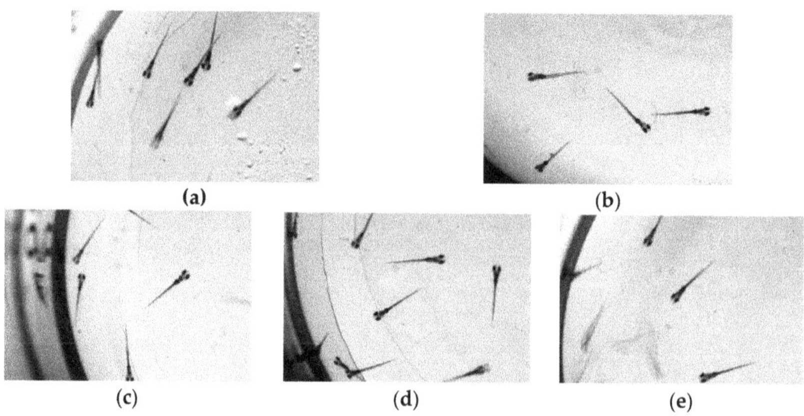

Figure 3. *Cont.*

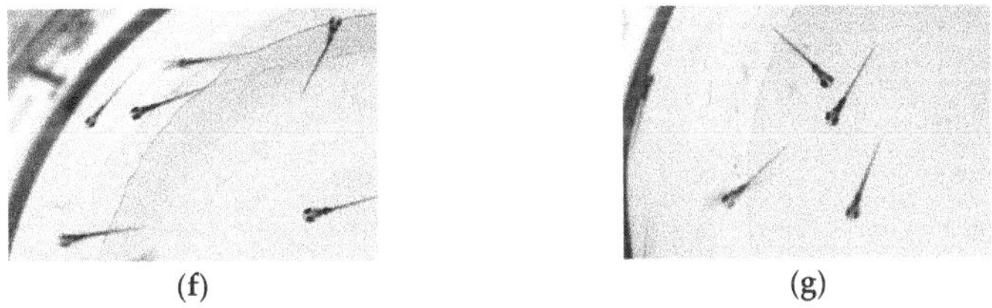

Figure 3. The phenotypic appearance of zebrafish fed with different concentration of FEAK. (a) Normal control group. (b) Model control group. (c) 31.2 μg/mL of FEAK group. (d) 62.5 μg/mL of FEAK group. (e) 125 μg/mL of FEAK group. (f) 250 μg/mL of FEAK group. (g) 500 μg/mL of FEAK group.

As shown in Table 4, the blood glucose level of zebrafish in the hyperglycemia group was 2.31 ± 0.129 mmol/L, which is about 2.5 times that of zebrafish in the normal control group (0.920 ± 0.044 mmol/L, $p < 0.001$), demonstrating that the hyperglycemia model was successful. As a short-term measurement of blood glucose, blood glucose levels can be directly used for blood glucose monitoring. Blood glucose levels were measured after zebrafish in the hyperglycemia group were treated with 125, 250, or 500 μg/mL of the FEAK for 2 days. Our results show that the hypoglycemic effect improved with increasing concentrations of the FEAK, indicating that the hypoglycemic effect of FEAK is dose-dependent. The blood glucose levels of the zebrafish dropped by 57.7% upon treatment with 500 μg/mL of FEAK ($p < 0.001$), and this hypoglycemic effect was similar to that of the positive control drug pioglitazone hydrochloride.

Table 4. Experimental result of FEAK-assisted hypoglycemic effect (n = 10).

Group	Concentration (μg/mL)	Blood Glucose Value (mmol/L, Mean ± SE)
Normal control group	-	0.92 ± 0.04 ***
Model control group	-	2.31 ± 0.13
Positive control group	20	1.07 ± 0.04 ***
Sample group	125	2.00 ± 0.13
	250	1.65 ± 0.10 *
	500	1.31 ± 0.05 ***

Note: the normal control group was prepared with pure water; the model control group was prepared with 0.15% yolk powder solution in the daytime, and 3% glucose solution was added in the evening after the removal of the yolk powder solution; the positive control group was prepared with pioglitazone hydrochloride tablets based on the model control group; and the sample groups were prepared with different concentrations of FEAK based on the model control group. Statistical differences between groups were obtained by comparing the data of each group with the model control group, * $p < 0.05$, *** $p < 0.001$.

2.6. Evaluation of Hypolipidemic Efficacy of FEAK In Vivo in a C. elegans Model

C. elegans is a good model organism for studies of lipid storage because it possesses similar regulatory factors and metabolic pathways that regulate adipose deposition and related metabolic diseases to mammals. Many mutants of lipid deposition have been generated in *C. elegans*. The dhs-3::gfp mutant exhibits green fluorescence on the surface of lipid droplets, allowing changes in lipid deposition in *C. elegans* to be easily observed. Therefore, we used dhs-3::gfp mutants to observe the effect of the FEAK on lipid accumulation in *C. elegans*. The fluorescence images were processed using ImageJ. The results in Figure 4 show that the mean fluorescence intensity of the DMSO group was 15.10 ± 0.086, while the mean fluorescence intensity of the 500 μg/mL FEAK group was 12.56 ± 0.108 ($p < 0.001$), a reduction of almost 17%. This result indicates that the FEAK decreases lipid accumulation in *C. elegans* and suggests that the FEAK lowers adipose deposition in vivo.

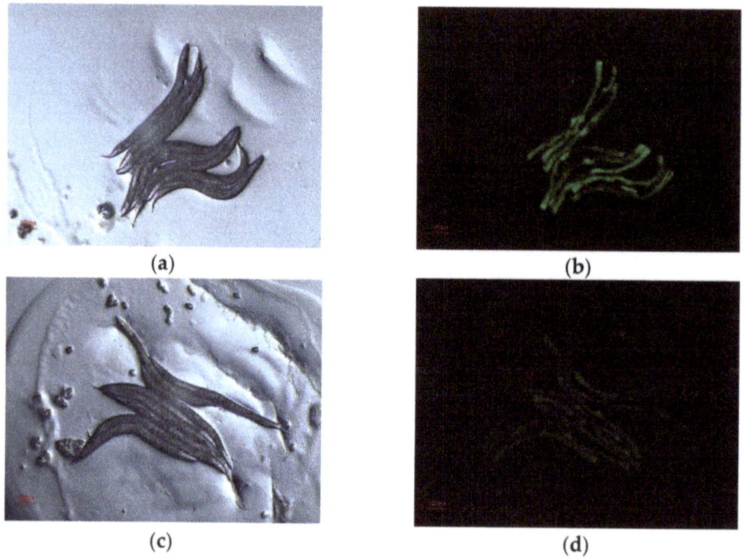

Figure 4. Representative images of dhs-3::gfp fluorescence in worms induced by DMSO and FEAK. (**a**) DMSO group (bright field, 100 px). (**b**) DMSO group (dark field,100 px). (**c**) FEAK group (bright field). (**d**) FEAK group (dark field).

3. Materials and Methods

3.1. Materials and Chemicals

Angelica keiskei (Xiancao Health Management Group Co., Ltd., Shandong, China); human hepatoma cells (HepG2) (Yongchuan Biotechnology, Shanghai, China); sodium oleate (Sigma, St. Louis, MO, USA); sodium palmitate (Sigma, St. Louis, MO, USA)); Dulbecco's modified Eagle medium (DMEM) (Hyclone, Beijing, China); TC and TG assay kits (Nanjing Jiancheng Biology Co., Ltd. Nanjing, China); α-Amylase (Yuanye Biology Co., Ltd. Shanghai, China); cholesterol esterase (Yuanye biology Co., Ltd. Shanghai, China); zebrafish (AB wild-type, Shanghai Feixi Biotechnology Co., Ltd. Shanghai, China); Caenorhabditis elegans (presented by the Wu Ziyun research group of Shanghai Jiaotong University, Shanghai, China); M9 buffer: $Na_2HPO_4 \cdot 12H_2O$ 0.15 g, KH_2PO_4 0.03 g, NaCl 0.05 g, $MgSO_4$ 0.0025 g; and pure water to a final volume of 100 mL.

3.2. Preparation of Flavonoid-Rich Extract from Angelica keiskei

Angelica keiskei samples were shattered after vacuum freeze-drying for 24 h. The dried samples were milled to a powder using a 100-mesh sieve. The powder was mixed with 50% ethanol at a mass/volume ratio of 1:10 (g:mL) and was heated and stirred in a water bath at 37 °C for 3 h. After cooling to room temperature, the extraction was collected by centrifuging at 8000 rpm/min at 4 °C. The ethanol extract process was repeated, and all extracts were placed in a rotary evaporator to remove trace amounts of ethanol. After lyophilization, the flavonoid-rich extract from *Angelica keiskei* (FEAK) was obtained and stored at −20 °C [24].

3.3. Analysis of the Flavonoid Composition of FEAK

FEAK samples were ground to a powder in a laboratory mill (30 Hz, 1.5 min). Then, 100 mg of powder was weighed and dissolved in 1.2 mL 70% methanol, stirred every 30 min for 30 s 6 times. The extract sample was placed in a refrigerator at 4 °C overnight. After centrifugation (12,000 rpm, 10 min), the supernatant was obtained, and the sample was filtered through a 0.22 μm microporous membrane and stored in the sample bottle for UPLC-MS/MS [25].

The UPLC-MS/MS analytical conditions were as follows. The UPLC was equipped with an Agilent SB-C18 column (1.8 μm, 2.1 mm × 100 mm). The mobile phase consisted of solvent A, pure water with 0.1% formic acid, and solvent B, acetonitrile with 0.1% formic acid. Sample measurements were performed with a gradient program that employed the starting conditions of 95% A, 5% B. Within 9 min, a linear gradient to 5% A, 95% B was programmed, and a composition of 5% A, 95% B was kept for 1 min. Subsequently, 95% A, 5.0% B was reached within 1.1 min and kept for 2.9 min. The flow velocity was set to 0.35 mL per minute. The column oven was set to 40 °C. The injection volume was 2 μL. The effluent was alternatively connected to an ESI triple quadrupole linear ion trap (QTRAP)-MS.

The relative content of flavonoid compound above 1% in the extract was detected by a relative quantification method, namely, it was calculated by the ratio of the peak area of each flavonoid component divided by the peak areas of all flavonoid components.

3.4. *Evaluation of the Hypoglycemic and Hypolipidemic Activity of FEAK In Vitro*

3.4.1. Measurement of α-Amylase Inhibitory Activity

We used the method of Yuca et al. [26], with slight modifications. The α-amylase solution (1 U/mL) was mixed into samples at varying concentrations and preheated in 37 °C water bath for 5 min, followed by addition of 1% preheated starch solution. The solution was incubated in a 37 °C water bath for 5 min. Then, 0.25 mL DNS solution was immediately added for color responses, and the reaction mixtures were incubated in boiling water for 5 min. After cooling in an ice bath, the reaction mixtures were diluted to 5 mL with distilled water. The absorbance was recorded at 540 nm on a microplate reader. The enzyme solution and starch solution were prepared in PBS buffer at pH 6.8, 0.1 mol/L. The additive and concentration of each tube are shown in Table 5. Equation (1) was used to calculate percent inhibition.

Table 5. Inhibition system of α-amylase activity.

Tube	α-Amylase Solution (μL)	Sample (μL)	Starch Solution (μL)	PBS (μL)
Blank 1	300	-	300	150
Blank control 2	-	-	300	450
Sample 3	300	150	300	-
Sample control 4	-	150	300	300

$$\text{Inhibitory activity} = 1 - \frac{A_3 - A_4}{A_1 - A_2} \times 100\% \quad (1)$$

A_1 is the absorbance of the blank group (equal volume buffer replaces the sample solution); A_2 is the absorbance of blank control group (equal volume buffer replaces sample solution and enzyme solution); A_3 is the absorbance of the sample group; and A_4 is the absorbance of the sample control group (equal volume buffer instead of enzyme solution).

3.4.2. Measurement of Cholesterol Esterase Inhibitory Activity in Porcine Pancreas

We used the method of Su et al. [27], with slight modifications. In brief, the measurement was performed in 0.1 mol/L sodium phosphate (pH 7.04) containing 0.1 mol/L NaCl, 0.2 mmol/L *p*-nitrophenyl butyrate (*p*-NPB) and 5.16 mmol/L sodium taurocholate buffer (STC). First, 10 U/mL of porcine pancreatic cholesterol esterase and PNPB were predissolved in acetonitrile and stored at −20 °C. Porcine pancreatic cholesterol esterase was added to the reaction tube, and samples were incubated at 25 °C for 5 min. Then, the absorbance of the solutions was measured at 450 nm using an ultraviolet-visible spectrophotometer. The additives and corresponding concentrations of each tube are shown in Table 6. Equation (1) was used to calculate percent inhibition.

Table 6. The amount of additive agents for cholesterol esterase inhibitory activity measurement.

Tube	Cholesterol Esterase Solution (μL)	Sample (μL)	PNPB (μL)	Buffer (mL)
Blank 1	50	-	10	1
Blank control 2	-	-	10	1
Sample 3	50	25	10	1
Sample control 4	-	25	10	1

3.4.3. Glucose Consumption Assay in HepG-2 Cells

The effect of FEAK on glucose consumption was investigated in insulin-resistant HepG-2 cells. After thawing, subculturing, and plating, HepG-2 cells were processed as follows. First, 100 nmol/L of recombinant human insulin was added, and cells were incubated at 37 °C for 30 min. Then, 0, 10, 25, 50 to 100 mg/mL FEAK was added, and cells were incubated at 37 °C for 1 h. Finally, 50 μM 2-NbDG was added, and cells were incubated at 37 °C for 1 h. After the incubation, the culture medium was removed, and cells were washed with PBS buffer twice. After digestion with 1 mL trypsin-EDTA solution, culture medium was added, and the resulting single cell suspension was placed in a centrifugal tube and centrifuged at 1000 RPM for 5 min. Afterward, the supernatant was removed and discarded, and the solution was centrifuged at 1000 rpm for 5 min. This process was repeated three times. HepG-2 cells were collected into flow cytometry tubes and placed on ice to maintain a low temperature for further determination of cell fluorescence intensity [28].

Fluorescence intensity was measured by flow cytometry in specific channels (10,000 cells per tube). The contrast value of fluorescence intensity was calculated according to the following Equation (2):

$$\delta = \frac{\text{Fluorescence intensity of samples} - \text{Fluorescence intensity of control group}}{\text{Fluorescence intensity of control group}} \times 100\% \quad (2)$$

The larger the value of δ, the stronger the ability of cells to absorb glucose and the stronger the hypoglycemic effect [20].

3.4.4. Total Cholesterol (TC) and Triglyceride (TG) Assay in HepG-2 Cells

The high lipid cell culture medium consisted of DMEM supplemented with 500 μmol/L sodium oleate and 250 μmol/L sodium palmitate [7]. HepG-2 cells were treated with 0.25% trypsin and transferred to petri dishes. The sample dose was set based on the cytotoxicity test. The blank control group and the model group were cultured in basal DMEM medium and high lipid medium at 37 °C and 5% CO_2 for 24 h, respectively. The sample intervention groups were first cultured in high lipid medium at 37 °C and 5% CO_2 for 24 h, then 50, 100, or 150 μg/mL FEAK was added for an additional 24 h [12].

After HepG-2 cells were cultured, the culture medium was discarded. The cells were washed with the PBS at 4 °C 3 times. Radio immunoprecipitation assay (RIPA)lysis buffer was added to each petri dish, and the cells were placed on ice for 30 min. The lysate was collected and centrifuged at 10,000 rpm for 10 min at 4 °C. The supernatant was obtained for the determination of TG and TC levels by using TC and TG assay kits.

3.5. Experimental Analysis of In Vivo Hypoglycemic Effect in a Zebrafish Model

3.5.1. Sample Preparation

Sample group: 50.0 mg/mL mother liquor was prepared using ethanol extract of FEAK and DMSO and stored at −20 °C.

Positive control group: 10.0 mg/mL mother liquor was prepared using pioglitazone hydrochloride tablets and DMSO and stored at −20 °C.

3.5.2. Evaluation of the Hypoglycemic Effect of FEAK in a Diabetic Zebrafish Model

Zebrafish of the AB wild-type strain were bred naturally in pairs. Zebrafish were all raised in fish culture water at 28 °C. Zebrafish at 5 days postfertilization (5 DPF) were used to evaluate the maximum tolerable concentration (MTC) and auxiliary hypoglycemic efficacy of FEAK.

The MTC of FEAK was determined. Zebrafish of the AB wild-type strain at 5 DPF were randomly selected and placed in 25 mL beakers, with 30 zebrafish in each beaker (experimental group). FEAK was dissolved in the water (31.2, 62.5, 125, 250, and 500 μg/mL). All experimental groups except for the normal control group were given 0.15% yolk powder solution in the daytime, and 3% glucose solution was given in the evening after the removal of the yolk powder solution to establish the zebrafish hyperglycemia model. After treatment at 28 °C for 48 h, the MTC of FEAK on zebrafish in the model control group was determined.

The auxiliary hypoglycemic efficacy of FEAK was further evaluated. Zebrafish of the AB wild-type strain at 5 DPF were randomly selected and placed in beakers, with 30 zebrafish in each beaker (experimental group). FEAK was given in aqueous solution (concentrations as shown in Tables 1 and 2), and 20.0 μg/mL of pioglitazone hydrochloride was used as a positive control. The normal control group and model control group were set up at the same time. After establishing the zebrafish hyperglycemia model and treating the zebrafish with FEAK at 28 °C for 48 h, the zebrafish were washed three times with fish culture water, and the glucose level of zebrafish was detected using a blood glucose meter. The auxiliary hypoglycemic efficacy was evaluated by statistical analysis of the results obtained using this indicator.

3.6. Evaluation of Hypolipidemic Effect of FEAK in C. elegans Model

First, *Escherichia coli* (*E. coli)* solution with FEAK was prepared as follows: 500.0 mg/mL mother liquor of FEAK was prepared with DMSO and stored at −20 °C; 10 μL of mother liquor was added to 10 mL concentrated *E. coli* solution and mixed, then the mixture was added to the nematode growth medium (NGM) agar plate and dried by airing for later use. The DMSO control group received 10 μL DMSO instead of FEAK.

The effect of FEAK on lipid deposition was observed by determining the green fluorescence of the dhs-3::gfp mutants. The animals were first synchronized at the L1 stage and then divided into groups. About 50 nematodes were cultured in each plate in a constant temperature incubator set to 20 °C for 6 days. The animals were then washed with M9 buffer, anesthetized with 40 mmol/L imidazole, observed, and photographed under fluorescence microscope. The average fluorescence intensity was calculated using ImageJ.

3.7. Statistical Analysis

Each experimental group was set up in three parallel groups. Statistical results were expressed as mean ± SE. SPSS 26.0 software (IBM, 2022, Shanghai, China). was used for statistical analysis, with $p < 0.05$ indicating a statistically significant difference. GraphPad Prism 9.0 software (GraphPad Software, 2021, Shanghai, China) was used for plotting.

4. Conclusions

This study aimed to investigate the effect of FEAK on regulating glucose and lipid metabolism. The hypoglycemic and hypolipidemic activities of FEAK were analyzed through a series of in vitro and in vivo experiments, including assay testing the inhibition of α-amylase and cholesterol esterase, the determination of the intracellular TC and TG levels and glucose uptake in HepG2 cells, and the evaluation of hypoglycemic and hypolipidemic efficacy in vivo using zebrafish and *C. elegans.* We found that FEAK is rich in flavonoids, including aureusidin, xanthoangelol, kaempferol, luteolin, and quercetin. We also found that 40 mg/mL of the FEAK showed an inhibition rate of 57.13 ± 1.88% against α-amylase and that 15 mg/mL of the FEAK showed a maximum inhibition rate of 72.11 ± 1.69% against cholesterol esterase. Further, 150 μg/mL of FEAK decreased intracellular TC levels in the HepG2 cells by 33.86% ($p < 0.001$), and 150 μg/mL of the FEAK decreased the intracellular TG levels in the HepG2 cells by 27.89% ($p < 0.001$). When the concentration of the FEAK reached 100 mg/mL, the contrast value indicating the glucose uptake reached its highest value of 68.12% ($p < 0.01$). Moreover, 500 μg/mL of FEAK decreased the blood glucose levels of zebrafish by 57.7% ($p < 0.001$), similar to the positive control drug pioglitazone hydrochloride. Finally, 500 μg/mL of the FEAK decreased the fluorescent intensity of *C. elegans* by 17% ($p < 0.001$) compared to that of the DMSO group. These findings provide strong evidence that FEAK has hypoglycemic and hypolipidemic activity and could be a promising natural product with potential value for the development of functional foods or drugs to prevent or treat type 2 diabetic or hyperlipidemia. However, the mechanism of the hypoglycemic and hypolipidemic effects of FEAK is not very clear, and further research is needed to understand the detailed action mechanism.

Author Contributions: Conceptualization, J.W. and Y.D.; Methodology, L.T. and R.W.; Software, L.T. and M.S.; Validation, L.T. and X.S.; Formal analysis, Z.F.; Data curation, L.T.; Writing—original draft preparation, L.T. and R.W.; Writing—review and editing, L.T.; Visualization, L.T. and J.L.; Supervision, J.W.; and project administration, J.W. All authors have read and agreed to the published version of the manuscript.

Funding: The authors are grateful to the financial fund of the Bayannaoer City National Industrial high-tech Industrial Demonstration Zone key project of "Science and technology to promote Inner mongolia development" (NMKJXM202209-2) and the National Natural Science Foundation of China (Grant No. 31972017).

Institutional Review Board Statement: Not applicable.

Informed Consent Statement: Not applicable.

Data Availability Statement: Not applicable.

Conflicts of Interest: The authors declare no conflict of interest.

Sample Availability: Samples of the plant material are available from the authors.

References

1. Zheng, Y.; Ley, S.H.; Hu, F.B. Global Aetiology and Epidemiology of Type 2 Diabetes Mellitus and Its Complications. *Nat. Rev. Endocrinol.* **2018**, *14*, 88–98. [CrossRef] [PubMed]
2. Sun, L.; Lu, E.; Wu, H.; Zhou, C. Study on diabetes complicated with hyperlipidemia. *Chin. J. Neuroanat.* **2013**, *29*, 346–350.
3. Diabetes Branch of Chinese Medical Association. Guidelines for the prevention and control of type 2 diabetes in China (2017 Edition). *Chin. J. Pract. Intern. Med.* **2018**, *38*, 292–344. [CrossRef]
4. Ma, L. Research progress of related adverse reactions and non-cardiovascular effects of statins. *Guide China Med.* **2018**, *16*, 8–9. [CrossRef]
5. Qin, S.; Wang, Y.X.; Chen, J.B.; Zhang, Y.Q.; Yu, X.W.; Tang, X.N. Research progress of hypoglycemic effect of traditional Chinese Medicine. *J. Pract. Gynecol. Endocrinol.* **2018**, *5*, 198. [CrossRef]
6. Rong, Y.Z.; Gu, X.Z.; Li, D.N.; Chen, L.H.; Zhang, Y.H.; Wang, Z.W. Characterization of aroma, sensory and taste properties of *Angelica keiskei* tea. *Eur. Food Res. Technol.* **2021**, *247*, 1665–1677. [CrossRef]
7. Gao, X.; Jiang, Y.; Xu, Q.; Liu, F.; Pang, X.; Wang, M.; Li, Q.; Li, Z. 4-Hydroxyderricin Promotes Apoptosis and Cell Cycle Arrest through Regulating PI3K/AKT/MTOR Pathway in Hepatocellular Cells. *Foods* **2021**, *10*, 2036. [CrossRef]
8. Amalia, R.; Aulifa, D.L.; Zain, D.N.; Pebiansyah, A.; Levita, J. The Cytotoxicity and Nephroprotective Activity of the Ethanol Extracts of *Angelica keiskei* Koidzumi Stems and Leaves against the NAPQI-Induced Human Embryonic Kidney (HEK293) Cell Line. *Evid. Based Complementary Altern. Med.* **2021**, *2021*, e6458265. [CrossRef]
9. Zhang, J. Effect of Ashitaba Chalcone on Lipid Metabolism of Rats with Type II Diabetes. *Food Sci.* **2015**, *36*, 250–254.
10. Zhang, W.; Jin, Q.; Luo, J.; Wu, J.H.; Wang, Z.W. Phytonutrient and anti-diabetic functional properties of flavonoid-rich ethanol extract from *Angelica keiskei* leaves. *J. Food Sci. Technol.* **2018**, *55*, 4406–4412. [CrossRef]
11. Liu, B.; Sun, J.P.; Zhao, Y.; Li, L.; Zhong, J.Y. Effect of ashitabe chalcones on the mRNA expression of PI3K and Akt in hepatocytes of rats with diabetes. *J. Hyg. Res.* **2013**, *42*, 466–469.
12. Bu, S.; Xue, Q.; Yuan, C.; Chen, Y.; Cao, F. Effects of Three Flavonoids Extracted from Ginkgo biloba on Lipid Metabolism in 3T3-L1 Cells. *J. Shandong Agric. Univ. Nat. Sci. Ed.* **2020**, *51*, 598–604.
13. Yuan, W.; Yan, M.X.; Wang, Y.T.; Liu, X.; Gong, Y.L. Optimized preparation of eugenol microcapsules and its effect on hepatic steatosis in HepG2 cells. *Drug Dev. Ind. Pharm.* **2020**, *47*, 225–234. [CrossRef] [PubMed]

14. Zhang, H.L.; Wu, Q.X.; Wei, X.; Qin, X.M. Pancreatic lipase and cholesterol esterase inhibitory effect of Camellia nitidissima Chi flower extracts in vitro and in vivo. *Food Biosci.* **2020**, *37*, 100682. [CrossRef]
15. Zhang, M.; Wang, L.M.; Chen, Z.; Zhao, Z.; Li, Y.C.; Deng, Q.; Huang, Z.J.; Zhang, X.; Li, C.; Zhou, M.G.; et al. Multilevel logistic regression analysis on hypercholesterolemia related risk factors among adults in China. *Chin. J. Prev. Med.* **2018**, *52*, 151–157.
16. Amanguli, A.; Ma, H.; Lan, W. Study on improvement effect of total flavonoids from *Matricaria recutita* on lipid deposition in HepG2 cells. *China Pharm.* **2022**, *33*, 1306–1312.
17. Matsumoto, K.; Yokoyama, S.; Gato, N. Bile Acid-Binding Activity of Young Persimmon (*Diospyros kaki*) Fruit and Its Hypo lipidemic Effect in Mice. *Phytother. Res.* **2010**, *24*, 205–210. [CrossRef]
18. Liu, C. Effects of Red Raspberry and Its Extracts on Fat Accumulation in HepG2 Cells. Master's Thesis, Northeast Forestry University, Harbin, China, 2019.
19. Lu, Y.C.; Liu, Y.X.; Wu, W.; Zhou, F.; Ji, B.P.; Su, C.Y. Preventive Effect of Blueberry Polyphenols on Oleic Acid-induced Fat Accumulation in HepG2 Cells. *Food Sci.* **2011**, *32*, 308–312.
20. Malle, E.K.; Zammit, N.W.; Walters, S.N.; Koay, Y.C.; Wu, J.; Tan, B.M.; Villanueva, J.E.; Brink, R.; Loudovaris, T.; Cantley, J.; et al. Nuclear Factor KB–Inducing Kinase Activation as a Mechanism of Pancreatic β Cell Failure in Obesity. *J. Experiment. Med.* **2015**, *212*, 1239–1254. [CrossRef]
21. Zong, H.; Bao, B. Construction of rict-1/daf-2&Pdhs-3::dhs-3::gfp in *C. elegans*. *J. HeFei Univ. Technol.* **2018**, *41*, 1695–1699.
22. Arneson, L.; Gleeson, M.; Connaughton, V. Induction of Hyperglycemia and Microvascular Retinal Complications in Zebrafish, Danio Rerio. *Investig. Ophthalmol. Vis. Sci.* **2006**, *47*, 1739.
23. Capiotti, K.M.; Antonioli, R.; Kist, L.W.; Bogo, M.R.; Bonan, C.D.; Da Silva, R.S. Persistent Impaired Glucose Metabolism in a Zebrafish Hyperglycemia Model. *Comp. Biochem. Physiol. Part B Biochem. Mol. Biol.* **2014**, *171*, 58–65. [CrossRef] [PubMed]
24. Deng, R.; Zou, Y.; Shan, H.; Zhao, Y.; Lv, H. Study on Optimum of Extraction and Determination of Total Flavonoids in *Angelica keiskei* Koidzumi. *Food Ind.* **2012**, *33*, 44–47.
25. Chen, W.; Gong, L.; Guo, Z.; Wang, W.; Zhang, H.; Liu, X.; Yu, S.; Xiong, L.; Luo, J. A Novel Integrated Method for Large-Scale Detection, Identification, and Quantification of Widely Targeted Metabolites: Application in the Study of Rice Metabolomics. *Mol. Plant* **2013**, *6*, 1769–1780. [CrossRef] [PubMed]
26. Yuca, H.; Özbek, H.; Demirezer, L.Ö.; Güvenalp, Z. Assessment of the α-glucosidase and α-amylase inhibitory potential of Paliurus spina-christi Mill. and its terpenic compounds. *Med. Chem. Res.* **2022**, *31*, 1393–1399. [CrossRef]
27. Su, J.; Ma, C.; Yang, L.; Wang, H.; Gao, C.; Nie, R. Inhibition of pancreatic cholesterol esterase activities and cholesterol micelle of EGCG and quercetin. *Sci. Technol. Food Ind.* **2015**, *36*, 346–349.
28. Shi, H. Effect of Purple Sweet Potato Anthocyanins on Lipid Metabolism in HepG2 Cells Induced by Oleic Acid. *Port Health Control* **2019**, *24*, 27–31, 35.

molecules

Article

Comparison of Different Volatile Extraction Methods for the Identification of Fishy Off-Odor in Fish By-Products

Yuanyuan Zhang [1], Long Tang [1], Yu Zhang [1,*,†], Huanlu Song [1,*,†], Ali Raza [1], Wenqing Pan [2], Lin Gong [2] and Can Jiang [3]

[1] Laboratory of Molecular Sensory Science, Beijing Technology and Business University, Beijing 100048, China
[2] Hunan Province Jiapinjiawei Biotechnology Co., Ltd, Changde 415401, China
[3] Wuzhou Testing Co., Ltd, Jining 273200, China
* Correspondence: bigzhyu@163.com (Y.Z.); songhl@th.btbu.edu.cn (H.S.)
† These authors contributed equally to this work.

Abstract: This study was conducted to analyze volatile odor compounds and key odor-active compounds in the fish soup using fish scarp and bone. Five extraction methods, including solid-phase microextraction (SPME), dynamic headspace sampling (DHS), solvent-assisted flavor evaporation (SAFE), stir bar sorptive extraction (SBSE), liquid-liquid extraction (LLE), were compared and SPME was finally selected as the best extraction method for further study. The volatile odor compounds were analyzed by gas chromatography-olfactometry-mass spectrometry (GC-O-MS) and comprehensive two-dimensional gas chromatography-olfactometry-mass spectrometry (GC × GC-O-MS) techniques, and the key odor-active compounds were identified via aroma extract dilution analysis (AEDA) and relative odor activity value (r-OAV) calculation. A total of 38 volatile compounds were identified by GC-O-MS, among which 10 were declared as odor-active compounds. Whereas 39 volatile compounds were identified by GC × GC-O-MS, among which 12 were declared as odor-active compounds. The study results revealed that 1-octen-3-one, 2-pentylfuran, (E)-2-octenal, 1-octen-3-one, hexanal, 1-octen-3-ol, 6-methylhept-5-en-2-one, (E,Z)-2,6-nondienal and 2-ethyl-3,5-dimethylpyrazine were the key odor-active compounds in the fish soup.

Keywords: fish scraps; fish soup; GC-O-MS; GC × GC-O-MS; AEDA; r-OAV

Citation: Zhang, Y.; Tang, L.; Zhang, Y.; Song, H.; Raza, A.; Pan, W.; Gong, L.; Jiang, C. Comparison of Different Volatile Extraction Methods for the Identification of Fishy Off-Odor in Fish By-Products. *Molecules* **2022**, *27*, 6177. https://doi.org/10.3390/molecules27196177

Academic Editors: Weiying Lu and Yanping Chen

Received: 27 August 2022
Accepted: 16 September 2022
Published: 21 September 2022

1. Introduction

China is a large producer of freshwater fish, and the production of fish products is steadily increasing year by year. China has been rich in fish products since ancient times, and Dongting Lake in Hunan Province is the main producer of freshwater fish, especially having large amounts of white chub [1]. During the processing of freshwater fish, especially in the processing of minced fish, large amounts of scraps such as fish heads, fish bones, fish tails, fish skins, and offal are produced [2]. Previously, it was found that the meat extraction rate of seven kinds of bulk freshwater fish, namely mackerel, grass carp, silver carp, bighead carp, crucian carp and bream, was the highest for mackerel (54.33%) and the lowest for bighead carp (32.80%), while the remaining 45% to 67% were fish head and fish bone scraps [3]. At present, these scraps are not utilized as value-added products, except a limited portion of them are used for processing feed fish meal, and the majority of them are directly discarded, which not only wastes a lot of resources but also pollutes the environment and increases the cost of environmental waste management [4]. An important issue faced by the aquatic food industry these days is the effective utilization of fish waste in order to generate economic benefits. The fish head is rich in nucleotides, amino acids and inorganic elements such as potassium, calcium, sodium and magnesium that contribute relatively to the flavor, and it is valuable to utilize it in seasoned soup, fish bone paste or aquatic seasoning base after proper treatment [5]. However, the removal of a fishy odor is a key step in the production of seasoning bases, which first requires the investigation of the nature of a fishy taste. In this study, fish scraps (fish heads and bones) were used as raw materials for boiling fish soup and then analyzing the off-flavor compounds in fish soup in order to provide a research background for the removal of fishy odor compounds as well as a scientific basis for studies such as the production of seasoning bases using fish scraps.

Recently, the odor compounds in fish have emerged as one of the most important studies for the development of fish seasonings, which are directly related to the sensory quality of the product. The fishy flavor is one of the main indicators used to determine the quality of fish flavoring, and it is formed by the joint action of a variety of odor compounds. Liu et al. [6] used SPME combined with the GC-MS technique to analyze fishy odor substances in the muscle (35 kinds), head (35 kinds) and skin (38 kinds) of tilapia, and nonanal, octanal and (E)-octenal were found to be the key fishy substances which were identified via AEDA analysis. Similarly, in the study of Li et al. [7], the DHS-GC-MS technique was employed to analyze the odor compounds produced during hydrolysis of grass carp, and 37 odor compounds were detected. In previous studies, a series of investigations were conducted on the volatile compounds of different varieties of fish, although the site, type and source of the study subject were not specified. However, these volatile compounds do not necessarily determine the key aroma compounds in the products. Therefore, the study of key odor-active compounds is also necessary. Although there have been some advances in the study of aroma-active substances in fish and boiled fish soup, most of the studies have been conducted with fish raw materials; it is interesting to think about the Maillard reaction, caramelization reaction and oxidation of fats and oils that may occur after heat treatment of foods, and these reactions produce various volatile compounds [8,9]; these types of reactions are also present in the boiling process of fish scraps, which helps to further investigate the odor compounds in them and provides a research basis for the preparation of seasonings using fish scraps.

GC-O-MS is the key approach to analyze odor compounds in fish soup. However, during the boiling process, the odor compounds in the fish soup may react in various ways, causing changes in the volatile components [10]. In this study, a new gas chromatographic technique, namely two-dimensional integrated gas chromatography-olfactometry-mass spectrometry (GC $\times$ GC-O-MS), was compared with conventional one-dimensional gas chromatography in order to analyze the sample; this two-dimensional gas chromatography allows the conversion between two modes (one- and two-dimensional analysis modes). The two-dimensional mode allows volatile compounds in the sample to be detected more quickly and efficiently [11]. With the continuous promotion of GC $\times$ GC-O-MS, this technique has been applied to the analysis of volatile odor compounds in various samples. For example, Yang et al. [12] used GC $\times$ GC-O-MS to identify the differences in volatile odorants of cinnamon tea leaves at different roasting temperatures and detected 97 aroma substances. Very recently, SPME combined with GC-O-MS and GC $\times$ GC-O-MS techniques were employed by Zhao et al. [13] to identify volatile odor compounds of pepper via AEDA and OAV analysis.

Therefore, in order to make the useful recovery of volatile compounds from fish by-products, the offal (fish head and bones) of silver carp from Dongting Lake, Hunan Province was utilized. The present study was mainly aimed to accomplish the following objectives: (1) to select the most suitable volatile extraction method from solid-phase microextraction (SPME), dynamic headspace sampling (DHS), solvent-assisted flavor evaporation (SAFE), stir bar sorptive extraction (SBSE) and liquid-liquid extraction (LLE); (2) the optimal extraction method from (1) was selected and combined with the aroma extraction dilution analysis to identify the key odor-active compounds of fish soup. The significance of this study includes the selection of an appropriate method to analyze the odor compounds in fish broth and to study the key odor compounds, in order to provide a research basis for futuristic studies dealing with the removal of fishy odor from fish soup, and to provide a theoretical basis for the production of flavors from fish broth boiled with fish scraps.

2. Materials and Methods

2.1. Samples and Chemicals

The fish heads and bones used in the study were supplied by Hunan Jiapinjiawei Biotechnology (Hunan, China). Firstly, 60 °C warm water was added to a heavy-bottomed pot having a fish head and bone; furthermore, the waste water was removed, and washing was performed. The samples were mixed with water in a ratio of 1:1 by volume, boiling was performed under high pressure at 121 °C for 2 h, and the dregs were filtered off to obtain fish soup. Finally, the samples were concentrated to about 25% of the total soluble solid content with a rotary evaporator at 55°C.

Ethyl ether, n-hexane, and anhydrous sodium sulfate, all having purities >99%, were purchased from Lab Gou e-mall (Beijing, China). 2-Methyl-3-heptanone and n-alkanes (C7–C30) were provided by Sigma-Aldrich (St. Louis, MO, USA.). Nitrogen gas (99.9992% purity) was obtained from Beijing AP BAIF Gases Industry Co., Ltd. (Beijing, China) and the liquid nitrogen was obtained from Xian Heyu Trading Co., Ltd. (Beijing, China).

2.2. Solid-phase Microextraction (SPME)

The SPME method was applied to extract the flavor compounds from fish soup according to the protocols mentioned by Li et al. [14]. Five grams of the sample and 0.3 µ L of 2-methyl-3-heptanone (0.816 µg/mL internal standard) were placed in a headspace vial (20 mL, Beijing Banxia Technology Development Co., Ltd., Beijing, China). The stock was equilibrated in a constant temperature water bath at 60 °C for 20 min, and then the volatile compounds were extracted with SPME needles having divinylbenzene/carboxyl/polydimethylsiloxane fibers (50/30 µm, Supelco, Bellefonte, PA, USA) at 60 °C for 40 min. Immediately after the extraction was completed, the extraction needle was inserted into the GC-O-MS and GC × GC-O-MS instruments for analysis and thermal desorption at 250 °C for 5 min. For accuracy, each sample was analyzed three times.

2.3. Solvent-Assisted Flavor Evaporation (SAFE)

The SAFE system is a compact combination of a distillation unit and a high vacuum pump; 40 g of fish soup and 5 µL of 2-methyl-3-heptanone (0.816 µg/µL, internal standard) was added to a Teflon bottle with diethyl ether/pentane (V1/V2 = 2:1). The resulting preparation was extracted by stirring for 8 h on a shaker (4 °C), and the volatile odor compounds were extracted using a SAFE apparatus (Deutschen Forschungsanstalt für Lebensmittelchemie, Freising, Free State of Bavaria, Germany). The distillation process was carried out by a molecular turbine pump (Edwards, Munich, Germany) for 2 h at 10-4 torr. The resulting fractions were collected in a trap that was submerged under liquid nitrogen. After collection, water was removed from the collected fraction by adding anhydrous Na2SO4 and the collected fraction was concentrated to 10 mL through a Vigreux column (50 cm × 1 cm I.D.; Beijing Banxia Technology Development Co., Ltd., Beijing, China). The concentrate was further reduced to 200 µL by a nitrogen stream (purity ≥ 99.999%) purging. Finally, one microliter of the sample was injected into the GC-O-MS and GC × GC-O-MS instruments for detection using a 5 µL syringe. All the samples were analyzed in triplicates for statistical accuracy.

2.4. Dynamic Headspace Sampling (DHS)

Fish soup (20 g) and 2 µL of 2-methyl-3-heptanone (0.816 µg/µL, internal standard) were added to a dynamic headspace flask, which was purged with 99.9992% high purity nitrogen at a flow rate of 150 mL/min at one end and inserted into a Tenax TA tube at the other end for adsorption. The temperature of water bath was established at 60 °C, equilibrated for 20 min, and the adsorption was performed for 40 min. After that, the Tenax TA tube was removed and placed in a thermal desorption unit (TDU) (Gerstel, Germany) for GC-O-MS or GC × GC-O-MS analysis. For statistical accuracy, all the samples were analyzed in triplicates.

2.5. *Stir Bar Sorptive Extraction (SBSE)*

Fish soup (15 g) and 1.5 μL of 2-methyl-3-heptanone (0. 816 μg/μL, internal standard) were added to a 40 ml headspace vial, and a Twister®(Gerstel, Germany) (such as PDMS, EG (ethylene glycol) or silica gel) was immersed inside the headspace vial (carrying the sample) and stirred at 60 °C for 40 min. After that, the stirring rod was picked out, rinsed with deodorized water, and placed in a hollow glass tube, which was further transferred to a TDU for thermal desorption in order to perform GC-O-MS and GC × GC-O-MS analysis. For accuracy, all the samples were analyzed three times.

2.6. *Liquid-Liquid Extraction (LLE)*

The volatile substances in the samples were extracted according to the conventional liquid-liquid extraction (LLE) method [15]. Fifty grams of sample was transferred to a triangular flask, after that, 50 mL distilled water, 50 mL dichloromethane, 50 mL diethyl ether and 5 μL of internal standard 2-methylhept-3-one (0. 816 μg/μL, internal standard) were added together. The mixture was stirred at 800 rpm for 10 min. After centrifugation at 8,000g for 40 minutes, the extracts were separated using a partition funnel. The resulting extract was added with anhydrous sodium sulfate and left to dry at 4 °C for 12 h. The dried extract was then concentrated to 100 μL using a nitrogen stream, and finally the concentrated extract was analyzed by GC-O-MS and GC × GC-O-MS. The samples were analyzed in triplicates for accuracy and statistical analysis.

2.7. *Gas Chromatography-Olfactometry-Mass Spectrometry (GC-O-MS) Analysis*

A GC-MS (7890A-7000, Agilent Technologies Inc., Santa Clara, CA, USA) instrument combined with an olfactory detection port (ODP4, Gerstel, Germany) was used to identify volatile odor compounds. Separation of odor-active substances in samples was performed on a polar DB-WAX capillary column (30 mm × 0.32 mm, 0.25 μm film thickness; J & W Scientific, Folsom, CA, USA). The gas chromatographic instrument condition includes an initial column temperature setting of 40 °C, holding for 3 min, followed by an increase in temperature up to 230 °C at 4 °C/min and holding for 3 min. Ultra-pure helium (99.999%, Beijing AP BAIF Gas Industry Co., Ltd., Beijing, China) was used as the carrier gas. The electron impact mass spectra were generated at an ionization energy of 70 eV with an m/z scan range of 25–370 amu. The temperatures of the mass spectrometer source and quadrupole were programed at 230 °C and 150 °C, respectively. Moisture gas was delivered to the olfactory detection port through a blank capillary column.

2.8. *GC × GC-O-MS Analysis*

A GC-MS instrument (8890A-5977B; Agilent Technologies Inc., Santa Clara, CA, USA) with an olfactory detector (OFD) Sniffer 9100 (Brechbühler, Schlieren, Switzerland) was used to identify volatile odor compounds in the samples. Two columns were used to separate the components, DB-WAX (polar, 30 m × 0.25 mm × 0.25 μm; Agilent Technologies) and DB-17 (mid polar, 2.22 m × 0.18 mm × 0.18 μm; Agilent Technologies). The initial column temperature was established at 40 °C and held for 3 min, followed by a ramp-up to 230 °C at 4 °C/min, held at 230 °C for 3 min. Ultra-pure helium (99.999%, Beijing APB Gas Industry Co., Ltd., Beijing, China) was used as the carrier gas. The electron impact mass spectra were generated at an ionization energy of 70 eV with an m/z scan range of 35–375 amu. The temperatures of the mass spectrometer source and quadrupole were programmed at 230 °C and 150 °C, respectively. A solid-state modulator SSM1800 (J&X Technologies, Shanghai, China) was placed between the two columns for the heating and cooling phases. The temperature of the cold zone was kept at −50 °C and the modulation cycle was set to 5 s.

The concentrated fractions in the instrument were analyzed at the sniffing port by three trained sensory panelists who were members of the Molecular Sensory Laboratory from Beijing University of Technology; they were trained for four weeks to analyze the effluent components at the sniffing port for 2 h per day. During the GC-O analysis, moisture was delivered to the sniffing port through a blank capillary column. The odor descriptors and label odor intensity values, as well as retention times, were recorded [16]. The compounds that were perceived by two or more panelists were tentatively identified as odor compounds.

2.9. Aroma Extraction Dilution Analysis (AEDA)

AEDA was analyzed by varying the fractionation ratio of GC-O. The importance of each volatile component was ranked by the average flavor dilution (FD) factor, which was determined at the sniffing port according to the AEDA procedure. According to the ratios of 1:3n, which were 0, 1:3, 1:9, 1:27, 1:81, etc. The corresponding flavor dilution (FD) factors were defined as 1, 3, 9, 27, 81, etc. The result of the FD factor analysis refers to the maximum value for which the compound can be detected. Consequently, the FD factor of the odor compound that was analyzed at the sniffing port was 81, which indicates that the odor-active substance was not analyzed by the panelist at the sniffing port when the GC-O dilution ratio was 81:1. Usually, the higher the FD factor obtained for the odor-active compound, the more critical it is to be analyzed.

2.10. Qualitative and Quantitative Analysis

Volatile odor compounds were identified by mass spectrometry (MS) and comparison of identified peaks with the NIST library mounted on GC-MS. The linear retention index (RI) was calculated for dual identification according to the following equation:

$$RI = 100N + 100n\{[t\,Ra - t\,RN] / [t\,R(N + n) - t\,RN]\} \quad (1)$$

where N represented the number of carbon atoms with peaks in front of the identified compound, n represented the numerical difference between the upper and lower n-alkanes in gas chromatography, and the variables t Ra, t RN, t R (N + n) denoted the retention time of the identified compounds and the upper alkane and lower alkanes, respectively. The unknown odor compounds were positively identified by the three methods: (1) comparison of RI and odor descriptions (O) with reference compounds; (2) comparison of MS data with NIST library; (3) finally by standard (STD) compounds verification [17].

2.11. Relative Odor Activity Value (R-OAV) Calculation

By adapting the method of Yang et al, the r-OAV was calculated using the following formula:

$$r\text{-}OAV = Ci/OTi \quad (2)$$

where Ci represented the relative concentration of a certain volatile odor compound in the fish soup and OTi denoted the odor threshold of the compound [18]. The threshold of the compound used in this study was the actual threshold in water [19].

2.12. Omission Experiment

An aroma model composed of the selected key odor-active compounds was prepared by reducing one of the 11 key aroma compounds at a time. A triangulation experiment assessed by a sensory panel was used to compare the differences between the blended quintile and the complete reconstituted fraction according to the PRC national standard. In other words, if 8, 9, and 10 or more of the 12 sensory members were able to identify differences between a compound when it was omitted and completed the reconstitution model, the results were considered significant ($\alpha \leq 0.05$), highly significant ($\alpha \leq 0.01$), or very highly significant ($\alpha \leq 0.001$), respectively. If fewer than 8 were identified, the result was considered "not significant". Sensory evaluation is a common method used in the food science area, which also does not involve informed consent and ethics.

2.13. Statistical Analysis

All experiments were performed in triplicate, and the data was expressed as mean $\pm$ standard deviation. Statistical analysis was performed using Microsoft Excel 2019 (Microsoft Corp., Redmond, WA, USA).

3. Results and discussion

3.1. Comparison of Different Extraction Methods for Analysis of Odor Compounds

Based on the efficiencies of the five extraction methods including SPME, SAFE, DHS, SBSE and LLE the best extraction method was selected and the odor substances in the fish soup were analyzed by GC-O-MS and GC × GC-O-MS. As shown in Table 1, in the 1D GC mode, a total number of 38 odor compounds were extracted by SPME, including 5 alcohols, 15 aldehydes, 7 ketones, 6 acids and 5 others. Ten odor compounds were detected from SAFE extract, including 3 alcohols, 2 aldehydes, 1 ketone, 1 acid and 3 others. Eighteen different types of odor compounds were extracted by DHS, including 8 aldehydes, 3 ketones, 2 esters, 2 acids and 3 others. Whereas, 11 odor compounds were extracted from SBSE, including 6 aldehydes, 2 ketones, 2 esters and 1 acid. While for LLE, only one aldehyde and one ketone were extracted. In the 2D GC mode. A total of 39 odor compounds were extracted by SPME, including 3 alcohols, 12 aldehydes, 9 ketones, 1 ester, 2 acids and 12 others; 33 odor compounds were extracted by SAFE, including alcohols (5), aldehydes (7), ketones (8), ester (1), acids (2) and others (10); by DHS, 21 volatiles were extracted including alcohols (2), aldehydes (8), ketones (4), esters (2), acids (3) and others (2). A total of 14 odor compounds were extracted from SBSE, including alcohol (1), aldehydes (4), ketones (2), ester (1), acids (3), and others (2); and only 10 volatile compounds were extracted through LLE, including alcohol (1), aldehydes (4), ketones (2) and esters (3). The results of the five extraction methods using 1D GC mode were compared, and the results are presented in Figures 1 and 2; it can be clearly concluded that SPME extracted a higher number of odor compounds.

To better determine the best method for the extraction of volatile compounds, OPLS-DA was chosen for further differentiation. As shown in Figure 3a,b, the OPLS-DA analysis showed that SPME was better differentiated from the other four extraction methods. In 1D mode, LLE could not be distinguished from SBSE. In 2D mode, three extraction methods cannot be distinguished, namely DHS, LLE and SBSE.

It has been shown that among the alcohols, 1-octen-3-ol is one of the potent compounds that contributed the most to the fishy odor [10,20], and it was predicted that lipoxygenases cause the breakdown of polyunsaturated fatty acids or the reduction of carbonyl compounds [21]. 1-Octen-3-ol was detected in SPME, SAFE and DHS, and the highest level (7.07 ± 0.98 μg/Kg) was detected when analyzed in the 2D mode in SPME; this was in agreement with the study of Xue et al. [10]. However, the aforementioned compound was not detected in both the SBSE and LLE extraction methods. Aldehydes are a relatively large group of compounds that contribute to fish odor [22]. Some of the aldehydes that can be easily sniffed were detected in SPME, but almost no aldehydes were extracted via LLE; this also confirms that LLE is not suitable for the analysis of odor compounds in fish. The extraction of odorous substances from fish by the LLE method has not been reported in recent studies. 1-Octen-3-one has the same mushroom flavor as 1-octen-3-ol, which has a low threshold and can be distinctly sniffed; it is worth noting that this substance is only detected in the 2D mode of SPME and is not detected by any other method. A few previous reports have shown 1-octen-3-one as a fishy substance, but to the best knowledge of the authors, this is the first time that 1-octen-3-one has been identified in fish. Pyridine was once considered as one of the main fishy substances in freshwater fish [23]. However, it has not been reported in recent studies [24–26]; this may be related to the detection method. In this study, only the 2D mode of SAFE was detectable at minimal levels and was not detected by olfaction. In addition, small amounts of pyrazines and furans were detected in SPME, DHS and SAFE, probably due to substances generated by prolonged high temperature heating during the boiling process [27], and the contents in SPME were higher than the other two methods. In summary, the best choice was concluded as SPME.

Table 1. Volatile compounds identified by five extraction methods in combination with GC-O-MS and GC×GC-O-MS.

No.	Component	CAS	Odor	RI [a]		Method of Identification [b]		Relative Content (ng/g) [c]									
								SPME		SAFE		DHS		SBSE		LLE	
				GC	GC × GC	GC	GC × GC	GC	GC × GC	GC	GC × GC	GC	GC × GC	GC	GC × GC	GC	GC × GC
	Alcohols																
1	1-Pentanol	71-41-0	balsam	1247	1239	MS/RI	MS/RI	50.54 ± 4.37	53.99 ± 6.18	ND	41.27 ± 12.73	ND	ND	ND	ND	ND	ND
2	2-Furanmethanol	98-00-0	musty	1702	1697	MS/RI	MS/RI	229.61 ± 16.49	288.19 ± 23.04	165.24 ± 26.33	ND	ND	203.15 ± 35.86	ND	ND	ND	75.09 ± 8.39
3	2-Thiophenemethanol	636-72-6	roasted	1885	ND	MS/RI	ND	23.83 ± 1.83	ND	ND	ND	ND	ND	ND	ND	ND	ND
4	1-Octen-3-ol	3391-86-4	mushroom	1430	1418	MS/RI	MS/RI/O	6.08 ± 1.08	7.07 ± 0.98	ND	5.49 ± 2.04	ND	6.57 ± 4.19	ND	ND	ND	ND
5	1-Butanol	71-36-3	oil	ND	1126	ND	MS/RI	ND	ND	ND	ND	ND	ND	ND	ND	ND	ND
6	Linalool	78-70-6	floral	ND	1509	ND	MS/RI	ND	ND	ND	12.48 ± 1.06	ND	ND	ND	ND	ND	ND
7	α-Cumyl alcohol	617-94-7	green	1743	ND	MS/RI	ND	ND	ND	2.05 ± 0.16	ND	ND	ND	ND	ND	0.16 ± 0.11	ND
8	2-Ethyl-1-hexanol	104-76-7	citrus	ND	1492	ND	MS/RI	ND	ND	ND	ND	ND	ND	ND	0.96 ± 0.06	ND	ND
	Aldehydes																
9	Furfural	98-01-1	sweet woody	1493	ND	MS/RI/O	ND	117.6 1± 5.40	ND	ND	ND	ND	ND	ND	ND	ND	ND
10	5-Methyl-2-thiophenecarboxaldehyde	13679-70-4	almond	1764	ND	MS/RI/O	ND	33.50 ± 1.57	ND	ND	ND	ND	ND	ND	ND	ND	ND
11	(*E*,*Z*)-2,6-Nonadienal	557-48-2	green fatty	1590	ND	MS/RI/O	ND	3.06 ± 0.05	ND	ND	ND	ND	ND	ND	ND	ND	ND
12	2-Methylbutyraldehyde	96-17-3	musty	987	ND	MS/RI	ND	25.10 ± 3.72	ND	ND	ND	ND	ND	ND	ND	ND	ND
13	3-Methylbutanal	590-86-3	cocoa chocolate	924	935	MS/RI	MS/RI	110.01 ± 7.46	134.51 ± 11.36	ND	ND	58.19 ± 2.63	108.32 ± 4.27	ND	ND	ND	ND
14	Pentanal	110-62-3	fermented	956	944	MS/RI	MS/RI	159.20 ± 4.02	201.26 ± 18.43	ND	ND	ND	ND	ND	ND	ND	ND
15	Hexanal	66-25-1	grass	1062	1058	MS/RI/O	MS/RI/O	92.66 ± 1.17	118.80 ± 7.34	ND	88.63 ± 9.33	66.04 ± 8.39	74.68 ± 6.62	26.59 ± 1.95	29.27 ± 2.29	ND	32.64 ± 5.15
16	Octanal	124-13-0	waxy citrus	1279	1265	MS/RI	MS/RI	2.49 ± 0.36	23.31 ± 1.26	ND	3.91 ± 0.73	ND	ND	0.27±0.06	3.08 ± 0.74	ND	ND
17	Nonanal	124-19-6	orange	1384	1391	MS/RI	MS/RI	235.93 ± 12.73	304.04 ± 25.62	175.56 ± 14.76	209.37 ± 12.94	104.27 ± 9.26	ND	ND	52.60 ± 2.46	ND	17.40 ± 0.95
18	(*E*)-2-Octenal	2548-87-0	cucumber	1437	1426	MS/RI	MS/RI/O	8.71 ± 0.76	9.56 ± 0.38	ND	ND	ND	ND	ND	ND	ND	ND
19	Benzaldehyde	100-52-7	sweet	1503	1509	MS/RI	MS/RI	103.81 ± 8.25	144.38 ± 12.34	ND	73.29 ± 8.37	68.53 ± 10.55	71.27 ± 9.27	34.17 ± 2.91	63.26 ± 5.72	ND	9.63 ± 0.94
20	5-Methyl-2-furancarboxaldehyde	620-02-0	spice caramel	1578	1554	MS/RI	MS/RI	352.37 ± 18.45	329.13 ± 18.23	ND	164.24 ± 24.53	228.72 ± 13.66	303.24 ± 36.81	73.02 ± 10.95	142.88 ± 11.91	ND	ND
21	Phenylacetaldehyde	122-78-1	green	1662	ND	MS/RI	ND	13.68 ± 2.33	ND	ND	ND	ND	ND	ND	ND	ND	ND
22	Pentadecanal	2765-11-9	fresh waxy	2019	ND	MS/RI	ND	10.82 ± 0.12	ND	ND	ND	ND	ND	ND	ND	ND	ND
23	(*Z*)-4-Heptenal	6728-31-0	oily fatty	ND	1287	ND	MS/RI	ND	8.46 ± 1.04	ND	ND	ND	ND	ND	ND	ND	ND
24	Methional	3268-49-3	musty	ND	1439	ND	MS/RI	ND	41.43 ± 6.12	ND	ND	ND	ND	ND	ND	ND	ND
25	Decanal	112-31-2	sweet waxy	ND	1538	ND	MS/RI/O	ND	25.18 ± 2.25	ND	9.73 ± 0.92	ND	11.04 ± 1.52	3.16 ± 0.06	3.95 ± 0.31	ND	ND
26	5-Methyl-2-thiophenecarboxaldehyde	13679-70-4	woody	ND	1783	ND	MS/RI/O	ND	20.34 ± 1.59	ND	ND	ND	ND	ND	ND	ND	ND
27	Dodecyl aldehyde	112-54-9	soapy waxy	1745	ND	MS/RI	ND	ND	ND	18.77 ± 0.35	ND	ND	ND	ND	ND	ND	ND
28	(*E*,*E*)-2,4-Heptadienal	4313-03-5	green	ND	1497	ND	MS/RI	ND	ND	ND	6.38 ± 0.74	ND	ND	ND	ND	ND	ND

Table 1. *Cont.*

No.	Component	CAS	Odor	RI [a]		Method of Identification [b]		Relative Content (ng/g) [c]									
								SPME		SAFE		DHS		SBSE		LLE	
				GC	GC × GC	GC	GC × GC	GC	GC × GC	GC	GC × GC	GC	GC × GC	GC	GC × GC	GC	GC × GC
	Alcohols																
	Aldehydes																
29	Heptaldehyde	111-71-7	fresh	1174	1165	MS/RI	MS/RI	ND	221.07 ± 14.28	ND	ND	84.93 ± 5.27	164.29 ± 12.63	43.28 ± 13.75	53.22 ± 2.76	ND	64.01 ± 6.37
30	3-Methylhexanal	19269-28-4	sweet green	1102	1116	MS/RI	MS/RI	ND	ND	ND	ND	5.37 ± 1.10	12.84 ± 2.12	ND	ND	ND	ND
31	2-Undecenal	2463-77-6	fruity	1793	1794	MS/RI	MS/RI	ND	ND	ND	ND	ND	ND	16.38 ± 0.97	9.96 ± 0.72	ND	ND
	Ketones																
32	Hydroxyacetone	116-09-6	caramel	1272	ND	MS/RI	ND	522.52 ± 48.29	ND	ND	ND	ND	ND	ND	ND	ND	ND
33	4-Hydroxy-2,5-dimethyl-3(2*H*)furanone	3658-77-3	sweet	2037	ND	MS/RI/O	ND	35.25 ± 1.45	ND	ND	ND	ND	ND	ND	ND	ND	ND
34	6-Methyl-5-hepten-2-one	110-93-0	citrus	1342	1332	MS/RI/O	MS/RI/O	4.34 ± 0.03	8.06 ± 1.24	ND	2.96 ± 0.91	7.47 ± 0.26	9.26 ± 0.27	0.38 ± 0.05	5.67 ± 0.11	ND	ND
35	2-Butanone	78-93-3	camphor	901	914	MS/RI	MS/RI	22.05 ± 9.25	29.37 ± 1.96	ND	19.62 ± 2.28	ND	ND	ND	ND	ND	ND
36	2,3-Pentanedione	600-14-6	pungent	1047	1058	MS/RI	MS/RI	7.17 ± 0.45	16.54 ± 0.52	ND	ND	ND	ND	ND	ND	ND	ND
37	Acetoin	513-86-0	sweet	1284	1273	MS/RI	MS/RI	31.42 ± 3.72	35.21 ± 0.74	ND	17.82 ± 3.32	27.37 ± 3.56	29.52 ± 2.17	ND	ND	ND	5.83 ± 0.64
38	2-Dodecanone	6175-49-1	fruity	1669	ND	MS/RI	ND	0.42 ± 0.07	ND	ND	ND	ND	ND	ND	ND	ND	ND
39	5-Ethyl-2(5*H*)-furanone	2407-43-4	spice	1756	ND	MS/RI	ND	5.69 ± 1.65	ND	ND	ND	ND	ND	ND	ND	ND	ND
40	2,3-Butanedione	431-03-8	butter	ND	961	ND	MS/RI	ND	46.93 ± 4.38	ND	ND	ND	ND	ND	ND	ND	ND
41	1-Penten-3-one	1629-58-9	peppery	ND	1018	ND	MS/RI	ND	6.14 ± 0.47	ND	ND	ND	ND	ND	ND	ND	ND
42	2-Heptanone	110-43-0	fruity	ND	1165	ND	MS/RI	ND	2.17 ± 0.02	ND	ND	ND	ND	ND	ND	ND	ND
43	1-Hydroxy-2-propanone	116-09-6	pungent	1310	1307	MS/RI/O	MS/RI	ND	492.86 ± 42.74	ND	318.74 ± 41.17	148.02 ± 16.14	358.60 ± 22.61	ND	ND	74.26 ± 12.37	169.60 ± 19.95
44	1-Octen-3-one	4312-99-6	mushroom	ND	1208	O	MS/RI/O	ND	0.12 ± 0.01	ND	ND	ND	ND	ND	ND	ND	ND
45	Acetophenone	98-86-2	sweet	1643	1637	MS/RI	MS/RI	ND	ND	0.97 ± 0.13	3.02 ± 1.05	ND	ND	ND	ND	ND	ND
46	2-Decanone	693-54-9	orange	ND	1498	ND	MS/RI	ND	ND	ND	1.34 ± 0.07	ND	ND	ND	ND	ND	ND
47	1-Hydroxy-2-butanone	5077-67-8	coffee	ND	1397	ND	MS/RI	ND	ND	ND	0.94 ± 0.19	ND	ND	ND	ND	ND	ND
48	2-Tridecanone	593-08-8	fatty	ND	1809	ND	MS/RI	ND	ND	ND	9.48 ± 0.17	ND	ND	ND	ND	ND	ND
49	Geranylacetone	3796-70-1	green	ND	1853	ND	MS/RI	ND	ND	ND	ND	ND	7.47 ± 0.65	ND	ND	ND	ND
50	4-Hexen-3-one	2497-21-4	metallic	1187	1164	MS/RI/O	MS/RI/0	ND	ND	ND	ND	ND	ND	16.33 ± 3.61	15.27 ± 4.21	ND	ND
51	5-Ethyldihydro-2(3*H*)-furanone	695-06-7	herbal	1692	1687	MS/RI	MS/RI	ND	9.11 ± 1.03	ND	ND	ND	ND	2.09 ± 0.95	7.14 ± 1.27	ND	ND
	Esters																
52	Triethyl phosphate	78-40-0	cider	ND	1674	ND	MS/RI	ND	ND	ND	0.14 ± 0.03	ND	ND	ND	ND	ND	ND
53	Allyl 2-ethylbutyrate	7493-69-8	nut	1252	1246	MS/RI/O	MS/RI/O	ND	ND	ND	ND	9.27 ± 1.15	9.83 ± 1.63	ND	ND	ND	ND

Table 1. *Cont.*

No.	Component	CAS	Odor	RI [a]		Method of Identification [b]		Relative Content (ng/g) [c]									
								SPME		SAFE		DHS		SBSE		LLE	
				GC	GC × GC	GC	GC × GC	GC	GC × GC	GC	GC × GC	GC	GC × GC	GC	GC × GC	GC	GC × GC
	Alcohols																
54	geranyl isovalerate	109-20-6	green	1934	1948	MS/RI/O	MS/RI/O	ND	ND	ND	ND	14.38 ± 1.91	47.26 ± 3.25	21.22 ± 1.57	27.81 ± 4.38	ND	ND
55	Nitrous acid, ethyl ester	109-95-5	sweet	ND	1417	ND	MS/RI	ND	ND	ND	ND	ND	ND	ND	ND	ND	0.56 ± 0.27
56	Acetic acid, butyl ester	123-86-4	banana	ND	1049	ND	MS/RI	ND	ND	ND	ND	ND	ND	ND	ND	ND	19.47 ± 3.26
57	Butyrolactone	96-48-0	oily	ND	1634	ND	MS/RI	ND	ND	ND	ND	ND	ND	ND	ND	ND	5.22 ± 0.84
	Acids																
58	Propionic acid	79-09-4	pungent	1509	1528	MS/RI/O	MS/RI	20.387 ± 2.65	ND	ND	ND	ND	ND	ND	ND	ND	ND
59	3-Methylpentanoic Acid	105-43-1	sweaty	1761	ND	MS/RI/O	ND	49.50 ± 2.73	ND	ND	15.63 ± 3.68	ND	ND	ND	ND	ND	ND
60	Acetic acid	64-19-7	sour	1452	1433	MS/RI/O	MS/RI/O	487.86 ± 36.94	461.79 ± 54.27	ND	ND	268.38 ± 34.76	322.47 ± 29.81	173.28 ± 14.25	236.73 ± 25.77	ND	ND
61	Butyric acid	107-92-6	cheese	1631	1642	MS/RI	MS/RI	38.72 ± 6.28	41.76 ± 3.14	ND	ND	ND	ND	ND	ND	ND	ND
62	Hexanoic acid	142-62-1	fatty	1859	1846	MS/RI	MS/RI	69.13 ± 4.92	ND	ND	ND	27.48 ± 3.25	36.91 ± 2.44	ND	ND	ND	ND
63	Octanoic acid	124-07-2	waxy	2056	ND	MS/RI	ND	27.44 ± 1.74	ND	ND	ND	ND	ND	ND	ND	ND	ND
64	3-Methylbutanoic acid	503-74-2	sweaty	1646	ND	MS/RI/O	ND	ND	ND	ND	7.30 ± 1.93	ND	ND	ND	ND	ND	ND
65	Valeric acid	109-52-4	sickening	ND	1850	ND	MS/RI	ND	ND	ND	ND	ND	0.78 ± 0.15	ND	ND	ND	ND
	Others																
66	Styrene	100-42-5	sweet	1264	1239	MS/RI	MS/RI	9.00 ± 0.37	ND	5.46 ± 1.58	17.29 ± 2.73	ND	ND	ND	ND	ND	ND
67	2-Ethenyl-6-methylpyrazine	13925-09-2	hazelnut	1486	1502	MS/RI	MS/RI	21.00 ± 1.25	28.09 ± 3.38	ND	8.63 ± 1.35	ND	ND	ND	ND	ND	ND
68	2-Pentylfuran	3777-69-3	fruity	1231	1226	MS/RI/O	MS/RI/O	19.93 ± 0.05	19.46 ± 2.77	ND	ND	ND	ND	ND	ND	ND	ND
69	2,6-Di-tert-butyl-4-methylphenol	128-37-0	camphor	1905	ND	MS/RI	ND	6.33 ± 0.42	ND	ND	ND	ND	ND	ND	ND	ND	ND
70	2,6-Dimethylpyrazine	108-50-9	cocoa	1309	ND	MS/RI	ND	51.84 ± 10.47	ND	ND	ND	ND	ND	ND	ND	ND	ND
71	Pyrrole	109-97-7	sweet	ND	1528	ND	MS/RI	ND	11.89 ± 0.99	ND	ND	ND	ND	ND	ND	ND	ND
72	Dimethyl disulfide	624-92-0	vegetable	ND	1082	ND	MS/RI	ND	33.50 ± 1.16	ND	18.43 ± 2.55	ND	ND	ND	ND	ND	ND
73	Dimethyl trisulfide	3658-80-8	sulfurous	ND	1365	ND	MS/RI/O	ND	6.04 ± 0.53	ND	0.58 ± 0.17	ND	ND	ND	ND	ND	ND
74	Benzene	71-43-2	aromatic	ND	955	ND	MS/RI	ND	7.39 ± 0.56	ND	7.73 ± 0.36	ND	ND	ND	ND	ND	ND
75	2-Ethylfuran	3208-16-0	burnt	ND	945	ND	MS/RI	ND	11.25 ± 0.97	ND	ND	ND	ND	ND	ND	ND	ND
76	2,5-Dimethylpyrazine	123-32-0	roasted	ND	1329	ND	MS/RI/O	ND	63.21 ± 3.24	ND	ND	ND	ND	ND	ND	ND	ND
77	2,6-Dimethylpyrazine	108-50-9	cocoa	1352	1351	MS/RI	MS/RI	ND	1.36 ± 0.17	ND	0.06 ± 0.02	1.63 ± 0.16	5.37 ± 0.68	ND	ND	ND	ND
78	3-Ethyl-2,5-dimethylpyrazine	13360-65-1	potato	ND	1430	ND	MS/RI/O	ND	34.18 ± 1.25	ND	ND	ND	ND	ND	ND	ND	ND
80	Toluene	108-88-3	sweet	1039	1034	MS/RI	MS/RI	ND	ND	3.28 ± 0.53	7.58 ± 0.95	ND	ND	ND	ND	ND	ND
81	1,2-Xylene	95-47-6	geranium	1171	1164	MS/RI	ND	ND	ND	7.94 ± 1.27	ND	ND	ND	ND	ND	ND	ND
82	Pyridine	110-86-1	sour	ND	1169	ND	MS/RI	ND	ND	ND	0.93 ± 0.03	ND	ND	ND	ND	ND	ND
84	Naphthalene	91-20-3	pungent	ND	1701	ND	MS/RI	ND	ND	ND	16.39 ± 1.55	ND	ND	ND	ND	ND	ND
85	Benzothiazole	95-16-9	sulfury	1933	1930	MS/RI	MS/RI	ND	ND	ND	ND	0.27 ± 0.06	3.05 ± 0.74	ND	ND	ND	ND
86	Dimethyl trisulfide	624-92-0	onion	ND	1365	ND	MS/RI/O	ND	1.69 ± 0.03	ND	ND	ND	ND	ND	ND	ND	ND

a: RI, The Retention index on capillaries DB-WAX and DB-5. b: The different identification methods, including MS, RI, and O. In particular, for those compounds which were not able to identify by all three methods, they would be regarded as temporarily identified. ND, not detected. c: Relative concentration stated as the mean ± SD and the unit is ppt (part per trillion, ng/g).

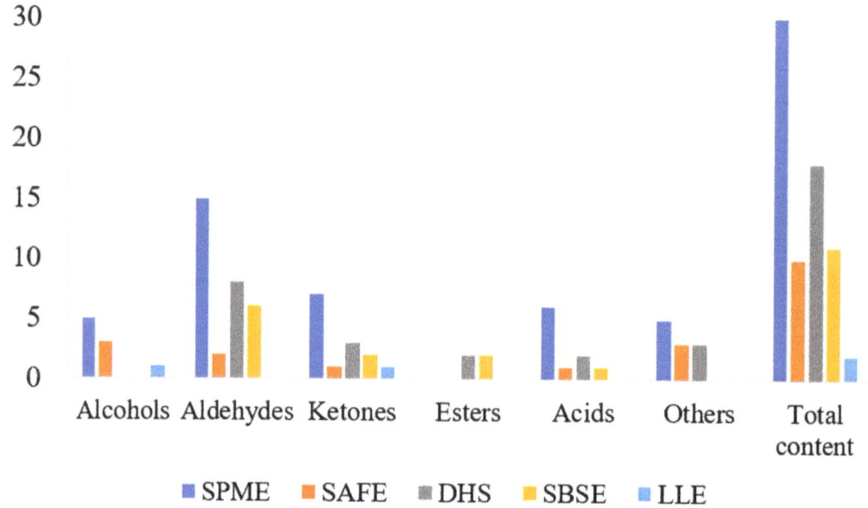

Figure 1. The number of volatile compounds measured by GC-O-MS for five extraction methods.

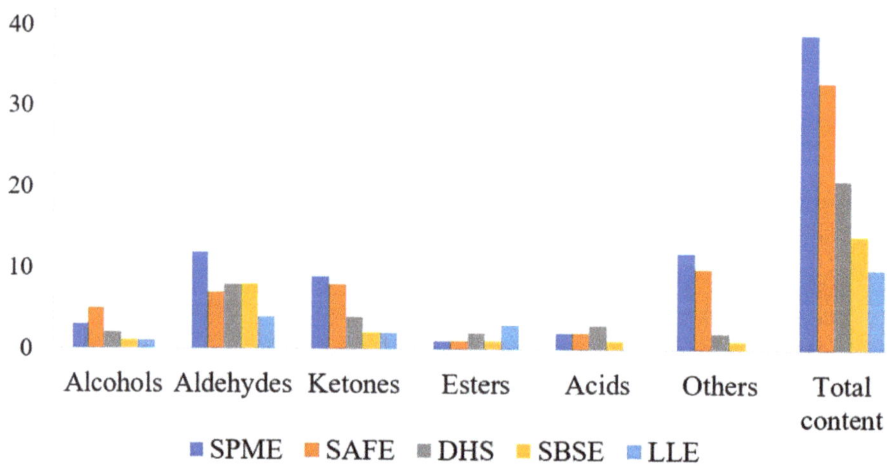

Figure 2. Number of volatile compounds measured by GC × GC-O-MS for five extraction methods.

3.2. Comparison of Aroma Compounds between GC-O-MS and GC × GC-O-MS Analysis

For a clearer comparison of the methods of extracting volatile compounds and analysis of odor-active substances in fish soup, GC-O-MS (1D, one dimension) and GC × GC-O-MS (2D, two dimension) were employed. For this, the results of the two-dimensional mode (GC × GC-O-MS) were compared with the analysis of all volatile compounds detected through the one-dimensional column (first column). Compared with GC-O-MS, GC × GC-O-MS can analyze the odor compounds in the sample more accurately based on its high resolution and high sensitivity. In other words, the GC × GC-O-MS analysis can be used to confirm whether each odor region is shared by multiple odor compounds; and it can also be used to quickly distinguish the compounds in multiple odor regions. Therefore, the GC × GC-O-MS method can be used to analyze volatile odor compounds more efficiently than GC-O-MS, although the analytical results of both instruments show a high degree of agreement.

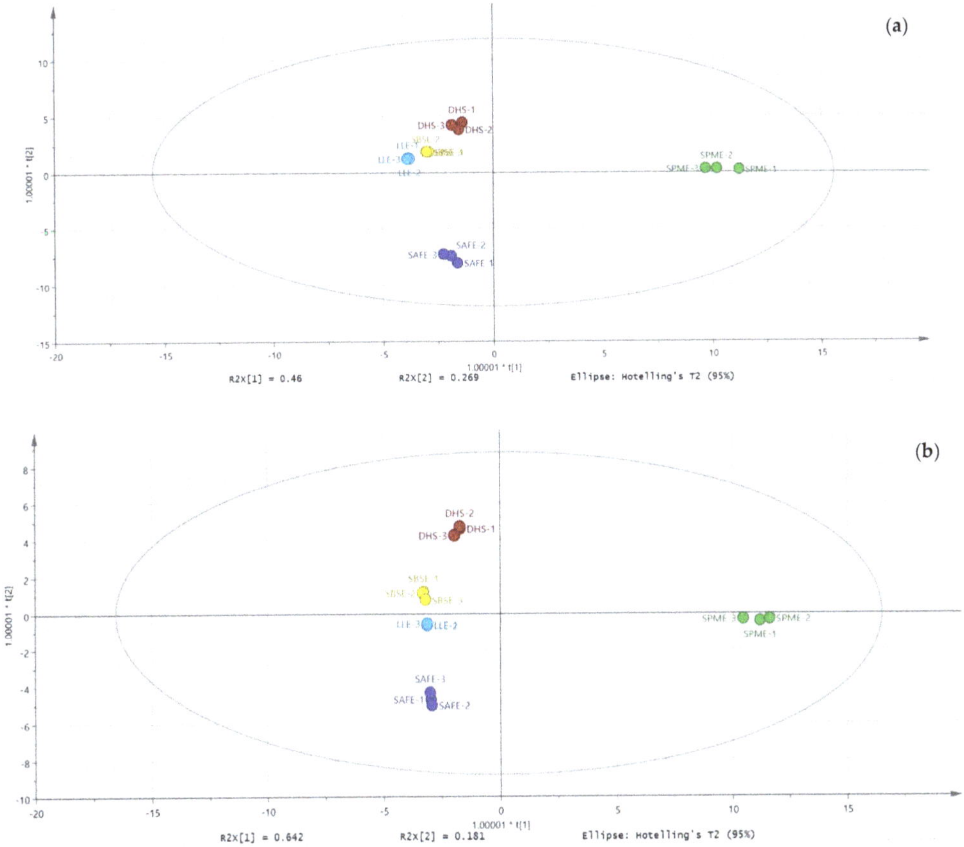

Figure 3. OPLS-DA analysis chart of five extraction methods. (**a**) OPLS-DA analysis of five extraction methods in 1D mode. (**b**) OPLS-DA analysis of five extraction methods in 2D mode.

As shown in Figure 4a–e, among the five extraction methods, more compounds were detected in 2D mode than in 1D mode. As shown in Table 1, the relative content of the compounds analyzed by the 2D model was also higher. The 3D view of the 2D mode was presented in Figure 5a,b. Specifically, 9 compounds, such as decanal, dimethyl trisulfide, and 2,5-dimethylpyrazine, were not analyzed in the 1D mode, but several of these substances could be identified in the 2D mode and could be analyzed in the sniffing port. Interestingly, it was observed that 1-octen-3-one can be smelled but not identified in the 1D mode, which does not reach the detection limit of the instrument, while it can be both analyzed and smelled in the 2D mode, which further confirms the high resolution of the 2D mode. In addition, other substances, such as 1-octen-3-ol, acetic acid, and (E)-2-octenal, can be analyzed in the 1D mode but cannot be identified in the sniffing port, while these compounds can be easily analyzed and sniffed in the 2D mode. Some compounds such as furfural, (E,Z)-2,6-nonadienal, and 4-hydroxy-2,5-dimethyl-3(2H) furanone, which were not identified in 2D mode, could be analyzed or smelled in the 1D mode. Clearly, the above results revealed that the use of both analytical methods can analyze the odor compounds in the samples more comprehensively.

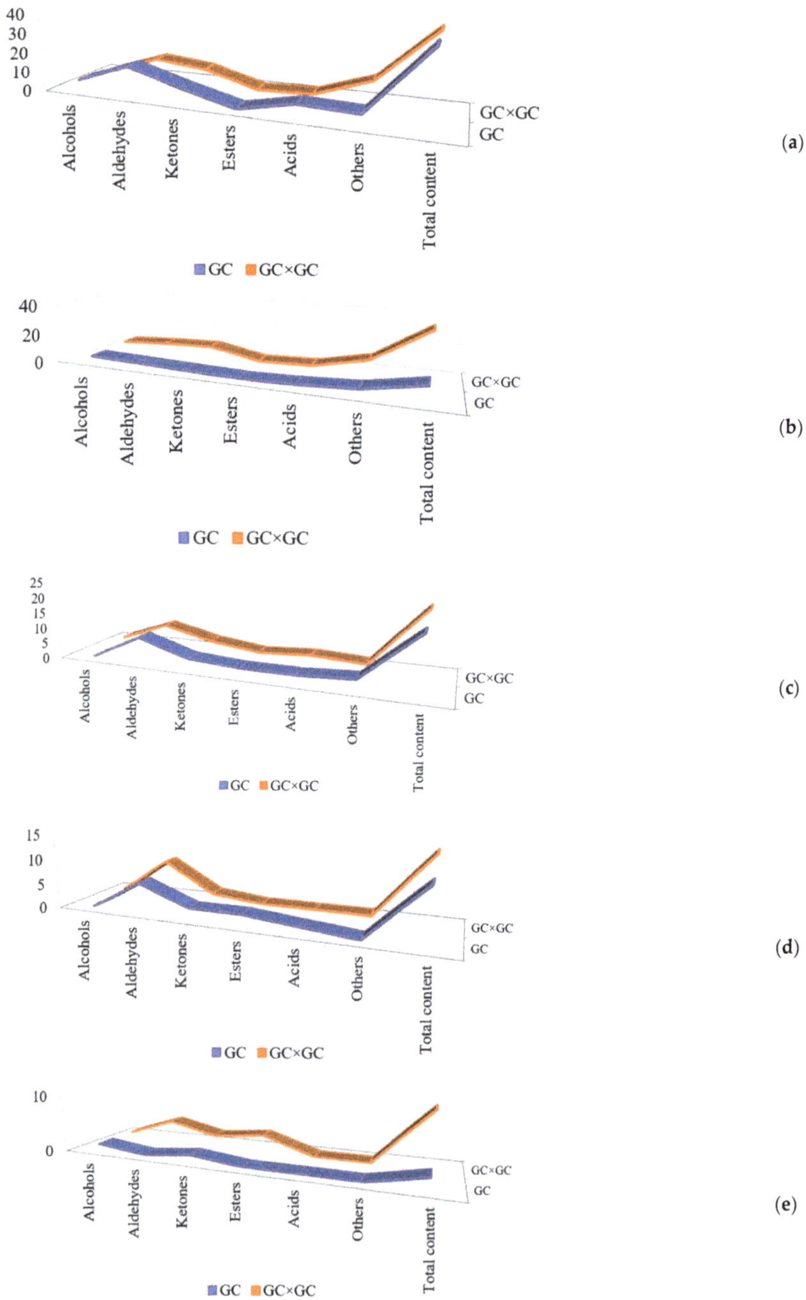

Figure 4. Five extraction methods combined with GC and GC × GC to measure the amount of volatile compounds. (**a**) SPME combined with GC and GC × GC to measure the amount of volatile compounds; (**b**) SAFE combined with GC and GC × GC to measure the amount of volatile compounds; (**c**) DHS combined with GC and GC × GC to measure the amount of volatile compounds; (**d**) SBSE combined with GC and GC × GC to measure the amount of volatile compounds; (**e**) LLE combined with GC and GC × GC to measure the amount of volatile compounds.

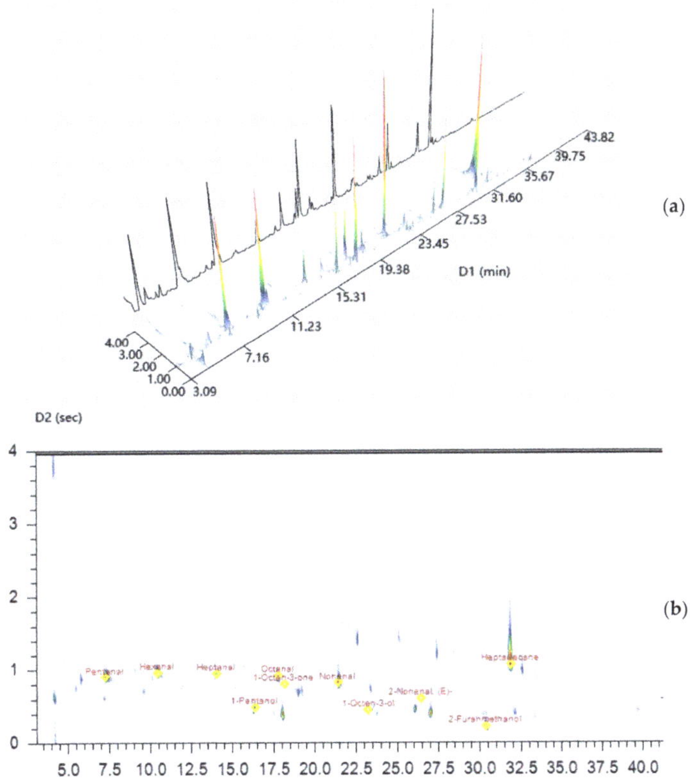

Figure 5. (**a**) The 3D analytical images of GC × GC-O-MS in sample; (**b**) The 2D analytical images of GC × GC-O-MS in sample.

3.3. Key Aroma-Active Compounds Identified by AEDA and r-OAV

It was found that all the odor compounds in the sample contributed to the overall odor profile. To analyze the compounds that contributed more, a dilution analysis was performed in order to identify the key odor compounds in the sample. The critical level of the odor-active compounds was determined by the FD factor obtained from the dilution analysis. In other words, the FD factor is positively correlated with the contribution of the odor compounds to the overall odor profile [28].

By using SPME combined with GC-O-MS and GC × GC-O-MS, 55 odor compounds were identified, including 5 alcohols, 19 aldehydes, 12 ketones, 1 ester, 6 acids, 5 pyrazines, 3 furans, and 6 others (Table 1), and their FD factors were obtained by AEDA, as shown in Table 2. Among these, 2-pentylfuran, and 1-octen-3-one showed the highest FD factor (FD = 729), followed by 4-hydroxy-2,5-dimethylfuran-3(2H)-one, (E)-2-octenal, 1-octen-3-ol and hexanal with FD factors of 243. However, hexanal, 2-pentylfuran, 1-octen-3-one, (E)-2-octenal, and 1-octen-3-ol contributed significantly to the odor characteristics of the fish soup; these results were consistent with previous studies [10]; these compounds delivered the odor of mushroom, beany, grass and cucumber, creating a characteristic odor in fish soup. The FD factors of these odor-active compounds were then obtained, including 6-methylhept-5-en-2-one (81), (E,Z)-2,6-nonadienal (27), 2-ethyl-3,5-dimethylpyrazine (27) and 3-methylpentanoic acid (27). Since these compounds showed a high FD factor, they were considered to be the key odorants in fish soup. However, some typical substances that usually affect the fish odor, such as heptanal, nonanal and trimethylamine, did not contribute to the overall odor profile and were not even detected in this study, probably due to their changes in content after boiling; this was for the first time that 3-methylpentanoic acid was detected as a compound with sweaty odor with an FD factor of 9, and also as a contributing compound to the fish odor. Importantly, the advantages of GC × GC-O-MS in the detection of odor compounds were also confirmed. In addition, the FD factor for acetic acid, 2,6-dimethylpyrazine and 5-methyl-2-thiophenecarboxaldehyde was 9, while furfural and dimethyl trisulfide had an FD factor of 3. Although the FD factors of these compounds are relatively small compared to other substances, since they were also smelled at the sniffing port, so it was predicted that they also made contributions to the overall odor profile of the fish soup.

Table 2. FD factors and relative odor activity values (r-OAV) of key aroma compounds.

No.	RI [a]		Odor	Compound	FD Factors [b]	Odor Threshold in Water (mg/kg) [c]	r-OAV
	DB-WAX	DB-5					
1	1279	974	mushroom	1-Octen-3-one	729	0.000016	7500
2	1209	992	fruity	2-Pentylfuran	729	0.0058	3436
3	1440	1032	cucumber	(*E*)-2-Octenal	243	0.006	1452
4	1062	793	grass	Hexanal	243	0.098	1212
5	1960	1112	sweet	4-Hydroxy-2,5-dimethylfuran-3*(2H)*-one	243	0.039	904
6	1424	957	mushroom	1-Octen-3-ol	243	0.007	869
7	1430	1092	potato	3-Ethyl-2,5-dimethylpyrazine	27	0.0086	486
8	1582	1178	green fatty	*(E,Z)*-2,6-nonadienal	27	0.008	383
9	1325	954	citrus	6-Methylhept-5-en-2-one	81	0.025	202
10	1761	951	sweaty	3-Methylpentanoic acid	27	0.28	177
11	1428	668	sour	Acetic acid	9	22	40
12	1309	944	cocoa	2,6-Dimethylpyrazine	9	1.5	35
13	1493	802	sweet woody	Furfural	3	23	5
14	1764	1149	almond	5-Methyl-2-thiophenecarboxaldehyde	9	-	-
15	1365	972	sulfurous	Dimethyl trisulfide	3	-	-

a: RI, The Retention index on capillaries DB-WAX and DB-5. b: The FD factors of all aroma compounds in three samples were calculated by varying the split ratio of the GC inlet. c: Odor thresholds were referenced from a book named Odor thresholds compilations of odor threshold values in air, water and other media (second enlarged and revised edition).

The concentrations of the compounds after AEDA analysis were calculated by a semi-quantitative method. Table 2 summarizes the concentrations as well as the results of r-OAV calculations. Among the odor compounds, 1-octen-3-one had the highest r-OAV of 7500, attributed to the fishy odor; this result was also consistent with the previous results [29]. Secondly, 2-pentylfuran had a higher r-OAV of 3436, and its odor attribute was beany; this compound has not been previously reported in the literature as a fishy odorant and may be generated from fatty acids during the boiling of fish scraps. (E)-2-Octenal, hexanal, 4-hydroxy-2,5-dimethylfuran-3(2H)-one, 1-octen-3-ol, and 6-methylhept-5-en-2-one had the r-OAVs of 1452, 1212, 904, 869 and 602, respectively. The r-OAVs of these compounds were consistent with the AEDA results. Additionally, these compounds were previously reported as fishy substances [30]. However, there was one exception, such as 6-methylhept-5-en-2-one with an FD factor of 81, but it showed a lower FD factor relative to other substances with an FD factor of 27 (2-ethyl-3,5-dimethylpyrazine and (E,Z)-2,6-nonadienal). The results of the above analysis indirectly suggested the existence of some synergistic effect between odor compounds, rather than a single relationship, which could potentially enhance or diminish the overall aroma; these results better illustrate the importance of olfactory detection when analyzing the odor compounds.

3.4. Omission Experiment

Aroma model was constructed by selecting the top 11 key odor compounds in order of r-OAV value. Triangle tests were performed for the deletion of the 11 key odor compounds one by one and the results are presented in Table 3. The absence of 2-pentylfuran, (E)-2-octenal, 1-octen-3-one and hexanal was detected by all panelists, which indicates the importance of these compounds in contributing to the overall odor profile. While in the absence of acetic acid, 4-hydroxy-2,5-dimethylfuran-3(2H)-one, and 3-methylpentanoic acid, the results showed insignificant, i.e., less impact on the overall aroma. Thus, the omission experiments further confirmed that 2-pentylfuran, (E)-2-octenal, 1-octen-3-one, hexanal, 1-octen-3-ol, 6-methylhept-5-en-2-one, (E,Z)-2,6-nonadienal and 2-ethyl-3,5-dimethylpyrazine were the key odor compounds in the fish soup.

Table 3. Omission experiment.

No.	Compounds [a]	N [b]	Significant [c]
1	2-Pentylfuran	12	***
2	1-Octen-3-one	11	***
3	Acetic acid	6	-
4	4-Hydroxy-2,5-dimethylfuran-3*(2H)*-one	4	-
5	(E)-2-Octenal	10	***
6	1-Octen-3-ol	11	**
7	Hexanal	12	***
8	6-Methylhept-5-en-2-one	10	**
9	*(E,Z)*-2,6-Nonadienal	8	*
10	2-Ethyl-3,5-dimethylpyrazine	8	*
11	3-Methylpentanoic acid	6	-

a: The compounds which r-OAV in Table 3 showing a significant difference among OF, PC and AO. b: Number of correct judgments from 12 assessors by the triangle test. c: Significance: "*", significant ($\alpha \leq 0.05$); "**", highly significant ($\alpha \leq 0.01$); "***", very highly significant ($\alpha \leq 0.001$); "-", non-significant.

4. Conclusions

In this study, fish scraps were used as raw materials for the preparation of fish soup. Five volatile extraction methods were compared in order to select the best method for extracting odor compounds from fish soup. Finally, SPME combined with GC-O-MS and GC $\times$ GC-O-MS were selected to analyze odor-active compounds together. A total of 57 volatile compounds were identified. Among these, 38 volatile compounds were detected by GC-O-MS, of which 10 were declared as odor-active compounds, i.e., substances that could be smelled at the sniffing port; while 39 volatile compounds were detected by GC $\times$ GC-O-MS, of which 12 were odor-active. The above results revealed that the combination of these two analytical methods can provide a more comprehensive analysis of the volatile aroma compounds in fish soup. The results of AEDA analysis showed that 1-octen-3-one and 2-pentylfuran contributed most significantly to the odor of fish soup, followed by (E)-2-octenal, hexanal, 4-hydroxy-2,5-dimethylfuran-3(2H)-one, 1-octen-3-ol, 2-ethyl-3,5-dimethylpyrazine, (2E,6Z)-nona-2,6-dienal, 6-methylhept-5-en-2-one and 3-methylvaleric acid. From the r-OAV results, it was concluded that the compound with the highest r-OAV was 1-octen-3-one, followed by 2-pentylfuran. In summary, 2-pentylfuran, (E)-2-octenal, 1-octen-3-one, hexanal, 1-octen-3-ol, 6-methylhept-5-en-2-one, (E,Z)-2,6-nonadienal and 2-ethyl-3,5-dimethylpyrazine were the most important odor-active compounds in the fish soup as confirmed by omission experiment and triangle test. In the subsequent studies, based on the fishy compounds in this study, enzymatic digestion of fish soup using protease was performed to remove the fishy compounds and enhance the umami taste of the fish soup, contributing to the production of a full-flavored and distinctive seasoning; this approach not only improves the comprehensive utilization value of fish scraps, but also avoids the environmental pollution caused by discarding fish scraps.

Author Contributions: Writing—original draft, formal analysis, Y.Z. (Yuanyuan Zhang); conceptualization, investigation, validation, L.T.; conceptualization, methodology, Y.Z. (Yu Zhang); writing—review & editing, H.S., Y.Z. (Yu Zhang) and A.R.; supervision, resources, H.S., W.P., L.G. and C.J.; methodology, A.R. All authors have read and agreed to the published version of the manuscript.

Funding: This research was financially supported by Hunan Province Jiapinjiawei Biotechnology Co., Ltd. Funder: Huanlu Song. No funding number is available.

Institutional Review Board Statement: Not applicable.

Informed Consent Statement: Not applicable.

Data Availability Statement: The data will be available at request.

Acknowledgments: Authors gratefully acknowledge the panelists of Laboratory of Molecular Sensory Science, Beijing Technology and Business University.

Conflicts of Interest: The authors declare no conflict of interest. The methods used in the manuscript are those commonly used in the field of food science and do not require ethical proof.

Sample Availability: Samples of the compounds are not available from the authors.

References

1. Coppola, D.; Lauritano, C.; Esposito, F.; Riccio, G.; Pascale, D. Fish Waste: From Problem to Valuable Resource. *Mar. Drugs* **2021**, *19*, 116. [CrossRef] [PubMed]
2. Shanthi, G.; Premalatha, M.; Anantharaman, N. Potential utilization of fish waste for the sustainable production of microalgae rich in renewable protein and phycocyanin-Arthrospira platensis/Spirulina. *J. Clean. Prod.* **2021**, *294*, 106–126. [CrossRef]
3. Jiang, X.M.; Wang, J.; Pan, B.Z.; Li, D.B.; Wang, Y.Z.; Liu, X.Y. Assessment of heavy metal accumulation in freshwater fish of Dongting Lake, China: Effects of feeding habits, habitat preferences and body size. *J. Environ. Sci.* **2022**, *112*, 355–365. [CrossRef] [PubMed]
4. Alfio, V.G.; Manzo, C.; Micillo, R. From Fish Waste to Value: An Overview of the Sustainable Recovery of Omega-3 for Food Supplements. *Molecules* **2021**, *26*, 1002. [CrossRef]
5. Yasin, N.H.M.; Ahmad, N.A.N.; Hanapi, M.F.M. Extraction of FAME from fish waste by using modified soxhlet method. *IOP Conf. Ser. Mater. Sci. Eng.* **2021**, *1092*, 012015. [CrossRef]
6. Liu, H.H.; Hu, F.; Li, P.F. Volatile Compositions Analysis of Tilapia (Oreochram niloticus) and Fishy Odour Development in the Muscle. *Adv. Mater. Res.* **2014**, *1033*, 767–776. [CrossRef]
7. Li, Y.L.; Wang, X.J.; Xue, Y.; Ruan, S.Y.; Zhou, A.Q.; Huang, S.F.; Ma, H.L. The Preparation and Identification of Characteristic Flavour Compounds of Maillard Reaction Products of Protein Hydrolysate from Grass Carp (Ctenopharyngodon idella) Bone. *J. Food Qual.* **2021**, *14*, 8394152. [CrossRef]
8. Cui, H.P.; Yu, J.Y.; Xia, S.Q.; Duhoranimana, E.; Huang, Q.R.; Zhang, X.M. Improved controlled flavor formation during heat-treatment with a stable Maillard reaction intermediate derived from xylose-phenylalanine. *Food Chem.* **2019**, *271*, 47–53. [CrossRef]
9. Buyukkurt, O.K.; Selli, S. Factors affecting on the release of aroma compounds. *Gida* **2020**, *45*, 204–216. [CrossRef]
10. Xue, D.D.; He, T.P.; You, M.C.; Song, H.L.; Gong, L.; Pan, W. Effects of different treatments on fishy odor of fish soups. *J. Aquat. Food Prod. Technol.* **2018**, *27*, 722–732. [CrossRef]
11. Wang, X.J.; Guo, M.Y.; Song, H.L.; Meng, Q.; Guan, X. Characterization of key odor-active compounds in commercial high-salt liquid-state soy sauce by switchable GC/GC x GC-olfactometry-MS and sensory evaluation. *Food Chem.* **2021**, *342*, 128224. [CrossRef]
12. Yang, P.; Yu, M.G.; Song, H.L.; Xu, Y.Q.; Lin, Y.P.; Granvogl, M. Characterization of Key Aroma-Active Compounds in Rough and Moderate Fire Rougui Wuyi Rock Tea (Camellia sinensis) by Sensory-Directed Flavor Analysis and Elucidation of the Influences of Roasting on Aroma. *J. Agric. Food Chem.* **2021**, *70*, 267–278. [CrossRef]
13. Zhao, M.; Li, T.; Yang, F.; Cui, X.Y.; Zou, T.T.; Song, H.L.; Liu, Y. Characterization of key aroma-active compounds in Hanyuan Zanthoxylum bungeanum by GC-O-MS and switchable GC × GC-O-MS. *Food Chem.* **2022**, *385*, 132659. [CrossRef]
14. Li, J.L.; Tu, Z.C.; Zhang, L.; Sha, X.M.; Wang, H.; Pang, J.J.; Tang, P.P. SPME-GC-MS Analysis of Changes in Volatile Compounds during Preparation of Grass Carp Soup. *Food Sci.* **2016**, *37*, 149–154. [CrossRef]
15. Xi, Y.R.; Tang, K.; Xu, Y.; Wang, D.; Wang, Q.W.; Zhang, H.N. Aroma characterization of Cabernet sauvignon wine from Loess Plateau with quantitative description analysis and GC-O/MS. *Food Ferment. Ind.* **2016**, *42*, 192–197. [CrossRef]
16. Zhang, Y.; Liu, Y.P.; Yang, W.X.; Huang, J.; Liu, Y.Q.; Huang, M.Q.; Sun, B.G.; Li, C.L. Characterization of potent aroma compounds in preserved egg yolk by gas chromatography−olfactometry, quantitative measurements, and odor activity value. *J. Agric. Food Chem.* **2018**, *66*, 6132–6141. [CrossRef]
17. Zhang, Y.; Song, H.L.; Li, P.; Yao, J.; Xiong, J. Determination of potential off-flavour in yeast extract. *LWT-Food Sci. Technol.* **2017**, *82*, 184–191. [CrossRef]

18. Yang, P.; Song, H.L.; Lin, Y.P.; Guo, T.Y.; Wang, L.J.; Granvogl, M.; Xu, Y.Q. Differences of characteristic aroma compounds in Rougui tea leaves with different roasting temperatures analyzed by switchable GC-O-MS and GC × GC-O-MS and sensory evaluation. *Food Funct.* **2021**, *12*, 4797–4807. [CrossRef]
19. Gemert, L.J.V. *Odour Thresholds-Compilations of Odour Threshold Values in Air, Water and Other Media*, 2nd ed.; Oliemans Punter & Partners BV: Zeist, The Netherlands, 2011.
20. Xu, Y.X.; Jiang, Z.D.; Du, X.P.; Zheng, M.J.; Yuan, F.Y.; Ni, H.; Chen, F. The identification of biotransformation pathways for removing fishy malodor from Bangia fusco-purpurea using fermentation with Saccharomyces cerevisiae. *Food Chem.* **2022**, *380*, 132103. [CrossRef]
21. Liu, Y.; Huang, Y.Z.; Wang, Z.M.; Cai, S.H.; Zhu, B.W.; Dong, X.P. Recent advances in fishy odour in aquatic fish products, from formation to control. *Int. J. Food Sci. Technol.* **2021**, *56*, 4959–4969. [CrossRef]
22. Gu, S.Q.; Dai, D.L.; Chong, Y.Q.; Ly, F.; Zhou, X.X.; Ding, Y.T. The binding of key fishy off-flavor compounds to silver carp proteins: A thermodynamic analysis. *RSC Adv.* **2020**, *10*, 11292–11299. [CrossRef]
23. Duan, Z.H.; Zhang, M.; Hao, J. Preparation process of crispy bighead carp fillets. *Food Ind. Technol.* **2003**, *24*, 44–47.
24. An, Y.Q.; Qian, Y.P.L.; Alcazar Magana, A.; Xiong, S.B.; Qian, M.C. Comparative characterization of aroma compounds in silver carp (Hypophthalmichthysmolitrix), pacific whiting (Merluccius productus), and alaska pollock (Theragra chalcogramma) surimi by aroma extract dilution analysis, odor activity value, and aroma recombination studies. *J. Agric. Food Chem.* **2020**, *68*, 10403–10413. [CrossRef]
25. Chen, J.; Tao, L.; Zhang, T.; Zhang, J.; Wu, T.; Luan, D.L.; Ni, L.; Wang., X.C.; Zhong, J. Effect of four types of thermal processing methods on the aroma profiles of acidity regulator-treated tilapia muscles using E-nose, HS-SPME-GC-MS, and HS-GC-IMS. *LWT-Food Sci. Technol.* **2021**, *147*, 111585. [CrossRef]
26. Liang, S.; Zhang, T.; Fu, X.; Zhu, C.; Mou, H. Partially degraded chitosan-based flocculation to achieve effective deodorization of oyster (Crassostrea gigas) hydrolysates. *Carbohydr. Polym.* **2020**, *234*, 115948. [CrossRef]
27. Li, X.F.; Xie, W.; Bai, F.; Wang, J.L.; Zhou, X.D.; Gao, R.C.; Xu, X.X.; Zhao, Y.H. Influence of thermal processing on flavor and sensory profile of sturgeon meat. *Food Chem.* **2022**, *347*, 131689. [CrossRef]
28. Liu, Y.; He, C.C.; Song, H.L. Comparison of SPME Versus SAFE Processes for the Analysis of Flavor Compounds in Watermelon Juice. *Food Anal. Methods* **2018**, *11*, 1677–1689. [CrossRef]
29. Mahmoud, M.; Buettner, A. Characterisation of aroma-active and off-odour compounds in German rainbow trout (Oncorhynchus mykiss). Part II: Case of fish meat and skin from earthen-ponds farming. *Food Chem.* **2016**, *232*, 841–849. [CrossRef]
30. Guo, Q.Y.; Yu, J.W.; Zhao, Y.Y.; Liu, T.T.; Su, M.; Jia, Z.Y.; Zhao, Y.; Mu, Z.; Yang, M. Identification of fishy odor causing compounds produced by Ochromonas sp. and Cryptomonas ovate with gas chromatography-olfactometry and comprehensive two-dimensional gas chromatography. *Sci. Total Environ.* **2019**, *671*, 149–156. [CrossRef]

molecules

Review

Recent Progress on Techniques in the Detection of Aflatoxin B_1 in Edible Oil: A Mini Review

Shipeng Yin [1], Liqiong Niu [2] and Yuanfa Liu [1,*]

[1] School of Food Science and Technology, Jiangnan University, No. 1800 Lihu Road, Binhu District, Wuxi 214122, China
[2] School of Life Sciences, Guangzhou University, Guangzhou 510006, China
* Correspondence: yfliu@jiangnan.edu.cn; Tel.: 86-510-8587-6799

Abstract: Contamination of agricultural products and foods by aflatoxin B_1 (AFB_1) is becoming a serious global problem, and the presence of AFB_1 in edible oil is frequent and has become inevitable, especially in underdeveloped countries and regions. As AFB_1 results from a possible degradation of aflatoxins and the interaction of the resulting toxic compound with food components, it could cause chronic disease or severe cancers, increasing morbidity and mortality. Therefore, rapid and reliable detection methods are essential for checking AFB_1 occurrence in foodstuffs to ensure food safety. Recently, new biosensor technologies have become a research hotspot due to their characteristics of speed and accuracy. This review describes various technologies such as chromatographic and spectroscopic techniques, ELISA techniques, and biosensing techniques, along with their advantages and weaknesses, for AFB_1 control in edible oil and provides new insight into AFB_1 detection for future work. Although compared with other technologies, biosensor technology involves the cross integration of multiple technologies, such as spectral technology and new nano materials, and has great potential, some challenges regarding their stability, cost, etc., need further studies.

Keywords: aflatoxin B_1; edible oil; chromatographic technology; spectroscopic technology; biosensor technology; recognition elements

Citation: Yin, S.; Niu, L.; Liu, Y. Recent Progress on Techniques in the Detection of Aflatoxin B_1 in Edible Oil: A Mini Review. *Molecules* **2022**, 27, 6141. https://doi.org/10.3390/molecules27196141

Academic Editors: Yanping Chen and Weiying Lu

Received: 10 August 2022
Accepted: 15 September 2022
Published: 20 September 2022

1. Introduction

Food security has always been an issue of concern in the international community, and, in recent years, food contamination has become a major factor affecting food security. Contaminated food can not only adversely influence human health (poisoning events, chronic diseases, etc.) but also affect and slow down the economy. When people consume contaminated food, they need to spend a lot of money and time on treatment. There are many factors causing food contamination, such as biological, chemical, and physical factors. Among these, microbial contamination is common and mainly includes contamination by bacteria, fungi, molds, viruses, or their toxins and by-products [1,2]. Mycotoxins are common food contaminants, which can cause changes in the appearance, flavor, smell, and other characteristics of food [3–7]. Mycotoxins are secondary metabolites produced by fungi (e.g., *Fusarium*, *Aspergillus*, and *Penicillium*) that have multiple toxic effects on organisms and contaminate agricultural products (cereals, milk, etc.). More than 400 kinds of mycotoxins have been identified. Among them, aflatoxins (AFs) have become one of the major concerns due to their high toxicity and carcinogenicity, causing approximately 25% of animal deaths [8–12].

Edible vegetable oil plays an irreplaceable role in the human diet. The world oil crop output has increased year by year and had reached 635.5 million tons by 2021 [13]. From the growth of oil crops to the final product, i.e., oil, each link may be affected by external factors (such as mycotoxins), which may affect the quality and safety of edible vegetable oil [14]. This is because most oil crops, such as corn, peanut, soybean, rapeseed, sunflower seeds, olives, and nuts, are seasonal. During the growth process, they will be affected by climate, pests, and other factors and can be easily be infected by *Aspergillus* flavus. After harvest, the oil may deteriorate or be affected by mildew due to storage conditions (such as temperature and humidity, etc.) and storage methods [15]. At the same time, during the production of edible oil, fresh-pressed edible oil is vulnerable to contamination of raw materials infected with *Aspergillus* by aflatoxin B_1 (AFB_1) [16–22]. Therefore, contamination of edible vegetable oil products by AFB_1 is a serious food safety problem (Figure 1) [20,23–25].

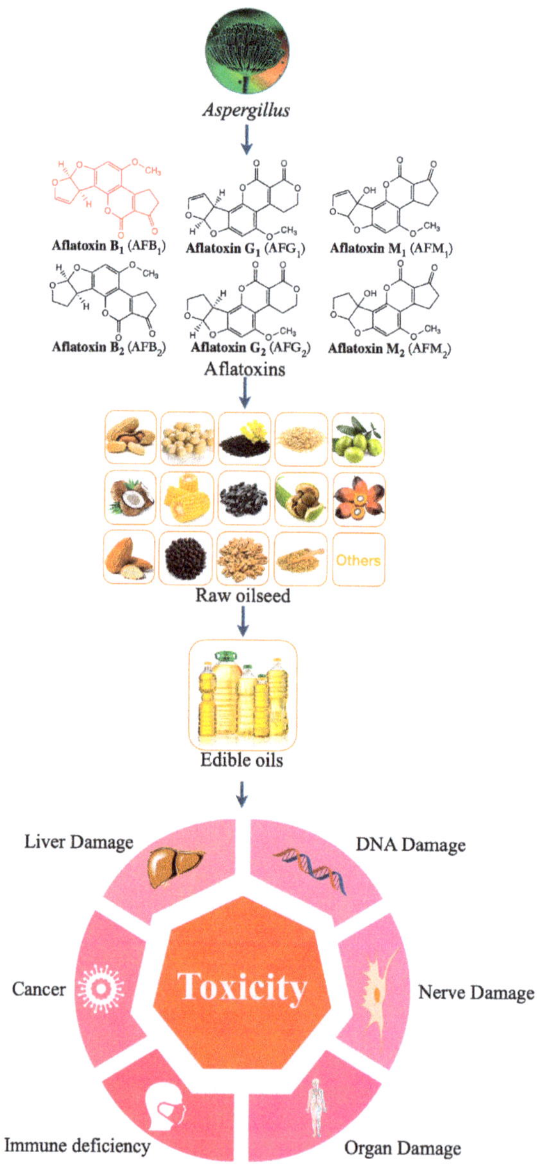

Figure 1. Harmful effects of different types aflatoxins contaminated edible oil.

The presence of aflatoxin is usually detected by using precision instruments, such as high-performance liquid chromatography–mass spectrometry (HPLC–MS), high-performance liquid chromatography–fluorescence detection (HPLC–FD), or other molecular techniques, while rapid detection is mainly realized by enzymatic immunoassay ELISA [26,27]. Although different methods are available for the detection of AFB_1 toxicity, these methods require expensive equipment and complex sample pretreatment or can only be performed at relatively high concentrations [28]. Therefore, simple, sensitive, efficient, economical, rapid, and stable AFB_1 detection methods are required. Recently, new technologies, such as biosensors, have been applied in many fields, such as health care and food detection. Because of their key advantages, such as convenient operation, rapid response, and excellent portability, these technologies can detect harmful substances in food sensitively and accurately, helping effectively avoid their harmful effects. They have attracted increasing attention of researchers and also promoted the rapid development of biosensors. With progress in nanotechnology, scientists are paying special attention to biosensors based on nanomaterials. These new biosensors or detection systems are sensitive, rapid, consistent, and cost-effective and can be used to detect AFB_1 in food [29–33].

Regarding the increased importance of biosensors for accurate detection of AFB_1 in edible oil, we have summarized the recent advances in biosensors for AFB_1 analysis, specifically from the points of view of the development of novel bioinspired recognition elements and nanomaterials-based electrochemical biosensors.

Therefore, we searched PubMed and web of science for publications describing the detection technology of aflatoxin B1 in edible oil. Search terms were as follows: aflatoxin B_1 OR AFB_1 OR *Aspergillus* OR mycotoxins OR AFB_2 OR AFG_1 OR AFG_2 OR AFM_1 OR AFM_2 OR AFs OR AFBO OR CYP450 OR edible oil OR vegetable oil OR corn oil OR peanut oil OR soybean oil OR sesame oil OR rapeseed oil OR sunflower seeds oil OR olives oil OR nuts oil OR maize oil OR canola oil OR blend oil OR coconut oil OR almond oil OR rice oil OR palm oil OR tea oil OR chromatographic technology OR spectroscopic technology OR immunological technology OR biosensor technology OR QuEChERS OR Fluorescence spectrophotometry OR Infrared spectroscopy OR Terahertz spectroscopy OR surface-enhanced raman spectroscopy (SERS) OR enzyme-linked immunosorbent assay (ELISA) OR amperometric OR impedometric OR electrochemical impedance spectroscopy (EIS) OR voltammetry (potentiometric) OR Conductometric OR LOD OR chromogenic OR Luminogenic OR Chemiluminescence OR Gravimetric OR Piezoelectric OR Magnetoelastic OR Acoustic OR electrodes (SPEs)OR SRP OR biosensors OR Nanomaterial-based biosensors OR electrochemical biosensors OR bioinspired recognition elements OR antibodies OR aptamers OR molecularly imprinted polymers OR Phylogenetic Evolution of Ligands for Exponential Enrichment (SELEX) OR fluorescence resonance energy transfer (FRET).

Publications until 29 August 2022 were included. This review only had the detection technology targeted at aflatoxin B1 in edible oil, and that had not included other types of toxins or other food carriers. After 4692 publications were searched, 596 full-text articles were reviewed and 132 articles were finally identified to meet our requirements.

2. Importance of Aflatoxins

Aflatoxins are a type of mycotoxins. They are highly toxic metabolites of fungi, produced in food and agricultural products. They have severe toxic effects, such as immunosuppressive, nephrotoxic, teratogenic, carcinogenic, and mutagenic, on human and animal health [34–38].

Aflatoxins can be divided into aflatoxin B_1 (AFB_1), aflatoxin B_2 (AFB_2), aflatoxin G_1 (AFG_1), and aflatoxin G_2 (AFG_2) according to their fluorescence properties and chromatographic mobility (Figure 1) [39–41]. Aflatoxin M_1 (AFM_1) and aflatoxin M_2 (AFM_2) are hydroxylated metabolites of AFB_1 and AFB_2, respectively. AFB_1 is the most toxic among all AF species, with a high incidence rate and the most complex detection mechanism (Figure 2) [42].

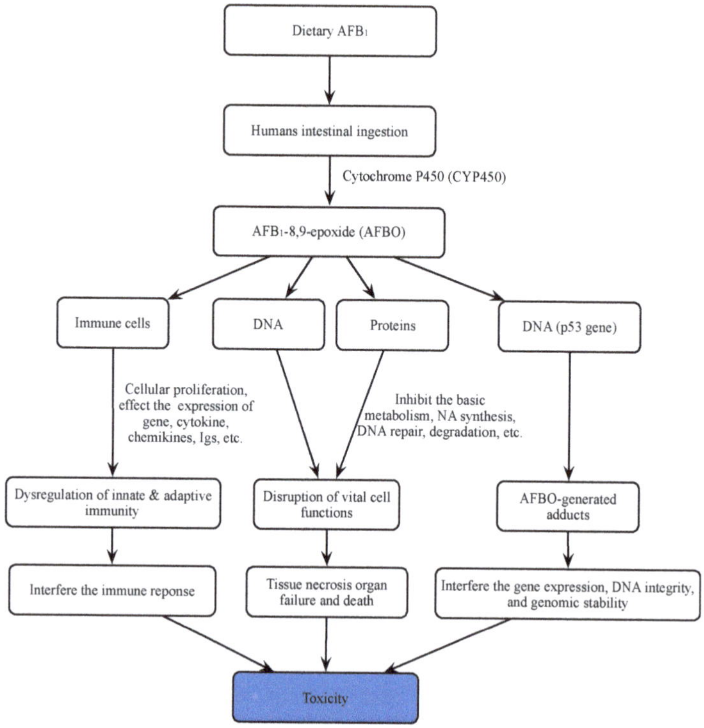

Figure 2. Main mechanisms of toxicity of aflatoxin B_1 for humans.

AFB_1 is a powerful carcinogenic, teratogenic, mutagenic, immunotoxic, hepatotoxic, and reproducible poison. Previous studies have shown that the toxicity of AFB_1 is 10, 68, and 416 times that of KCN, arsenic and melamine, respectively [43,44] (Figure 2). Therefore, AFB_1 has been classified as a class 1 carcinogen by many international authoritative organizations or institutions [45,46]. Due to the structural double bonds in the furan ring, AFB_1 has high carcinogenicity and toxicity [17,47]. The lipophilic structure of atrial fibrillation promotes its entry into the blood through gastrointestinal and respiratory tracts [48,49]. Once AFB_1 enters blood, it is distributed in various tissues and accumulates in the liver or other organs, resulting in liver cancer (Figure 3). In the liver, AFB_1 produces a variety of metabolites through the hydroxylation and demethylation of the first-stage drug metabolism enzymes (for example, cytochrome P450 oxidase and CYP450 superfamily members, such as CYP1A2, CYP3A4, and CYP2A6) [50]. Metabolic reaction (internal and external) activates the final carcinogen AFB_1 -8,9-epoxy metabolite, which covalently binds to cellular macromolecules (DNA, RNA, or protein) and plays a key role in acute and chronic poisoning. AFB_1 residues also destroy the function of tumor suppressor genes (p53 and Rb) in the liver, which affects normal cells and leads to liver injury, increasing the probability of tumor and liver cirrhosis [51–55]. It is estimated that about 30% of liver cancers in the world are caused by AFB_1. Its toxicity increases the infection rate of hepatitis B virus (HBV) and the risk of liver cancer [56]. A recent study found that the synergistic effect of AFB_1 and HBV leads to liver cancer [50]. The reason is that HBV infection directly or indirectly exposes hepatocytes to AFB_1 sensitive to tumors. The toxic effect of AFB_1 is also related to dose, age, sex, nutrition, exposure time, and type [57]. In addition, AFB_1 can be transmitted to the fetus through the placenta and affect the health of infants [58]. AFB_1 exposure also inhibits immunity, thereby increasing the susceptibility to immunodeficiency virus attack and the probability of infection with other infectious diseases [59–63].

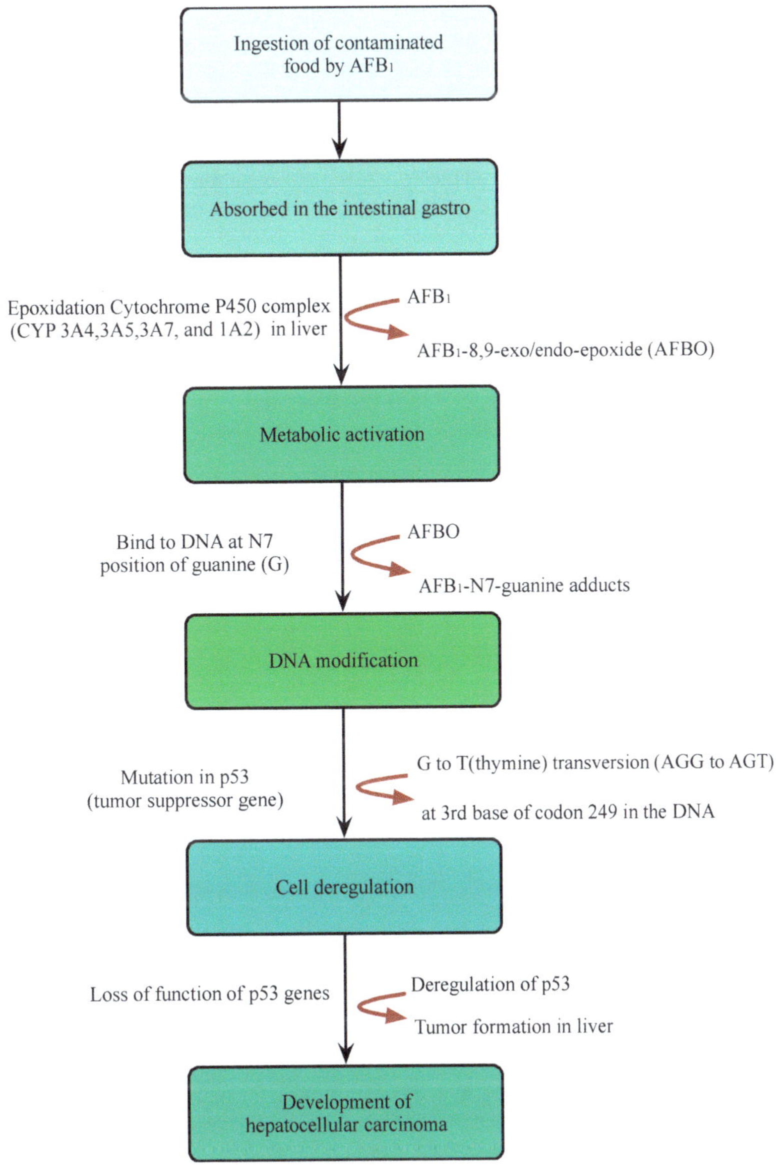

Figure 3. Illustration of the mechanism of hepatocellular carcinoma caused by ingestion of AFB_1-contaminated foods.

3. AFB_1 Regulations on Edible Oil

Because AFB_1 poses many hazards to the human body, many governments and international research institutions have made many efforts to control AFB_1 pollution in different foods. For example, the FAO and the European Commission and Codex Alimentarius Commission have formulated regulations regarding the content of AFB_1 in various foods to ensure consumer safety [64–69].

As for edible oil, most countries have no legislative restrictions and only a few countries, such as China, have effective regulations, laws, and standards for the highest level of

AFB_1 in different edible oils (Table 1). Due to some adverse conditions in the traditional oil processing process, AFB_1 is usually degraded to the normal level in the extraction and refining process [17,70]. The EU has strict regulatory norms. The total amount of AFB_1 and AF allowed in oilseeds is restricted to 2 and 4 μg kg^{-1}, respectively. However, the maximum limit of AFs in oils has not been determined. The corresponding regulations in China, the United States, Kenya, and Thailand clearly stipulate the maximum level of total AFs in all edible oils, but the maximum level required is different. It is worth mentioning that in China, the AFB_1 limit in corn and peanut oil is stipulated to be 20 μg kg^{-1}, which may be because corn and peanut are most vulnerable to aflatoxin pollution [71,72]. See Table 1 for specific differences.

Table 1. The maximum limits (μg kg^{-1}) established for major AFB_1 in some countries/regions for edible oils.

Countries/Agencies	Food Products	Edible Vegetable Oil	Total of AFs (μg kg^{-1})	AFB_1 (μg kg^{-1})	Refs.
EU	Oil seeds	-	15	8	[32,69]
EU	-	Peanut oil Others oil	4	2	[32,69]
China	-	Maize oil Peanut oil Others oils	- - -	20 20 10	[73]
Greece	-	Olive oil	-	-	[69]
Russia	-	Vegetable oil	-	-	[64]
France	-	Vegetable oil	-	5	[64]
Kenya	-	Vegetable oil	20	-	[64,71]
Taiwan	-	Edible oil	10	-	[65]
Morocco	-	Vegetable oil	-	5	[72]
Thailand	All foods	Oil and fats	20	-	[64,65,74]
USA	All foods	-	20	-	[64,71,75]
Brazil	All foods	-	-	15	[76]
India	All foods	-	-	30	[65]
Chile	All foods	-	-	-	[76]
Indonesia	All foods	-	35	20	[65]
Singapore Australia	All foods	-	5	-	[64,65]
Malaysia	All foods	-	35	-	[65]
Japan Vietnam	All foods	-	10	-	[65]
Sri Lanka	All foods	-	30	-	[65]

4. Methods for Detecting AFB_1 in Edible Oil

The matrix is too complex for edible oil, and the mycotoxin content is relatively low, making it difficult to detect AFB_1. Therefore, researchers have developed various traditional and modern methods to detect AFB_1 in oil. AFB_1 detection technology is mainly divided into chromatographic technology, spectroscopic technology, immunological technology, and biosensor technology [16,77].

Figure 4 briefly summarizes the LOD timelines for AFB_1 detection in edible oils published from 2007 to 2022 mentioned in this review. It can be seen from the figure that with the advancement of time, no matter what type of detection technology or which specific detection method is used, the LOD of AFB_1 in edible oil tends to be lower. This shows that people have a great interest in the detection of AFB1 in edible oil. At the same time, the wide use of new materials represented by nanomaterials highlights the interdisciplinary characteristics of new sensors. Next is a brief introduction of the identification method of AFB_1, including its advantages and disadvantages, combined with actual cases.

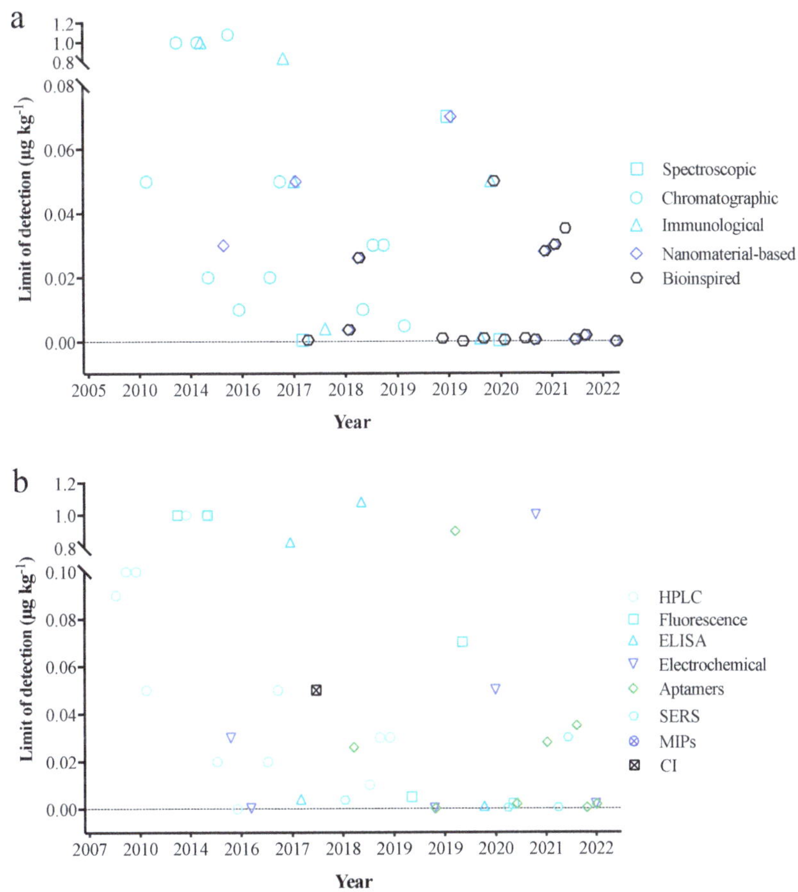

Figure 4. Timeline of the limit of detection on AFB_1 in edible oil by different (**a**) detection technology and (**b**) detection method. CI: Chemiluminescence immunoassay.

4.1. Chromatographic Technology

4.1.1. High Performance Liquid Chromatography (HPLC)

High-performance liquid chromatography is a common official detection method. Many countries and institutions have used it, such as China's national standard, the European Committee for Standardization (CEN), and the association of analytical organizations (AOAC). One characteristic of the HPLC method is that it can measure multiple targets with high sensitivity [78]. In recent years, researchers have developed new detection strategies combining HPLC with other sensors, such as fluorescence detection (FLD), ultraviolet (UV) detection, diode array detection, and mass spectrometry (MS) [79,80]. Compared to traditional HPLC, this further improves the reliability, sensitivity, and accuracy of target analytes and is widely used to detect harmful substances in food. For example, HPLC combined with FLD is the standard method for detecting AFB_1 in edible vegetable oil [81–86]. HPLC–FLD was able to detect AFB_1 levels as low as 0.01–0.04 μg kg^{-1} [81] and 0.005–0.03 μg L^{-1} [82].

Recently, liquid chromatography–tandem mass spectrometry (LC–MS–MS) methods are being increasingly used for the analysis of mycotoxins [85]. They have the advantages of not having a sample purification limitation during extraction, high resolution, high sensitivity, and suitability for various edible vegetable oils [19,87–99]. GC analysis is mostly used for volatile substances, and most mycotoxins are non-volatile, further limiting the application of GC in mycotoxin detection. A similar procedure to HPLC, UHPLC or UPLC is also used on the column to improve the resolution of AFB_1. Hidalgo et al. [100] developed a new analytical method by coupling UHPLC to a triple quadrupole analyzer (UHPLC–QqQ–MS/MS), which was well validated and applied to monitor mycotoxins, including AFB_1, in 194 samples of edible vegetable oil.

Many commonly used methods require sample preparation due to the different matrices of edible oil. Currently, a variety of methods are available for the extraction and isolation of mycotoxins from oil, such as liquid–liquid extraction or partitioning (LLE), frequently reported in the literature [101–105]; solid–phase extraction (SPE) [105–109]; immune affinity columns (IACs) [81,94]; IAC combined with dispersive liquid–liquid microextraction (DLLME) [91]; multifunctional cleanup columns [110]; the QuEChERS system [90]; gel permeation chromatography (GPC) [111]; immune assay extraction; and low-temperature cleanup (LTC) [112–115]. However, each method has its advantages and limitations. Thus, which method to choose still depends on the type of food matrix, mycotoxin characterization, and detection techniques [116].

4.1.2. Thin-Layer Chromatography (TLC)

Thin-layer chromatography (TLC) is an adsorption thin-layer chromatographic separation method suitable for complex mixed samples [117,118]. Since its development in the 1950s, thin-layer chromatography has been widely used in, for example, biology, medicine, and the chemical industry. It has recently been used in food analysis and quality control and has become a conventional technology in laboratories. Many reports have shown that TLC can be applied to all stages of the food industry, such as the stage of traditional substances, represented by food raw materials, ingredients, and additives, and the stage of unconventional substances, represented by harmful substances and pollutants. The detection and determination of compounds cover almost all substance categories [119–122].

Thin-layer chromatography uses the different adsorption capacities of each component to the same adsorbent so that when the mobile phase (the solvent) is flowing through the stationary phase (the adsorbent), there is continuous adsorption, desorption, readsorption, and redesorption to achieve the mutual separation of each component [123].

Although the TLC method has matured, it still has shortcomings, such as a low detection accuracy, volatility during the experiment being harmful to the experimental operators and the environment, and complex sample pretreatment [124,125]. In recent years, an interdisciplinary approach, such as the combination of TLC with image analysis and with new technologies, such as surface-enhanced Raman spectroscopy, mass spectrometry, and nuclear magnetic resonance, has further promoted the development of thin-layer chromatography and enhanced the practicability of this method in food analysis [126–129]. TLC is used to detect harmful substances in various foods, such as AF in edible oil, making it an effective analytical tool in food science methods [124,130].

4.2. Spectroscopic Technology

4.2.1. Fluorescence Spectrophotometry

Spectrum-based sensing technology has been developed and used to assess AFs contamination in food [131]. Among many spectral techniques, fluorescence spectrometry shows certain potential in determining AFs in a variety of agricultural products and foods [125,132,133]. Fluorescence spectrometry uses the target molecules in the sample to absorb ultraviolet or visible light to produce fluorescence and determine its molecular structure. It has excellent detection sensitivity and specificity in the study of AFs and other chemical components [134,135]. The study found that the fluorescence phenomenon is conducive to the characterization and monitoring of target detection objects. For example, AFB_1 can emit a specific range of fluorescence (425–500 nm) under the excitation of UV light source (340–400 nm), which provides the possibility of using fluorescence spectroscopy to analyze AFB_1 in different foods [135,136]. In recent years, laser-induced fluorescence (LIF) technology has developed rapidly and attracted more attention because it uses a certain wavelength of laser light source and has better specificity and detection sensitivity. The advantage of LIF is that it can realize online, rapid and nondestructive direct detection according to the characteristic fluorescence peak of AFB_1. Researchers have developed a detection model based on LIF, which can quickly and accurately screen AFB_1 in different edible oils. The information and conclusions obtained in the study further show that LIF technology can be used for rapid and nondestructive detection of AFB1 in different edible oils [19,137]. However, LIF is also vulnerable to the interference of external factors, such as the power and accuracy of the instrument, the environmental factors of temperature and humidity, and the physical and chemical index factors of the detected object. Although this limits the wide application of LIF technology, researchers are still trying and exploring.

4.2.2. Infrared (IR) Spectroscopy

Infrared spectroscopy (IRs) has the characteristics of rapid detection, simple sample preparation process and strong adaptability. It has been widely proven to be an effective food safety detection and control technology. Because IR covers a wide range of electromagnetic spectra (780 to 2500 nm), IR can be applied to the detection of a variety of foods including edible oil, meat, aquatic products, fruits and vegetables [138–144]. When IRs radiation penetrates the sample, the radiation is reflected, absorbed or transmitted by molecular bonds, resulting in the energy change of light, which can reflect some characteristic chemical bonds, thus reflecting the characteristics of the tested product [145,146]. In the application of edible oil, IR shows many abilities, such as distinguishing different kinds of oil, grading the quality of oil, detecting harmful substances in oil, etc. [138,143,147–150]. Using near infrared (NIR) technology to detect mold in edible oil has also been a research hotspot in recent years. Researchers have promoted the further application and development of IR technology by establishing qualitative and quantitative analysis models for AFB_1 pollution in edible oil [151–153].

4.2.3. Terahertz (THz) Spectroscopy

With the development of optical and electronic technology, terahertz spectroscopy (THz) has been a revolutionary development, and shows great potential as a new technology tool for nondestructive food testing [154–157]. As a technical information link between microwave spectroscopy and infrared spectroscopy, THz has the characteristics of both, making it widely used in basic research and industrial practice [158,159]. Like other spectral technologies, thanks to the development of chemometrics methods, THz has become a powerful technical tool in the food industry, due to its strong detection and quantification capabilities [156,157,160]. Through the combination of THz and chemometrics methods, researchers have constructed a rapid nondestructive detection model for AFB1 in edible oil. Although the accuracy is slightly lower than other conventional analysis methods, it provides a possibility for THz in food safety detection [161]. In a recent study, researchers further improved the accuracy of THz in detecting AFB1 in edible oil by adding pretreatment and other methods on the basis of predecessors, and reduced the LOD of AFB_1 to 1 $\mu g\ kg^{-1}$, and the accuracy is improved to more than 90% [161,162]. The cross integration of THz and chemometrics and other disciplines is conducive to promoting its application and development in the detection of AF_S in the edible oil industry. At the same time, the limitations of THz should also be clear, such as the low detection limit and sensitivity advantage are not obvious, the penetration of the detected object is limited, there is scattering effect, the technology is expensive, the database is lack, etc. [163].

4.2.4. Surface-Enhanced Raman Spectroscopy (SERS)

As a complementary analysis technology of IR, the Raman spectroscopy (RS) technology is sensitive to the symmetrical vibration of covalent bonds of non-polar groups (such as C=O, C-C and S-S) [164–166]. Therefore, RS has the advantages of being fast, sensitive and simple in the detection and evaluation system of food [165,167,168]. However, traditional RS has some limitations, such as Raman scattering. Therefore, researchers have developed SERS signal enhancement technology represented by electromagnetic field enhancement and chemical enhancement [165]. At present, the application of SERS technology in the detection of AFs is still challenging, and the intersection of technology development and multidisciplinarity (such as materials science, stoichiometry, etc.) is the focus of researchers [165]. In recent years, researchers have reported a variety of SERS schemes for AFB_1 detection in edible oil, such as SERS tag detection using antibodies and aptamers, sandwich immunoassay based on SERS, etc. [169–174]. The growing research results show that SERS technology is becoming a powerful tool to ensure the safety development of the food industry, especially in the safety supervision of AFs. However, it cannot be denied that challenges still exist, such as the development of targeted new materials, the optimization of key core technologies, and the practical application of research results [175–177].

4.3. Immunological Technology

Enzyme-Linked Immunosorbent Assay (ELISA)

In recent years, researchers have often used immunochemical methods to determine mycotoxins in food, in addition to traditional chromatographic techniques. The core of immunochemistry is the specific interaction between immunoglobulin (Igs) and antigen (Ag). Several immunochemical methods have been applied to detect mycotoxins in edible vegetable oils, such as enzyme-linked immunosorbent assay (ELISA) and biosensors based on immunoassay.

ELISA is one of the most commonly used methods for detecting mycotoxins [24]. It has been designed and developed on the basis of the principle of specific immune responses between Igs and Ags. The specificity of this immunoassay is due to the use of enzyme-labeled Igs or Ags and solid-matrix-restricted immunoglobulins to capture unlabeled silver in the analyte and detect it with labeled immunoglobulins. Although ELISA is well developed and widely used in food analysis, clinical practice, biotechnology, environmental, chemical, and other industries, it still has several deficiencies, such as excessive dependence on the matrix caused by the interaction between the target antigen and matrix components. The standard ELISA is composed of four main parts (immune-recognition element, sorbent substrate, enzyme label, and chromogenic reagent), and the deficiency of the central part is the root cause of the limitation of ELISA. In recent years, researchers have used the cross-fusion of multiple technologies to drive the performance of one of the components or the whole ELISA, especially in terms of sensitivity, accuracy, and stability [27].

For mycotoxins, due to the high singularity of ELISA, the developed kit has specific recognition ability and has been widely used in the detection of mycotoxins [70]. For example, Qi et al. [20] used ELISA and UPLC–MS/MS to detect AFB_1 in peanut oil, although the LOD was only 1.08 $\mu g\ kg^{-1}$, much higher than the LOD of UPLC–MS/MS (the LOD is 0.01 $\mu g\ kg^{-1}$) [20]. It has been affirmed because of its accuracy, rapidity, and other advantages. For the actual detection of other harmful substances, such as AFB_1, AFB_2, AFG_1 and AFG_2, in different edible oils (oils of soybean, coconut, peanut, fennel, melon, and palm kernel), ELISA showed satisfactory results and the concentration was lower than the legislative limit [178–180]. On this basis, the researchers developed a commercial ELISA kit that can detect AFB_1, which can be applied to a variety of samples including edible oil, and the detection limit can be as low as 3 ppb. Although the current ELISA technology or kit still has problems such being as time-consuming, high cost, and cumbersome operation, with the advancement of technology, ELISA technology shows strong application potential [27].

4.4. Electrochemical Biosensing Technology

Due to rapidity, small footprint, economy, sensitivity, and unique capabilities, electrochemical biosensing devices have received particular attention in assessing food quality, mainly reflecting AFB_1 levels in food samples [181]. The AFB_1 electrochemical biosensor can produce various types of analytical signals, such as voltage, current, and impedance [182,183].

The standard transduction methods are amperometric, electrochemical impedance spectroscopy (EIS), and voltammetry (potentiometry).

4.4.1. Amperometric Biosensors

The amperometric biosensor is an electrochemical device with high selectivity and sensitivity that takes the change in the measuring current as the analysis signal. Because the change in the current is closely related to the concentration of AFB_1 in food samples and the change can be achieved by maintaining a stable potential, an amperometric biosensor is relatively perfect. A typical amperometric biosensor consists of two- or three-electrode systems (containing a functional electrode, a reference electrode, and an auxiliary electrode), and the analytical performance of the latter is significantly higher than that of the former (Figure 5) [181]. This is because the additional auxiliary electrode not only increases the area of the detection surface but also increases the current between it and the functional electrode, as well as the operating potential between the functional electrode and the reference electrode, thereby enhancing the changes in the detection process of AFB_1 in food in electronic dynamics. On the contrary, the dual-electrode system does not include auxiliary electrodes, which may lose their function at high temperatures. Therefore, amperometric biosensors with dual-electrode systems are not used to analyze the quality of food samples [181].

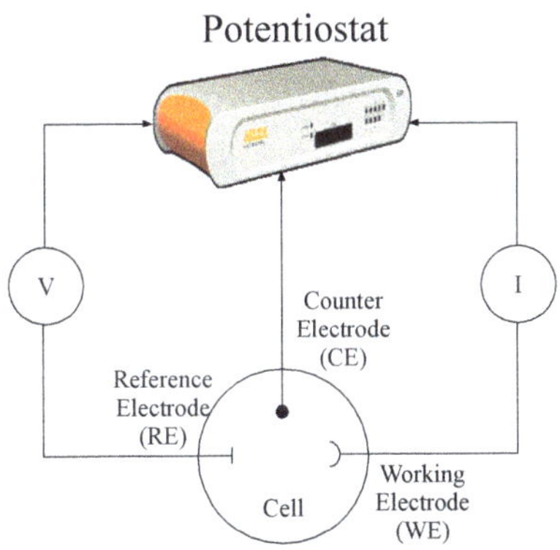

Figure 5. Scheme of the two or three-electrode setup used in electrochemical methods.

Even functional electrodes are usually made of inert metal materials (such as platinum, gold) or carbon (graphite, glassy carbon). The main drawback is reproducibility of measurements. Currently, printed electrodes have become a good substitute because their cost and mass production can be controlled [70,78].

Researchers used two kinds of nanomaterials with different charges to deposit on the electrode alternately, obtaining a multilayer electrode with a sandwich structure with excellent conductivity and rich electrochemical active sites [184]. Such a biosensor has good selectivity, reproducibility, and stability. In the subsequent optimization test, the optimized electrochemical biosensor was found to have significant stability and even after being placed for a period of time, it showed good LOD (0.002 ng mL^{-1}). This sensor is believed to be one of the best biosensors for detecting mycotoxins. Researchers applied the electrochemical biosensor to detect AFB_1 in real oil samples and found that it has a good recovery rate (98.11–103.36%).

Xuan et al. [185] developed an integrated AFB_1 detection platform that uses disposable screen-printed electrodes (SPEs), allowing routine detection without electrode modification. According to the SPE used, the platform can simplify the tedious sample processing process through high-throughput processing, reduce operating errors, and improve experimental reproducibility, which can benefit large-scale sample processing. The detectable concentration range of AFB_1 was 0.08–800 μg kg^{-1} with a LOD of 0.05 μg kg^{-1}. Analysis of real samples and verification of the method showed the results of the new sensor to be consistent with those of the classical method (LC–MS/MS), indicating that the developed method has the potential to monitor AFB_1 in peanut oil.

Another study reported a new aflatoxin biosensor based on the AFB_1 inhibition of acetylcholinesterase (AChE) [186]. The core of this method is to immobilize choline oxidase on a screen-printed electrode modified with Prussian blue (PB). The electrode used in the biosensor can detect H_2O_2 at low potential. As per the results, the linear operating range of the biosensor is estimated to be 10–60 ppb and the LOD is 2 ppb. On using real olive oil samples to evaluate the sensor, the recovery rate was found to be 78 ± 9% at 10 ppb.

4.4.2. Electrochemical Impedance Spectroscopy (EIS)

Electrochemical impedance spectroscopy (EIS) technology is an effective monitoring tool for identifying and monitoring changes in mycotoxins at the interface between electrode surface modifications. When the target analyte is combined with a biometric element, it generates an electrochemical response by changing conductivity and capacitance through an impedance biosensor [187]. These biosensors monitor the impedance changes caused by the interaction between the target detection object, such as AFB_1, and the biometric element fixed on the working electrode, and display the detection results in the form changed electron flow on the working electrode [188,189]. Typical potentiometric sensors are also suitable for the three-electrode system.

The main parameter of the EIS is the charge transfer resistance value (RCT), which is closely related to the reaction between immobilized mycotoxins and the antibody antigen and is also proportional to the target detector/concentration of the target [65,66]. For determining AFB_1, Yu et al. [190] reported a sensitive and convenient EIS method involving MWCNT/RTIL/Ab-modified electrodes coated in bare GCE. The experimental results show that the resistance of the MWCNT/RTIL/Ab-modified electrode (605.6 Ω) is higher than that of bare GCE (151.9 Ω). When AFB_1 was immobilized on MWCNT/RTIL/Ab-modified electrodes, the increase in the electron transfer resistance (Ret) value was found to be directly related to the AFB_1 amount. The specific interaction between AFB_1 and Ab causes an increase in the Ret value, which leads to the production of electrically insulating biological conjugates, which will prevent the electron transfer process of redox probes. Therefore, the EIS measurement results are consistent with the above cyclic voltammetry results. Because of its simple characteristics, this method can be widely used to detect various agricultural products and edible oils.

For many researchers exploring mycotoxin detection methods, aptamer-based EIS has become a hot research topic. Aptamer-based impedance biosensors have achieved satisfactory results in detecting mycotoxins in food and have great potential for practical application in edible oils.

4.4.3. Voltammetry Biosensors

Voltammetric biosensors solve the problem of obtaining analytical data using ion-selective electrodes. Similar to amperometric biosensors, voltammetry also requires a two- or three-electrode system. When the current is constant, it can detect target analytes, such as AFB_1, in food samples by evaluating the change in circuit potential between the functional electrode and the reference electrode [60,191].

Biosensors have also shown promising results in detecting the AFB_1 content in edible oils. For example, Wang et al. [192] developed a new disposable electrochemical biosensor based on stripping voltammetry to detect copper ions released from copper apatite. The biosensor uses copper ions as a signal label to immobilize AFB_1 antibody on a screen-printed carbon electrode (SPCE) modified by gold nanoparticles. The detection is performed by the voltammetric signal of the dissolution of copper ions released from acid hydrolysis of copper apatite, and copper apatite increases the number of loaded copper ions. The electrochemical signal is further amplified. Peanut oil was used to evaluate the reliability and application potential of biosensors. Researchers believe that this new method will be applied to many fields in the near future because of its many excellent characteristics (low cost, rapidity, accuracy, and high sensitivity).

4.4.4. Nanomaterial-Based Biosensors

Recently, different nanomaterials, such as carbon and metal, have been used to modify the active surfaces of macroelectrodes and microelectrodes to design electrochemical biosensors for the detection of AFB_1 [193–195]. This is because new biosensors directly use nanomaterials or other materials containing nanoparticles that show significant characteristics, such as high sensitivity and specificity for detecting targets, reliability, and consistency of products [181,196,197]. Nanomaterials significantly increase the effective surface area of biosensors and further improve the analytical performance [60,194]. Nanomaterials also enhance some characteristics of biometric elements in biosensor devices in terms of electrical, catalytic, optical, and thermal properties [198]. According to previous studies, some of the key functional enhancements are the enhanced immobilization of biomolecules, generation and expansion of analytical signals, and enhanced usability of fluorescent labels.

Characteristic of Nanomaterials Based Electrochemical Biosensors

The role of nanomaterials in biosensors is mainly reflected in the immobilization of biomolecules, signal generator, fluorescent labeling, and signal amplification.

Nanomaterials not only immobilize biomolecules but also increase the interaction between different molecular materials. In addition, nanomaterials enhance the stability of biomolecular immobilization, thereby increasing the signal strength of the immunoassay [31]. Metal nanomaterial particles, such as AgNPs and MOFs, can increase the surface area and biocompatibility of biomolecules bound to the detection target. However, non-metallic nanomaterials show negatively charged functional groups, which can be used as an effective carrier to bind and fix with positively charged targets.

Signal Generator

Xue et al. [31] reported that, when the photoelectric signal changes, nanoparticles such as gold and silver can act as a signal generator. By adjusting the fluorescence signal generated by nanomaterials, a new AFB_1 nanoprobe can be constructed. In addition, because these nanoparticles can be prepared in different sizes according to need, they have good functionality, stability, and scalability [133,199].

Fluorescent Label

Nanomaterials have unique optical properties that enable them to be widely used in a variety of disciplines, especially in the detection of hazardous substances in food. Nanomaterials can detect AFB_1 by sensing optical signals (absorbance, chemiluminescence, fluorescence, etc.) [31]. Some nanomaterials, such as metal nano-ions and quantum dots, have been used as fluorescence quenching agents because of the ability of AFB_1 to directly quench or reduce the fluorescence intensity. In addition, quantum dots have transformed fluorescein into a marker element that binds to aptamers or antibodies.

Signal Amplification

Nanomaterials can also be used as functional materials for various electrodes, signal components, etc., to amplify signals in various ways. For example, on the electrode surface of electrochemical sensors, nanomaterials such as gold and silver can amplify the analytical signal by enhancing the redox reaction. Some metal nanoparticles, such as gold, can amplify signals related to their characteristics, such as unique catalytic activity, biocompatibility, and multiple absorption sites. Carbon, graphene, and other non-metallic nanomaterials improve the analytical performance by increasing the surface area.

4.5. Bioinspired Recognition Elements for Biosensors

A biosensor is independent quantitative analysis equipment used to study the analytes required in different types of food samples. A biosensor consists of many parts [29,30] (Figure 6). Biometric elements are the core components of biosensors and can detect specific target analytes. The quality of biometric elements usually determines the specificity and sensitivity of analysis [200,201]. Biorecognition elements, including antibodies, aptamers, molecularly imprinted polymers, and enzymes, have been used to manufacture biosensors [34,64,202]. These elements show increased sensitivity and selectivity for target analytes. Critical biometric elements for developing biosensors to detect AFB_1 in edible vegetable oil are elaborated below.

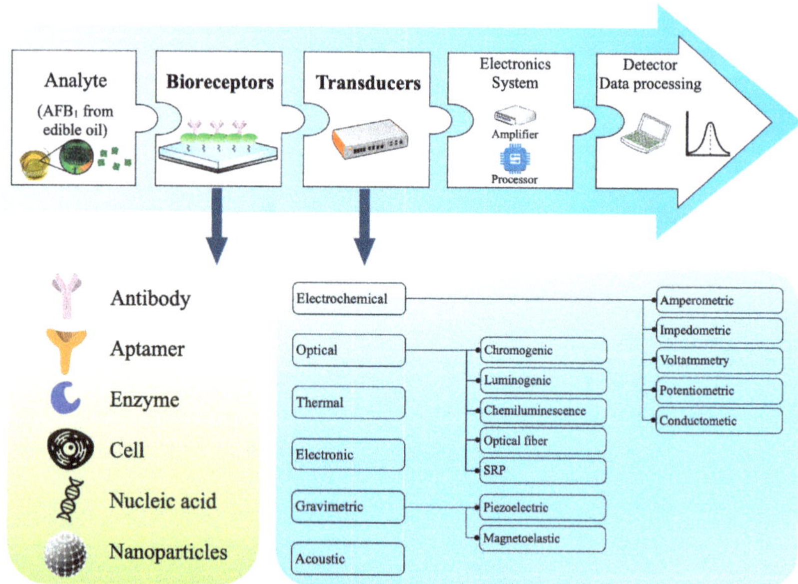

Figure 6. Schematic diagram of typical biosensor. Including analyzer, bioreceptor, transducer, electronic system (amplifier and processor), detector (for data processing).

4.5.1. Antibody

Antibodies have been used as recognition elements for developing biosensors because of their specificity and sensitivity [200]. Biosensors that use antibodies as recognition elements are called immunosensors, and their mechanism relies on the specific recognition of aflatoxin epitopes by antibodies.

The first batch of polyclonal antibodies, developed in 1976, became the basis for most mycotoxin detection methods. In the following decades, polyclonal and monoclonal antibodies were the basis for most mycotoxin detection methods [192,200,203]. Today, in addition to monoclonal and polyclonal antibodies, various other types of antibodies are being used to detect target analytes. Researchers have developed an antibody-based immunosensor that can directly recognize AFB_1 and is used in peanut oil with a concentration range of 0.001 to 100 ng mL^{-1}, with a detection limit of 0.2 pg mL^{-1} [192].

However, the production of monoclonal antibodies and polyclonal antibodies is complex and the antibodies degrade, denature, and aggregate easily [204,205]. In recent years, with the development of protein- and DNA-based new engineering technology, it has become possible to develop modified and recombinant antibodies (RAbs). RAbs integrate many advantages of biosensors, such as simple operation and a high degree of automation, high throughput screening, low requirements for configuration attributes, and the trend of more miniaturization [200]. Zhao et al. [206] developed a novel method of MB-dcELISA for AFB_1 based on the mimotope of an RAb and nanobody. This study effectively proved that compared with monoclonal antibodies, an RAb is more economical and easier to prepare. Compared with chemically synthesized toxic antigens, immunoassay is safer and performs better in validation studies. In real samples (corn germ oil and peanut oil), the LOD of AFB_1 is as low as 0.13 ng mL^{-1}. Other researchers also designed an RAb with increased sensitivity to low-molecular-weight haptens, and this RAb was validated in olive oil with a lower LOD (0.03 ng mL^{-1}) for AFB_1 [190].

Researchers recently found that, by increasing the immobilization of antibodies and giving full play to the characteristics of specific antibodies, the performance of the sensor can be effectively improved, and on this basis, some immunosensors have been developed for detecting AFB_1 in edible oil [133,196]. For example, to determine AFB_1, Shi et al. [207] proposed a novel immobilized immunosensor based on graphene supported with hybrid gold nanoparti-cles-poly4-aminobenzoic acid. In the study, after the reduction in graphene oxide by PABA via an epoxy ring opening reaction, the nanocomposite PABA-r-GO was obtained. Then, gold nanoparticles (AuNPs) were prepared on this basis to form a Au-PPABA-r-GO nanohybrid. The final sensor was obtained by the covalent binding of the COOH group of functional nanocomposites with an AFB_1-specific antibody. The sensor has good performance (linear range 0.01–25 ng mL^{-1} and LOD 0.001 ng mL^{-1}) and has been successfully applied to detect real vegetable oil. This sensor also has good reproducibility and selectivity, especially stability, and can be stored at a low temperature for a long time.

4.5.2. Aptamers

Aptamers are single-stranded RNA or DNA (20–90 oligonucleotide sequences with specific sequences) that can bind to various targets, such as ions, antibodies, proteins, cells, and organic molecules [208]. The particular recognition ability of aptamers relies on the three-dimensional structure of a high-affinity target-induced DNA three-dimensional structure. The researcher procured specific targets for aptamers by screening oligonucleotides using the Phylogenetic Evolution of Ligands for Exponential Enrichment (SELEX) program. Aptamer sensors are biosensors integrated with aptamers developed in the 1990s [200,208–211].

Recently, aptamers have attracted significant attention in food contamination analysis and are used for various sensing applications due to their inherent benefits: (1) aptamers are obtained from in vitro synthesis, so animals are not necessary; (2) aptamers have lower toxicity, immuno-genicity, and production cost; (3) aptamers have enhanced chemical and thermal stability; (4) aptamers have excellent batch-to-batch reproducibility; (5) aptamers have a smaller size and show a remarkable ability to penetrate the tissue and adhere to target molecules; and (6) it is possible to change their structure [193,212,213].

Notably, the immobilization of aptamers is a critical step in biosensor design because it can affect the affinity of aptamers for their targets and their long-term stability in fundamental sample analysis.

Therefore, researchers have developed many strategies to immobilize aptamers: (1) adsorption or π–π stacking interactions between DNA bases and modified graphene oxide (GO) interface [214], (2) aptamers with carboxylic acids on surfaces or nanomaterial covalent bonding of groups [184,215], (3) binding of sulfide aptamer with CdTe quantum dots (QDs) or gold-based materials [216], (4) binding with avidin or other affinity interactions based on biotin streptomycin affinity [200,217,218], and (5) hybridization with partially complementary single-stranded DNA previously fixed on the surface of nanoparticles [219–222].

Table 2 describes some examples of aptasensors recently reported for the detection of AFB_1 in edible oils. About half of the previous reports have been based on fluorescent mycotoxin aptamer sensors. Some of them use metal or nanostructured materials, such as gold nanoparticles (AuNPs), GO, single-walled carbon nanotubes, or TiO_2 tubes, and are used to prepare aptamer sensors.

Nanometer material has always been the focus of research, and its applications in biosensors are also diverse. Black phosphorus nanosheets (BPNSs) have great application prospects in biosensors due to their unique characteristics [223]. Wu et al. [224] developed a highly specific and sensitive aptamer sensor (UCNPs-BPNSs) based on the team's research on upconversion nanoparticles (UCNPs) [196]. The research team attached UCNPs to the surface of BPNSs at a very small space distance (less than 10 nm) through glutaraldehyde crosslinking method and π-π stacking effect method, and then constructed the fluorescence resonance energy transfer (FRET) system. This aptamer sensor can effectively detect AFB_1 in peanut oil and other foods quantitatively with good linear range (0.2–500 ng mL^{-1}) and LOD (0.028 ng mL^{-1}).

Xia et al. [225] proposed a label-free, single-tube, homogeneous, and inexpensive assay for AFB_1 based on fine-tunable double-ended stem aptamer beacons (DS) and the effect of aggregation-induced emission (AIE). The structure of the DS aptamer beacon can provide end protection against exonuclease I (EXO I) to the aptamer probe and endow it with specificity and a rapid response to the target AFB_1. Compared with the traditional molecular beacon structure, the stability of the DS aptamer beacon can be adjusted by adjusting its two terminal stems so that the affinity and selectivity of the probe can be precisely optimized. Using an AIE-active fluorophore, which is illuminated by the aggregation of negatively charged DNA, AFB_1 can be measured label-free. The method has been successfully applied to the analysis of AFB_1 in peanut oil, with a total recovery of 93.59–109.30%. Therefore, beacon-based DS assays may help in real-time monitoring and control of AFB_1 contamination.

Yang et al. [226] first devised a selection method based on rational truncation and post-splicing and developed a bivalent anti-AFB_1 chimeric aptamer (B72) that was measured by micro-thermophoresis (MST) compared to the initial selection. The affinity of the anti-AFB_1 aptamer (B50) increased by 188-fold, and the study also found that B72 has a dual binding site for AFB_1, which is consistent with the experimental results obtained by isothermal titration calorimetry (ITC) and molecular docking simulations. Therefore, on the basis of the peroxidase-like activity of gold nanoparticles catalyzing 3,3,5,5-tetramethylbenzidine (TMB), an aptamer sensor of gold nanoparticles (AuNPs) was developed by the colorimetric detection of AFB_1. The assay further validates the practical applicability of the chimeric aptamers. The aptasensor could identify AFB_1 with an excellent linear range (5–5120 nM) and detection limit (1.88 nM) in the corn oil environmental test of H_2O_2. Therefore, this study can be called a general selection method for designing high-affinity aptamers and constructing novel aptamer-based biosensing platforms for high-sensitivity and specificity analysis of other targets.

Zhong et al. [227] manufactured an electrochemical aptamer sensor in a similar way for the sensitive detection of AFB_1. The researchers used electrodeposited AuNPs to prepare AuNPs/ZIF-8 nanocomposites on glassy carbon electrodes (GCEs) decorated with the eight zeolite imidazolate framework (ZIF-8), which increased the surface area of the electro desorption molecular load. Compared with other previously reported sensors, the aptasensor developed under optimized conditions shows a more comprehensive linear range (10.0–1.0 $\times$ 10^5 pg mL^{-1}) and a lower detection limit (1.82 pg mL^{-1}). In addition, the constructed aptasensor possesses excellent selectivity, reproducibility, and stability. Moreover, the aptamer sensor has been successfully used to detect AFB_1 in corn oil and peanut oil samples, and the recovery was between 93.49% and 106.9%, which proves the potential application value of this method. Researchers are very interested in this kind of electrochemical aptamer sensor. Wang et al. [228] also developed an AFB_1 electrochemical aptamer sensor for detecting peanut oil in a similar way. The difference lies in the use of different composite materials (zinc and nickel bimetallic organic skeleton materials).

The hybridization chain reaction (HCR) is a commonly used isothermal nucleic acid amplification technique, and due to the characteristics such as no enzyme, high amplification efficiency etc., HCR is usually used as a new synthetic material technology and is widely used in various sensors. Researchers have fully combined the characteristics of HCR to build a signal amplification strategy, which has been successfully applied to the sensitive detection of AFB_1 [229–235]. Wang et al. [236] proposed a fluorescent aptamer sensor based on DNA walker, DNA tetrahedral nanostructures (DTNs) and network HCR. Among them, DNA walker was used as the signal amplifier induced by AFB_1 target, and combined with self-assembled DTNs. Finally, based on network HCR, signal amplification is realized and sensitive detection of AFB_1 in peanut oil was realized with with LOD of 0.492 pg mL^{-1} and the linear range of 1–1000 pg mL^{-1}. In the other report, Zuo et al. [237] combined DNAzyme with substrate chain (Zn-Sub) and enzyme chain (Zn-Enz) with HCR products to form a Y-shaped structure, which can significantly enhance the fluorescence intensity of the detection target. The fluorescent aptamer sensor proposed by researchers shows excellent performance with LOD of 0.22 nmol L^{-1} and the linear range of 0.4–16 nmol L^{-1}.

The emerging quantum dots (QDs), represented by carbon quantum dots (CQD_S), graphene quantum dots (GOD_S) etc., have attracted great attention and are widely used in various sensor fields since their discovery because of their excellent optical properties, low toxicity, stability and low cost etc. [238–240]. QDs-based sensors can adopt different working mechanisms and be applied to detect different substances, including AF_S in edible oil [51,173,238]. According to the characteristics of QDs, Xuan et al. [185] and Ye et al. [241] developed and constructed different magnetic control pretreatment platforms, which were actually applied to the detection of AFB_1 in peanut oil and agricultural products, and both showed good detection characteristics. Other researchers used a quencher system composed of quantum dots and graphene oxide to detect AFB_1 in peanut oil, which also showed good detection characteristics [242].

SERS

As mentioned above, SERS is a promising analytical tool with many advantages over traditional AFB_1 detection methods, including high sensitivity, easy sample preprocessing, and non-destructive testing [166,174,243]. Compared with antibodies, aptamers have the advantages of low cost, easy synthesis, good stability and strong specificity to target molecules. With the mature development of aptamer manufacturing technology, they have gradually become one of the most potential recognition elements in SERS labeling detection.

Recently, several authors have combined advanced composite materials with SERS aptamer sensors to develop new procedures for AFB_1 detection. For example, on the basis of the combination of a multifunctional capture probe (Fe_3O_4@Au report the strong Raman signal of probe 1 (AU)-4MBA@AgNSs-Apt), an ultrasensitive assay was successfully developed for a high-performance SERS aptamer sensor of AFB_1. He et al. [174] reported that, in the presence of AFB_1, the probe was released from the capture probe, resulting in a decrease in SERS intensity, possibly due to the specific binding affinity between the aptamer and AFB_1. For AFB_1 detection, a wide linear range, from 0.0001 to 100 ng mL^{-1}, was obtained, with an R^2 of 0.9911, and the LOD was calculated as 0.40 pg mL^{-1}. Finally, after extracting AFB_1 from peanut oil samples, the SERS aptamer sensor was successfully applied to the analysis of AFB_1, and the recovery was between 96.6% and 115%. Therefore, the novel SERS aptamer sensor is a promising analytical tool for detecting AFB_1 in actual samples.

In the report by Yang et al. [169], with the help of the specific interaction between AFB_1 and aptamer, a novel SERS-based universal aptamer sensor platform was constructed to detect AFB_1. First, gold nanotriangle (GNT)-DTNB@Ag-DTNB nanotriangles (GDADNTs) were synthesized and used as SERS active substrates. These magnetic beads and amino-terminal-aptamer-conjugated magnetic beads (CS-Fe_3O_4) were then used as capturer and reporter of AFB_1, respectively. Finally, the platform showed excellent sensitivity under optimized assay conditions, with a lower LOD (0.54 pg mL^{-1}) and a more comprehensive linear range (0.001–10 ng mL^{-1}). In addition, the high stability of SERS substrate activity was maintained for at least three months, with an RSD of ~5%, which has good selectivity for general coexistence interference. The excellent sensitivity and selectivity of micro-AFB_1 detection are mainly due to the substantial Raman-enhancing effect of GNTs as the core of GDADNTs, which results from the bilayer of reporter molecules, aptamer specificity, and the super-paramagnetic CS-Fe_3O_4, respectively. The researchers also evaluated and confirmed that the established SERS aptamer sensor can be used to detect AFB_1 in peanut oil samples.

In a subsequent study, on the basis of previous research, another simple and sensitive SERS aptamer sensor was developed for detecting AFB_1 in peanut oil [170]. In this study, the researchers used an aminoterminal AFB_1 aptamer (NH2-DNA1) as a SERS aptamer sensor, magnetic beads conjugated to a thiol-terminal-complementary AFB_1 aptamer (SH-DNA2) (CS-Fe_3O_4) as enrichment nanoparticle probes, and AuNR@DNTB@Ag nanorods (ADANR) as reporter nanoprobes. 5,5′-Dithiobis (2-nitrobenzoic acid) (DNTB) is embedded in gold and silver core/shell nanorods as a Raman reporter molecule, which has a large Raman scattering cross section and no fluorescence interference. Furthermore, CS-Fe_3O_4 has good biocompatibility and superparamagnetism, which can quickly enrich signals. Therefore, NH2-DNA1-CS-Fe_3O_4 and SH-DNA2-ADANRs were prepared by a mixed reaction between aptamers and complementary aptamers. When present, AFB_1 will compete with NH2-DNA1-CS-Fe_3O_4 to induce SH-DNA2-ADANRs to dissociate from CS-Fe_3O_4, further reducing SERS signals. According to the SERS aptamer sensor, the lower detection limit of AFB_1 is 0.0036 ng mL^{-1} and the correlation coefficient is as high as 0.986. The effective linear detection range is 0.01–100 ng mL^{-1}, obtained with a correlation coefficient as high as 0.986. Finally, the specificity and accuracy of the SERS aptasensor were proved by detecting AFB_1 in natural peanut oil.

Similar research strategies are reflected in other reports. Jiao et al. [244] developed a gold-silver core-shell nanoparticles (Au@Ag CSNPs) SERS sensor decorated with 5-aminotetramethylrhodamine (NH_2-Rh). Based on the optimization of experimental conditions, the sensor can be combined with solid phase extracts of peanut oil, hazelnut, and other samples to achieve a quantitative analysis of AFB_1 with detection range and LOD 0.1–5.0 ng·mL^{-1} and 0.03 ng·mL^{-1}, respectively.

Various AFB_1 sensors are also identified in edible vegetable oil by electrochemical detection, which has some unique advantages, such as low cost, high sensitivity, and the possibility of micromachining. For example, Xiong et al. [245] revealed a highly innovative method based on dual-DNA-tweezer nanomachines to detect AFB_1 in olive and peanut oils. Wu et al. [246] presented a method based on ferrocene and β-cyclodextrin (β-simple electrochemical aptamer sensor for host–guest recognition between CD) to detect AFB_1 in peanut oil, with a low LOD (0.049 pg mL^{-1}).

4.5.3. Molecularly Imprinted Polymers (MIPs)

MIPs have been used as recognition elements to develop biosensors, and synthetic polymers have displayed precise target recognition [133,202,247]. These artificial materials can recognize specific targets in complex mixtures because of specific recognition sites for binding or catalysis and functional groups with shapes and geometries complementary to those of the template molecule. These polymers self-assemble with template molecules and active/functional monomers through the polymerization of cross-linking agents. Therefore, when the template molecule is removed, pores with multiple active sites appear in the polymer, which match the spatial configuration of the template molecule [248,249]. In recent years, traditional MIPs have been applied in many cross fields, such as chromatography, drug delivery, solid-phase extraction, controlled release, bioremediation, and sensors [200,250–256]. In AFB_1 detection studies, MIP-based biosensors have shown many advantages, such as unique selectivity, sensitivity, user-friendliness, and cost-effectiveness [32,42,200,202]. For instance, Li et al. [173] exploited MIPs by preparing an electrochemiluminescence (ECL) platform for AFB_1 detection with an ultra-low LOD, of 8.5 fg mL^{-1}, and a wide linear range (10^{-5} to 10 ng mL^{-1}). While the MIP–ECL platform was used, the recovery rate of corn oil samples was close to that obtained by HPLC, indicating the reliability of the sensor and its potential in food safety evaluation. It is worth mentioning that, as of the publication of this review, this is the lowest LOD of AFB_1 in edible oil.

However, MIP-based biosensors also have some disadvantages, such as generally poorer affinity and specificity than antibodies, slower binding kinetics than biological receptors, incomplete template elimination, and lower utilization of binding sites [133,200,202]. Therefore, there is increasing interest in developing improved MIPs [257–259]. The key to the success of the sensor of an MIP is whether the MIP is effectively attached to the transducer. Three commonly used immobilization methods are in situ polymerization, electropolymerization, and physical coating. Additionally, the number of applications of MIP sensors for detecting AFB_1 in edible vegetable oil is limited [200,202,260].

Table 2. Techniques used for the detection of AFB_1 in different types of edible oil.

Matrix	Analytical Method	Sample Preparation Method	Linear Range	Recovery	LOD	Ref.
Peanut oil	ELISA	Immunoaffinity column cleanup	-	84.40–92.60%	1.08 μg kg^{-1}	[20]
Coconut oil	HPLC-FLD	Immunoaffinity chromatography	-	-	0.01–0.04 μg kg^{-1}	[81]
Peanut oil Corn oil Soybean oil	HPLC-FLD	DLLME with in situ derivatization	0.1–100 ng mL^{-1}	106.90–121.50%	0.03 ng mL^{-1}	[82]
Sunflower oil Olive oil Canola oil Frying oil Blend oil	HPLC-FLD	Immunoaffinity column cleanup	0.04–0.16 ng g^{-1}	95.56–102.13%	0.16 ng g^{-1}	[83]
Vegetable oil	HPLC-FLD	Immunoaffinity chromatography - Solid phase extraction	-	95.20–99.00%	0.25 μg kg^{-1}	[84]
Coconut oil Almond oil Soybean oil Olive oil	HPLC -FLD	Immuno Affinity chromatography combined with DLLME	0.005–10.00 ng mL^{-1}	96.00–109.90%	830 ng mL^{-1}	[85]
Sesame oil Sunflower oil Peanut oil Mixed oil	HPLC-FLD	Liquid-Liquid Extraction	0.34–109.2 μg kg^{-1}	83.00–96.00%	0.09–1.5 μg kg^{-1}	[86]
Canola oil Corn oil Olive oil Peanut oil Soybean oil	LC-MS/MS	Liquid-Partitioning	-	101.00–111.00%	0.030 μg kg^{-1}	[87]
Soybean oil Corn oil Rice bran oil	LC-MS/MS	QuEChERS × DLLME	-	70.70–76.00%	–	[88]
Blend oil Peanut oil Maize oil	LC-MS/MS	Hollow fiber liquid phase microextraction	0.1–500 μg kg^{-1}	78.59–80.61%	0.02 μg kg^{-1}	[89]
Sunflower oil Palm oil Corn oil	LC-ESI-MS/MS	QuEChERS	0.04–2000 ng g^{-1}	87.90–106.60%	0.01 ng kg^{-1}	[90]
Soybean oil Corn oil Peanut oil Blended oil	LC–MS/MS	Immunoaffinity chromatography	0.16 μg kg^{-1}	87.40–97.30%	0.05 μg kg^{-1}	[96]
Olive oil Peanut oil Sesame oil	LC-MS/MS	Immunoaffinity column cleanup	2–20 mg kg^{-1}	87.70–102.20%	0.1 μg kg^{-1}	[97]
Olive oil	LC/ESI-MS/MS	Matrix Solid Phase Dispersion	0.2–0.4 (pg inj)	95.00–98.00%	0.2 pg inj	[261]
Olive oil Sunflower oil Soybean oil Corn oil	UHPLC-QqQ-MS/MS	QuEChERS	0.5–25 μg kg^{-1}	96.00–107.90%	0.5 μg kg^{-1}	[100]
Sesame oil Groundnut oil Cottonseed oil	HPLC	Liquid-Liquid extraction	0.2–0.8 μg kg^{-1}	-	0.1 μg kg^{-1}	[102]

Table 2. *Cont.*

Matrix	Analytical Method	Sample Preparation Method	Linear Range	Recovery	LOD	Ref.
Vegetable oils	GPC-HPLC-FLD	Liquid-Liquid Extraction	1.0–30.0 μg kg^{-1}	82.60–90.60%	1.0 μg kg^{-1}	[104]
Peanut oil Sunflower oil Olive oil	IAC-LC-ESI–MS/MS	Liquid-Liquid Extraction	0.02–10 μg kg^{-1}	84.00–99.00%	0.02 μg kg^{-1}	[105]
Virgin olive oil	HPLC-FLD	Solid Phase Extraction		65.50–87.50%	0.25 ng g^{-1}	[110]
Canola oil Soybean oil Corn oil Olive oil Peanut oil	FL	LTC-IMSPE	0.0048–0.0126 ng·g^{-1}	79.60–117.90%	0.0048 ng·g^{-1}	[115]
Rapeseed oil Peanut oil Blended oil Blended olive oil Sunflower oil Tea oil Rice oil Corn oil Sesame oil Soybean oil	HPLC-MS/MS	QuEChERS	0.2–20 ng mL^{-1}	87.80–98.60%	0.05 ng g^{-1}	[116]
Soya bean oil Groundnut oil Beniseed oil Palm kernel oil Melon oil Coconut oil	ELISA	Immunoaffinity column cleanup (226 Aflatoxin clean-up Column)	-	-	≤0.8352 μg L^{-1}	[179]
Peanut oil Virgin olive oil	ELISA& TSA-ELISA	Liquid-Liquid Extraction	-	81.40–118.80%	0.004 ng mL^{-1}	[180]
Olive oil	Amperometric biosensor coupled with AChE enzyme	Liquid-Liquid Extraction	10–60 ppb	76.00–78.00%	2 ppb	[186]
Olive oil	EIS based on MWCNTs/RTIL composite films	Liquid-partitioning	0.1–10 ng mL^{-1}	96.00–116.00%	0.03 ng mL^{-1}	[190]
Peanut oil	Disposable electrochemical immunosensor with Au NPs modified SPCE	-	0.001–100 ng mL^{-1}	90.00–102.00%	0.2 pg mL^{-1}	[192]
Peanut oil	Fluorescence spectroscopy based on UCNPs upconversion nanoparticles (UCNPs)	Liquid-partitioning	0.2–100 ng mL^{-1}	92.80–113.40%	0.2 ng mL^{-1}	[196]
Oil	Chemiluminescence immunoassay combined with the magnetic particles (MPCLIA)	Liquid-partitioning	0.1–100 ng mL^{-1}	85.67–108.67%	0.05 ng mL^{-1}	[197]
Corn germ oil Peanut oil	An immunoassay based on both recombinant antibody and its mimotope	Liquid-Liquid extraction	0.242.21 ng mL^{-1}	86.70–116.20%	0.13 ng mL^{-1}	[206]

Table 2. *Cont.*

Matrix	Analytical Method	Sample Preparation Method	Linear Range	Recovery	LOD	Ref.
Vegetable oil	Immobilized immunosensor based on the hybrid gold nanoparticles-poly 4-aminobenzoic acid supported graphene	-	0.01–25 ng mL^{-1}	-	0.001 ng mL^{-1}	[207]
Peanut oil	UCNPs-BPNSs aptamer	-	0.2–500 ng mL^{-1}	92.89–99.24%	0.028 ng mL^{-1}	[224]
Peanut oil	Dual-terminal stemmed aptamer beacon, aggregation-induced emission	Liquid-Liquid Extraction	40–300 ng mL^{-1}	93.59–109.30%	27.3 ng mL^{-1}	[225]
Corn oil	A chimeric aptamer-based gold nanoparticles aptasensor	-	5–5120 nM	91.50–117.60%	1.88 nM	[226]
Corn oil Peanut oil	An electrochemical aptasensor base on an AuNPs/ZIF-8 nanocomposite	-	10.0–1.0 × 10^5 pg mL^{-1}	93.49–106.90%	1.82 pg mL^{-1}	[227]
Peanut oil	An electrochemical aptasensor base on an AuNPs/Zn/Ni-ZIF-8–800@ graphene nanocomposite	-	0.18–100 ng mL^{-1}	80.26–109.60%	0.18 ng mL^{-1}	[228]
Oil	An aptasensor of hybridization chain reaction and Zn^{2+}-dependent DNAzyme catalyzed cleavage	-	0.4–16 nmol L^{-1}	92.20–107.80%	0.22 nmol L^{-1}	[237]
Oil	Fabricating electrochemical aptasensors	-	0.04–0.10 ng m L-1	94.5–103.3%	0.002 ng m L^{-1}	[184]
Peanut oil	Electrochemical immunosensor base on AFB_1-BSA-QDs	-	0.08–80 μg kg^{-1}	102.70–113.30%	0.05 μg kg^{-1}	[185]
Peanut oil	SERS aptasensor	-	0.0001–100 ng·mL^{-1}	96.60–115.00%	0.40 pg·mL^{-1}	[174]
Peanut oil	SERS aptasensor	-	0.001–10 ng mL^{-1}	94.70–109.00%	0.54 pg mL^{-1}	[169]
Peanut oil	Q-dots-aptamer-GO fluorescence quenching system	-	1.6–160 μM	-	1.4 nM	[242]
Peanut oil	Atomic absorption spectroscopy	-	2.5–240 μg kg^{-1}	-	0.04 μg kg^{-1}	[241]
Peanut oil	SERS aptasensor	–	0.01–100 ng mL^{-1}	91.09–105.73%	0.0036 ng mL^{-1}	[170]
Peanut oil	SERS aptasensor with NH_2-Rh-Au@Ag CSNPs	Solid Phase Extraction	0.1–5.0 ng mL^{-1}	-	0.03 ng mL^{-1}	[244]

Table 2. *Cont.*

Matrix	Analytical Method	Sample Preparation Method	Linear Range	Recovery	LOD	Ref.
Olive oil Peanut oil	Dual DNA tweezers nanomachine	-	0.08–10 ppb	90.00–110.00%	0.035 ppb	[245]
Peanut oil	Electrochemical aptasensor based on smart host-guest recognition of β-cyclodextrin polymer	-	0.1×10^{-4}–10 ng mL^{-1}	94.50–106.70%	0.049 pg mL^{-1}	[246]
Peanut oil	A dual signal amplified aptasensor based on DNA walker, (DTNs) and network (HCR)	-	1−1000 pg mL^{-1}	87.56–105.28%	0.492 pg mL^{-1}	[236]
Corn oil	A novel fluorescence aptasensor based on mesoporous silica nanoparticles	-	0.5–50 ng mL^{-1}	90.30–92.40%	0.13 ng mL^{-1}	[262]
Peanut oil	Dual-terminal proximity aptamer probes	-	1.0–200 ng mL^{-1}	90.30–102.91%	0.9 ng mL^{-1}	[263]
Sesame oil Olive oil Peanut oil Soybean oil	An aptamer-based MCE-LIF	-	0.05–5.0 ng mL^{-1}	95.29–109.19%	0.026 ng mL^{-1}	[264]
Peanut oil	A simple fluorescent AFB_1 sensor based on a humic acid/carbon dots system	-	0.1–0.8 ng mL^{-1}	103.80–108.00%	70 pg mL^{-1}	[171]
Peanut oil	SERS aptasensor	-	0.01–100 ng mL^{-1}	90.40–113.10%	5.0 ng mL^{-1}	[172]
Edible oil	Immunoaffinity chromatography fluorometer	Immunoaffinity column clean-up	1.0–32.2 μg kg^{-1}	-	1 μg kg^{-1}	[265]
Corn oil	An MIP-ECP-ECL sensing platform based on $CH_3NH_3PbBr_3$ quantum dots (MAPB QDs)@SiO_2	-	10^{-5}–10 ng mL^{-1}	102.00–110.00%	8.5 fg mL^{-1}	[173]
Soybean oil	Terahertz spectroscopy (photoelectric techniques)	-	-	-	2 μg kg^{-1}	[162]
Peanut oil Corn oil	ELC based on Escherichia coli	-	0.01–0.3 μg mL^{-1}	90.00–112.00%	1 μg mL^{-1}	[266]

5. Conclusions and Perspectives

Mycotoxin contamination, especially AFB_1 contamination in edible oil, is usually unavoidable. A more sensitive and rapid sensor-based early warning tool for AFB_1 detection would help to reduce risk. Various traditional, modern, and biosensing technologies have been used to detect toxins in contaminated food. Spectroscopic techniques, chromatographic techniques are general methods for the detection of AFB1 in edible oils. In recent years, based on the cross-integration of multiple disciplines, the innovation, progress and development of general methods have also been promoted. Although traditional chromatographic techniques can effectively detect mycotoxins, their performance in all aspects cannot achieve satisfactory results. Combined use with other sensor equipment can effectively improve reliability, sensitivity and accuracy. However, due to the high cost of equipment, on-site inspection cannot be performed, and sample pretreatment is required, which limits the use of chromatography technology in the detection of AFB1 in edible oil. The development of spectroscopic techniques has become increasingly diverse and can effectively detect mycotoxins, especially AFB1 in edible oils. However, these methods are not suitable for on-site detection, because they still have many shortcomings, such as low sensitivity and reliability, and the need for professional personnel to operate.

Unlike conventional detection techniques, novel biosensors show high accuracy, sensitivity, and specificity; better cost controllability and portability; and reliability and simplicity in operation.

This review also discusses the development of important recognition elements in sensors. The recognition element of the sensor should have sensitivity and specificity sufficient enough to detect small amounts of target toxins, even in samples with complex matrix systems. The development and use of nanomaterials further improve the efficiency of biosensor conversion systems, but these require further improvements in their sensitivity, selectivity, and reproducibility. Of course, the stability and cost will also affect the selection of identification elements, which can improve the practicability.

Despite significant progress in biosensors for the detection of AFB_1, there are some problems and challenges in the future. (1) The recognition elements of biosensors (such as metal nanoparticles, quantum dots, and graphene) improve the efficiency of sensing systems, but all these require further improvements in terms of sensitivity, selectivity, and reproducibility. (2) Future studies can perform AFB_1 toxicity measurements and develop advanced nanomaterial-integrated biosensors to improve the overall detection of harmful substances, such as AFB_1, in contaminated food samples. (3) When detecting AFB_1 in contaminated food samples, researchers can focus on combining biosensing systems with microarray technology to fabricate more portable devices. (4) Reagent-free, clean-free, calibration-free, or nonbiological contamination biosensors for aflatoxin analysis require more effort and will reduce the possible future hazards.

Author Contributions: S.Y. Conceptualization, investigation, writing, review and editing of the manuscript; L.N. conceptualization and investigation of the manuscript; Y.L. review of the manuscript. All authors have read and agreed to the published version of the manuscript.

Funding: This research was financially supported by Key R&D project of Jiangsu Province (BE2021306); and Development Project (BE2021306) and the Shandong Province Key Research and Development Program (No. 2021CXGC010808).

Conflicts of Interest: The authors declare no conflict of interest.

Abbreviations

AChE	Acetylcholinesterase
Adanrs	Aunr@DNTB@Ag Nanorods
AFB_1	Aflatoxin B_1
AFB_2	Aflatoxin B_2
AFBO	Aflatoxin B_1–8,9-Epoxide
AFG_1	Aflatoxin G_1
AFG_2	Aflatoxin G_2
AFM_1	Aflatoxin M_1
AFM_2	Aflatoxin M_2
Afs	Aflatoxins
AIE	Induced Emission
AOAC	Association of Analytical Communities
AuNPs	Gold Nanoparticles
BPNSs	Black phosphorus nanosheets

CBNs	Carbon-Based Nanomaterials
CEN	European Committee for Standardization
CODEX	European Commission and The Codex Alimentarius Commission
CV	Cyclic Voltammetry
CQD$_S$	Carbon Quantum Dots
CSNPs	Core-Shell Nanoparticles
CYP	Cytochrome P
DLLME Dispersive	Liquid-Liquid Microextraction
DNTB	5,5′-Dithiobis (2-Nitrobenzoic Acid)
DS	dual-terminal stemmed
DTNs	DNA tetrahedral nanostructures
ECL	Electrochemiluminescence
EIS	Electrochemical Impedance Spectroscopy
ELISA	Enzyme-Linked Immunoassay
EU	European Union
EXO I	Exonuclease I
FAO	Food and Agriculture Organization
FRET	Fluorescence Resonance Energy Transfer
GCEs	Glassy Carbon Electrodes
Gdadnts	Gold Nanotriangles (Gnts)-DTNB@Ag-DTNB Nanotriangles
GO	Graphene Oxide
GOD	Graphene Quantum Dots
GPC	Gel Permeation Chromatography
GST	Glutathione-S-Transferase
HCR	Hybridization Chain Reaction
HBV	Hepatitis B Virus
HPLC-FD	High-Performance Liquid Chromatography Coupled with A Fluorescence Detector
HPLC-MS	High-Performance Liquid Chromatography Coupled with Mass Spectrometry
Iacs	Immunoaffinity Columns
IMSPE	Immuno magnetic solid phase extraction
IRs	Infrared spectroscopy
ITC	Isothermal Titration Calorimetry
KCN	Editpotassium Cyanide
LC-MS-MS	Liquid Chromatography-Tandem Mass Spectrometry
LIF	Laser-Induced Fluorescence
LLE	Liquid-Liquid Extraction or Partitioning
LOD	Limits of Detection
LTC	Immune Assay Extraction and Low Temperature Cleanup
MIPs	Molecularly Imprinted Polymers
MNPs	Metal Nanoparticles
MST	Microthermophoresis
NH2-DNA1	Amino-Terminal AFB1 Aptamer
NH_2-Rh	5-aminotetramethylrhodamine
NIR	Near Infrared
PABA-R-GO	Hybrid 4-Aminobenzoic Acid-Reduced Graphene Oxide
PB	Prussian Blue
QDs	Quantum Dots
QuEChERS	Quick, Easy, Cheap, Effective, Rugged, and Safe
RAbs	Recombinant Antibodies
RS	Raman spectroscopy
SELEX	Systematic Evolution of Ligand by the Exponential Enrichment Process
SERS	Surface-Enhanced Raman Spectroscopy
SH-DNA2	Complementary Aptamer
SPCE	Screen-Printed Carbon Electrode

SPE	Solid Phase Extraction
THz	Terahertz
TLC	Thin-Layer Chromatography
TMB	3,3,5,5-Tetramethylbenzidine
UCNPs	Upconversion Nanoparticles
UHPLC/UPLC	Ultra-High Performance Liquid Chromatography
UHPLC-Qqq-MS/MS	Ultra-High Performance Liquid Chromatography Coupled to A Triple Quadrupole Analyzer
US-FDA	US Food and Drugs Administration
WHO	The World Health Organization
β-CD	β-cyclodextrin

References

1. Shabeer, S.; Asad, S.; Jamal, A.; Ali, A. Aflatoxin Contamination, Its Impact and Management Strategies: An Updated Review. *Toxins* **2022**, *14*, 307. [CrossRef] [PubMed]
2. Skrzydlewski, P.; Twaruzek, M.; Grajewski, J. Cytotoxicity of Mycotoxins and Their Combinations on Different Cell Lines: A Review. *Toxins* **2022**, *14*, 244. [CrossRef]
3. Holban, A.M.; Grumezescu, A.M. *Microbial Contamination and Food Degradation*; Academic Press: London, UK, 2018; Volume 10.
4. Pickova, D.; Ostry, V.; Malir, F. A Recent Overview of Producers and Important Dietary Sources of Aflatoxins. *Toxins* **2021**, *13*, 186. [CrossRef] [PubMed]
5. Anater, A.; Manyes, L.; Meca, G.; Ferrer, E.; Luciano, F.B.; Pimpão, C.T.; Font, G. Mycotoxins and their consequences in aquaculture: A review. *Aquaculture* **2016**, *451*, 1–10. [CrossRef]
6. Wang, Y.; Nie, J.; Yan, Z.; Li, Z.; Cheng, Y.; Chang, W. Occurrence and co-occurrence of mycotoxins in nuts and dried fruits from China. *Food Control* **2018**, *88*, 181–189. [CrossRef]
7. Lee, H.J.; Ryu, D. Worldwide Occurrence of Mycotoxins in Cereals and Cereal-Derived Food Products: Public Health Perspectives of Their Co-occurrence. *J. Agric. Food Chem.* **2017**, *65*, 7034–7051. [CrossRef]
8. Azam, M.S.; Ahmed, S.; Islam, M.N.; Maitra, P.; Islam, M.M.; Yu, D. Critical Assessment of Mycotoxins in Beverages and Their Control Measures. *Toxins* **2021**, *13*, 323. [CrossRef]
9. Mahato, D.K.; Lee, K.E.; Kamle, M.; Devi, S.; Dewangan, K.N.; Kumar, P.; Kang, S.G. Aflatoxins in Food and Feed: An Overview on Prevalence, Detection and Control Strategies. *Front. Microbiol.* **2019**, *10*, 2266. [CrossRef]
10. Ünüsan, N. Systematic review of mycotoxins in food and feeds in Turkey. *Food Control* **2019**, *97*, 1–14. [CrossRef]
11. Ahlberg, S.H.; Joutsjoki, V.; Korhonen, H.J. Potential of lactic acid bacteria in aflatoxin risk mitigation. *Int. J. Food. Microbiol.* **2015**, *207*, 87–102. [CrossRef]
12. Jard, G.; Liboz, T.; Mathieu, F.; Guyonvarc'h, A.; Lebrihi, A. Review of mycotoxin reduction in food and feed: From prevention in the field to detoxification by adsorption or transformation. *Food Addit. Contam. Part A* **2011**, *28*, 1590–1609. [CrossRef] [PubMed]
13. OECD/FAO. *OECD-FAO Agricultural Outlook 2022–2031*; OECD Publishing: Paris, France, 2022. [CrossRef]
14. Mao, X.; Yan, A.; Wan, Y.; Luo, D.; Yang, H. Dispersive Solid-Phase Extraction Using Microporous Sorbent UiO-66 Coupled to Gas Chromatography-Tandem Mass Spectrometry: A QuEChERS-Type Method for the Determination of Organophosphorus Pesticide Residues in Edible Vegetable Oils without Matrix Interference. *J. Agric. Food Chem.* **2019**, *67*, 1760–1770. [CrossRef] [PubMed]
15. Pitt, J.I.; Taniwaki, M.H.; Cole, M.B. Mycotoxin production in major crops as influenced by growing, harvesting, storage and processing, with emphasis on the achievement of Food Safety Objectives. *Food Control* **2013**, *32*, 205–215. [CrossRef]
16. Vasseghian, Y.; Moradi, M.; Dragoi, E.-N.; Khaneghah, A.M. A review on mycotoxins detection techniques in edible oils. *Int. J. Environ. Anal. Chem.* **2020**, *102*, 2125–2139. [CrossRef]
17. Javanmardi, F.; Khodaei, D.; Sheidaei, Z.; Bashiry, M.; Nayebzadeh, K.; Vasseghian, Y.; Mousavi Khaneghah, A. Decontamination of Aflatoxins in Edible Oils: A Comprehensive Review. *Food Rev. Int.* **2020**, *38*, 1–17. [CrossRef]
18. Shavakhi, F.; Rahmani, A.; Piravi-Vanak, Z. A global systematic review and meta-analysis on prevalence of the aflatoxin B1 contamination in olive oil. *J. Food. Sci. Technol.* **2022**, 1–10. [CrossRef]
19. He, X.; Zhang, Y.; Yang, X.; Chen, M.; Pang, Y.; Shen, F.; Fang, Y.; Liu, Q.; Hu, Q. Estimating bulk optical properties of AFB1 contaminated edible oils in 300-900 nm by combining double integrating spheres technique with laser induced fluorescence spectroscopy. *Food Chem.* **2022**, *375*, 131666. [CrossRef]
20. Qi, N.; Yu, H.; Yang, C.; Gong, X.; Liu, Y.; Zhu, Y. Aflatoxin B1 in peanut oil from Western Guangdong, China, during 2016-2017. *Food Addit. Contam. Part B* **2019**, *12*, 45–51. [CrossRef]
21. Li, S.; Li, X.; Zhang, Q. Advances in the development of detection techniques for mycotoxins in vegetable oil. *Chin. J. Chromatogr.* **2019**, *37*, 569–580. [CrossRef]
22. Ji, J.; Jiang, M.; Zhang, Y.; Hou, J.; Sun, S. Co-occurrence of aflatoxins in plant oil products from China. *Food Addit. Contam. Part B* **2022**, 1–8. [CrossRef]
23. Einolghozati, M.; Talebi-Ghane, E.; Ranjbar, A.; Mehri, F. Concentration of aflatoxins in edible vegetable oils: A systematic meta-analysis review. *Eur. Food Res. Technol.* **2021**, *247*, 2887–2897. [CrossRef]

24. Bordin, K.; Sawada, M.M.; Rodrigues, C.E.d.C.; da Fonseca, C.R.; Oliveira, C.A.F. Incidence of Aflatoxins in Oil Seeds and Possible Transfer to Oil: A Review. *Food Eng. Rev.* **2014**, *6*, 20–28. [CrossRef]
25. Shephard, G.S. Aflatoxins in peanut oil: Food safety concerns. *World Mycotoxin J.* **2018**, *11*, 149–158. [CrossRef]
26. Wu, L.X.; Ding, X.X.; Li, P.W.; Du, X.H.; Zhou, H.Y.; Bai, Y.Z.; Zhang, L.X. Aflatoxin contamination of peanuts at harvest in China from 2010 to 2013 and its relationship with climatic conditions. *Food Control* **2016**, *60*, 117–123. [CrossRef]
27. Wu, L.; Li, G.; Xu, X.; Zhu, L.; Huang, R.; Chen, X. Application of nano-ELISA in food analysis: Recent advances and challenges. *TrAC, Trends Anal. Chem.* **2019**, *113*, 140–156. [CrossRef]
28. Hayashi, Y.; Matsuda, R.; Maitani, T.; Imai, K.; Nishimura, W.; Ito, K.; Maeda, M. Precision, limit of detection and range of quantitation in competitive ELISA. *Anal. Chem.* **2004**, *76*, 1295–1301. [CrossRef]
29. Bhalla, N.; Jolly, P.; Formisano, N.; Estrela, P. Introduction to biosensors. *Essays Biochem.* **2016**, *60*, 1–8. [CrossRef]
30. Naresh, V.; Lee, N. A Review on Biosensors and Recent Development of Nanostructured Materials-Enabled Biosensors. *Sensors* **2021**, *21*, 1109. [CrossRef]
31. Xue, Z.; Zhang, Y.; Yu, W.; Zhang, J.; Wang, J.; Wan, F.; Kim, Y.; Liu, Y.; Kou, X. Recent advances in aflatoxin B1 detection based on nanotechnology and nanomaterials-A review. *Anal. Chim. Acta.* **2019**, *1069*, 1–27. [CrossRef]
32. Goud, K.Y.; Reddy, K.K.; Satyanarayana, M.; Kummari, S.; Gobi, K.V. A review on recent developments in optical and electrochemical aptamer-based assays for mycotoxins using advanced nanomaterials. *Mikrochim. Acta* **2019**, *187*, 29. [CrossRef]
33. Bhardwaj, H.; Sumana, G.; Marquette, C.A. Gold nanobipyramids integrated ultrasensitive optical and electrochemical biosensor for Aflatoxin B1 detection. *Talanta* **2021**, *222*, 121578. [CrossRef] [PubMed]
34. Danesh, N.M.; Bostan, H.B.; Abnous, K.; Ramezani, M.; Youssefi, K.; Taghdisi, S.M.; Karimi, G. Ultrasensitive detection of aflatoxin B1 and its major metabolite aflatoxin M1 using aptasensors: A review. *TrAC Trends Anal. Chem.* **2018**, *99*, 117–128. [CrossRef]
35. Castillo, G.; Poturnayová, A.; Šnejdárková, M.; Hianik, T.; Spinella, K.; Mosiello, L. Development of electrochemical aptasensor using dendrimers as an immobilization platform for detection of Aflatoxin B1 in food samples. In Proceedings of the 2015 XVIII AISEM Annual Conference, Trento, Italy, 3–5 February 2015; pp. 1–4.
36. Zheng, M.Z.; Richard, J.L.; Binder, J. A review of rapid methods for the analysis of mycotoxins. *Mycopathologia* **2006**, *161*, 261–273. [CrossRef]
37. Bennett, J.W.; Klich, M. Mycotoxins. *Clin. Microbiol. Rev.* **2003**, *16*, 497–516. [CrossRef] [PubMed]
38. Robbins, C.A.; Swenson, L.J.; Nealley, M.L.; Gots, R.E.; Kelman, B.J. Health effects of mycotoxins in indoor air: A critical review. *Appl. Occup. Environ. Hyg.* **2000**, *15*, 773–784. [CrossRef] [PubMed]
39. Catanante, G.; Rhouati, A.; Hayat, A.; Marty, J.L. An Overview of Recent Electrochemical Immunosensing Strategies for Mycotoxins Detection. *Electroanalysis* **2016**, *28*, 1750–1763. [CrossRef]
40. Yao, H.; Hruska, Z.; Di Mavungu, J.D. Developments in detection and determination of aflatoxins. *World Mycotoxin J.* **2015**, *8*, 181–191. [CrossRef]
41. Liu, D.; Li, W.; Zhu, C.; Li, Y.; Shen, X.; Li, L.; Yan, X.; You, T. Recent progress on electrochemical biosensing of aflatoxins: A review. *TrAC Trends Anal. Chem.* **2020**, *133*, 115966. [CrossRef]
42. Hui, Y.; Wang, B.; Ren, R.; Zhao, A.; Zhang, F.; Song, S.; He, Y. An electrochemical aptasensor based on DNA-AuNPs-HRP nanoprobes and exonuclease-assisted signal amplification for detection of aflatoxin B1. *Food Control* **2020**, *109*, 106902. [CrossRef]
43. Luan, Y.; Chen, Z.; Xie, G.; Chen, J.; Lu, A.; Li, C.; Fu, H.; Ma, Z.; Wang, J. Rapid Visual Detection of Aflatoxin B1 by Label-Free Aptasensor Using Unmodified Gold Nanoparticles. *J. Nanosci. Nanotechnol.* **2015**, *15*, 1357–1361. [CrossRef]
44. Nguyen, B.H.; Tran, L.D.; Do, Q.P.; Nguyen, H.L.; Tran, N.H.; Nguyen, P.X. Label-free detection of aflatoxin M1 with electrochemical Fe3O4/polyaniline-based aptasensor. *Mater. Sci. Eng. C Mater. Biol. Appl.* **2013**, *33*, 2229–2234. [CrossRef]
45. Taghdisi, S.M.; Danesh, N.M.; Ramezani, M.; Abnous, K. A new amplified fluorescent aptasensor based on hairpin structure of G-quadruplex oligonucleotide-Aptamer chimera and silica nanoparticles for sensitive detection of aflatoxin B1 in the grape juice. *Food Chem.* **2018**, *268*, 342–346. [CrossRef]
46. Henry, S.H.; Bosch, F.X.; Troxell, T.C.; Bolger, P.M. Reducing liver cancer–global control of aflatoxin. *Science* **1999**, *286*, 2453–2454. [CrossRef] [PubMed]
47. Mao, J.; He, B.; Zhang, L.; Li, P.; Zhang, Q.; Ding, X.; Zhang, W. A Structure Identification and Toxicity Assessment of the Degradation Products of Aflatoxin B(1) in Peanut Oil under UV Irradiation. *Toxins* **2016**, *8*, 332. [CrossRef]
48. Mousavi Khaneghah, A.; Eş, I.; Raeisi, S.; Fakhri, Y. Aflatoxins in cereals: State of the art. *J. Food Saf.* **2018**, *38*, 12532. [CrossRef]
49. Qureshi, H. Is Aflatoxin B1 A Biomarker for Pathogenic Potential of Aspergillus flavus? *J. Cell Sci.* **2014**, *5*, 1000188. [CrossRef]
50. Ali, N. Aflatoxins in rice: Worldwide occurrence and public health perspectives. *Toxicol. Rep.* **2019**, *6*, 1188–1197. [CrossRef]
51. Sun, C.; Liao, X.; Jia, B.; Shi, L.; Zhang, D.; Wang, R.; Zhou, L.; Kong, W. Development of a ZnCdS@ZnS quantum dots-based label-free electrochemiluminescence immunosensor for sensitive determination of aflatoxin B1 in lotus seed. *Mikrochim. Acta* **2020**, *187*, 236. [CrossRef]
52. Bhardwaj, H.; Sumana, G.; Marquette, C.A. A label-free ultrasensitive microfluidic surface Plasmon resonance biosensor for Aflatoxin B1 detection using nanoparticles integrated gold chip. *Food Chem.* **2020**, *307*, 125530. [CrossRef]
53. Ramalho, L.N.Z.; Porta, L.D.; Rosim, R.E.; Petta, T.; Augusto, M.J.; Silva, D.M.; Ramalho, F.S.; Oliveira, C.A.F. Aflatoxin B1 residues in human livers and their relationship with markers of hepatic carcinogenesis in Sao Paulo, Brazil. *Toxicol. Rep.* **2018**, *5*, 777–784. [CrossRef]

54. Kensler, T.W.; Roebuck, B.D.; Wogan, G.N.; Groopman, J.D. Aflatoxin: A 50-year odyssey of mechanistic and translational toxicology. *Toxicol. Sci.* **2011**, *120* (Suppl. 1), S28–S48. [CrossRef] [PubMed]
55. Rushing, B.R.; Selim, M.I. Aflatoxin B1: A review on metabolism, toxicity, occurrence in food, occupational exposure, and detoxification methods. *Food Chem. Toxicol.* **2019**, *124*, 81–100. [CrossRef] [PubMed]
56. Chu, Y.J.; Yang, H.I.; Wu, H.C.; Liu, J.; Wang, L.Y.; Lu, S.N.; Lee, M.H.; Jen, C.L.; You, S.L.; Santella, R.M.; et al. Aflatoxin B1 exposure increases the risk of cirrhosis and hepatocellular carcinoma in chronic hepatitis B virus carriers. *Int. J. Cancer* **2017**, *141*, 711–720. [CrossRef] [PubMed]
57. Marin, S.; Ramos, A.J.; Cano-Sancho, G.; Sanchis, V. Mycotoxins: Occurrence, toxicology, and exposure assessment. *Food Chem. Toxicol.* **2013**, *60*, 218–237. [CrossRef]
58. Wild, C.P.; Turner, P.C. The toxicology of aflatoxins as a basis for public health decisions. *Mutagenesis* **2002**, *17*, 471–481. [CrossRef]
59. Zhou, Q.; Tang, D. Recent advances in photoelectrochemical biosensors for analysis of mycotoxins in food. *TrAC Trends Anal. Chem.* **2020**, *124*, 115814. [CrossRef]
60. Eivazzadeh-Keihan, R.; Pashazadeh, P.; Hejazi, M.; de la Guardia, M.; Mokhtarzadeh, A. Recent advances in Nanomaterial-mediated Bio and immune sensors for detection of aflatoxin in food products. *TrAC Trends Anal. Chem.* **2017**, *87*, 112–128. [CrossRef]
61. Keenan, J.; Jolly, P.; Preko, P.; Baidoo, J.; Wang, J.S.; Phillips, T.D.; Williams, J.H.; McGwin, G., Jr. Association Between Aflatoxin B1 Albumin Adduct Levels and Tuberculosis Infection Among HIV+ Ghanaians. *Arch. Clin. Microbiol.* **2011**, *2*, 3.
62. Williams, R.; Airey, M.; Baxter, H.; Forrester, J.; Kennedy-Martin, T.; Girach, A. Epidemiology of diabetic retinopathy and macular oedema: A systematic review. *Eye* **2004**, *18*, 963–983. [CrossRef]
63. Gong, Y.; Hounsa, A.; Egal, S.; Turner, P.C.; Sutcliffe, A.E.; Hall, A.J.; Cardwell, K.; Wild, C.P. Postweaning exposure to aflatoxin results in impaired child growth: A longitudinal study in Benin, West Africa. *Environ. Health Persp.* **2004**, *112*, 1334–1338. [CrossRef]
64. Abdolmaleki, K.; Khedri, S.; Alizadeh, L.; Javanmardi, F.; Oliveira, C.A.F.; Mousavi Khaneghah, A. The mycotoxins in edible oils: An overview of prevalence, concentration, toxicity, detection and decontamination techniques. *Trends Food Sci. Technol.* **2021**, *115*, 500–511. [CrossRef]
65. Claeys, L.; Romano, C.; De Ruyck, K.; Wilson, H.; Fervers, B.; Korenjak, M.; Zavadil, J.; Gunter, M.J.; De Saeger, S.; De Boevre, M.; et al. Mycotoxin exposure and human cancer risk: A systematic review of epidemiological studies. *Compr. Rev. Food Sci. Food Saf.* **2020**, *19*, 1449–1464. [CrossRef] [PubMed]
66. Adeyeye, S.A.O.; Yildiz, F. Fungal mycotoxins in foods: A review. *Cogent Food Agric.* **2016**, *2*, 1213127. [CrossRef]
67. Selamat, J.; Iqbal, S.Z. *Food Safety: Basic Concepts, Recent Issues, and Future Challenges*; Springer: Berlin/Heidelberg, Germany, 2016.
68. Directorate-General for Health and Consumers. *The Rapid Alert System for Food and Feed (RASFF): Annual Report 2005*; Publications Office: Luxembourg, 2006.
69. Papachristou, A.; Markaki, P. Determination of ochratoxin A in virgin olive oils of Greek origin by immunoaffinity column clean-up and high-performance liquid chromatography. *Food Addit. Contam.* **2004**, *21*, 85–92. [CrossRef] [PubMed]
70. Bhat, R.; Reddy, K.R. Challenges and issues concerning mycotoxins contamination in oil seeds and their edible oils: Updates from last decade. *Food Chem.* **2017**, *215*, 425–437. [CrossRef] [PubMed]
71. Mazumder, P.M.; Sasmal, D. Mycotoxins–limits and regulations. *Anc. Sci. Life* **2001**, *20*, 1.
72. Zinedine, A.; Mañes, J. Occurrence and legislation of mycotoxins in food and feed from Morocco. *Food Control* **2009**, *20*, 334–344. [CrossRef]
73. Selvaraj, J.N.; Wang, Y.; Zhou, L.; Zhao, Y.; Xing, F.; Dai, X.; Liu, Y. Recent mycotoxin survey data and advanced mycotoxin detection techniques reported from China: A review. *Food Addit. Contam. Part A* **2015**, *32*, 440–452. [CrossRef]
74. Anukul, N.; Vangnai, K.; Mahakarnchanakul, W. Significance of regulation limits in mycotoxin contamination in Asia and risk management programs at the national level. *J. Food Drug Anal.* **2013**, *21*, 227–241. [CrossRef]
75. Ji, F.; He, D.; Olaniran, A.O.; Mokoena, M.P.; Xu, J.; Shi, J. Occurrence, toxicity, production and detection of Fusarium mycotoxin: A review. *Food Prod. Process. Nutr.* **2019**, *1*, 1–14. [CrossRef]
76. Aiko, V.; Mehta, A. Occurrence, detection and detoxification of mycotoxins. *J. Biosci.* **2015**, *40*, 943–954. [CrossRef] [PubMed]
77. Le, V.T.; Vasseghian, Y.; Dragoi, E.N.; Moradi, M.; Mousavi Khaneghah, A. A review on graphene-based electrochemical sensor for mycotoxins detection. *Food Chem. Toxicol.* **2021**, *148*, 111931. [CrossRef]
78. Turner, N.W.; Bramhmbhatt, H.; Szabo-Vezse, M.; Poma, A.; Coker, R.; Piletsky, S.A. Analytical methods for determination of mycotoxins: An update (2009–2014). *Anal. Chim. Acta.* **2015**, *901*, 12–33. [CrossRef]
79. Ingenbleek, L.; Sulyok, M.; Adegboye, A.; Hossou, S.E.; Kone, A.Z.; Oyedele, A.D.; Kisito, C.; Dembele, Y.K.; Eyangoh, S.; Verger, P.; et al. Regional Sub-Saharan Africa Total Diet Study in Benin, Cameroon, Mali and Nigeria Reveals the Presence of 164 Mycotoxins and Other Secondary Metabolites in Foods. *Toxins* **2019**, *11*, 54. [CrossRef] [PubMed]
80. Zhu, C.; Liu, D.; Li, Y.; Shen, X.; Ma, S.; Liu, Y.; You, T. Ratiometric electrochemical aptasensor for ultrasensitive detection of Ochratoxin A based on a dual signal amplification strategy: Engineering the binding of methylene blue to DNA. *Biosens. Bioelectron.* **2020**, *150*, 111814. [CrossRef] [PubMed]
81. Karunarathna, N.B.; Fernando, C.J.; Munasinghe, D.M.S.; Fernando, R. Occurrence of aflatoxins in edible vegetable oils in Sri Lanka. *Food Control* **2019**, *101*, 97–103. [CrossRef]

82. Wang, N.; Duan, C.; Geng, X.; Li, S.; Ding, K.; Guan, Y. One step rapid dispersive liquid-liquid micro-extraction with in-situ derivatization for determination of aflatoxins in vegetable oils based on high performance liquid chromatography fluorescence detection. *Food Chem.* **2019**, *287*, 333–337. [CrossRef]
83. Nabizadeh, S.; Shariatifar, N.; Shokoohi, E.; Shoeibi, S.; Gavahian, M.; Fakhri, Y.; Azari, A.; Mousavi Khaneghah, A. Prevalence and probabilistic health risk assessment of aflatoxins B1, B2, G1, and G2 in Iranian edible oils. *Environ. Sci. Pollut. Res. Int.* **2018**, *25*, 35562–35570. [CrossRef]
84. Ma, F.; Chen, R.; Li, P.; Zhang, Q.; Zhang, W.; Hu, X. Preparation of an immunoaffinity column with amino-silica gel microparticles and its application in sample cleanup for aflatoxin detection in agri-products. *Molecules* **2013**, *18*, 2222–2235. [CrossRef]
85. Afzali, D.; Ghanbarian, M.; Mostafavi, A.; Shamspur, T.; Ghaseminezhad, S. A novel method for high preconcentration of ultra trace amounts of B(1), B(2), G(1) and G(2) aflatoxins in edible oils by dispersive liquid-liquid microextraction after immunoaffinity column clean-up. *J. Chromatogr. A* **2012**, *1247*, 35–41. [CrossRef]
86. Elzupir, A.O.; Suliman, M.A.; Ibrahim, I.A.; Fadul, M.H.; Elhussein, A.M. Aflatoxins levels in vegetable oils in Khartoum State, Sudan. *Mycotoxin Res.* **2010**, *26*, 69–73. [CrossRef] [PubMed]
87. Zhang, K.; Xu, D. Application of Stable Isotope Dilution and Liquid Chromatography Tandem Mass Spectrometry for Multi-Mycotoxin Analysis in Edible Oils. *J. AOAC Int.* **2019**, *102*, 1651–1656. [CrossRef] [PubMed]
88. Eom, T.; Cho, H.D.; Kim, J.; Park, M.; An, J.; Kim, M.; Kim, S.H.; Han, S.B. Multiclass mycotoxin analysis in edible oils using a simple solvent extraction method and liquid chromatography with tandem mass spectrometry. *Food Addit. Contam. Part A* **2017**, *34*, 2011–2022. [CrossRef] [PubMed]
89. Huang, S.; Chen, X.; Wang, Y.; Zhu, F.; Jiang, R.; Ouyang, G. High enrichment and ultra-trace analysis of aflatoxins in edible oils by a modified hollow-fiber liquid-phase microextraction technique. *Chem. Commun.* **2017**, *53*, 8988–8991. [CrossRef] [PubMed]
90. Sharmili, K.; Jinap, S.; Sukor, R. Development, optimization and validation of QuEChERS based liquid chromatography tandem mass spectrometry method for determination of multimycotoxin in vegetable oil. *Food Control* **2016**, *70*, 152–160. [CrossRef]
91. Pereira, V.L.; Fernandes, J.O.; Cunha, S.C. Comparative assessment of three cleanup procedures after QuEChERS extraction for determination of trichothecenes (type A and type B) in processed cereal-based baby foods by GC-MS. *Food Chem.* **2015**, *182*, 143–149. [CrossRef] [PubMed]
92. Nathanail, A.V.; Syvahuoko, J.; Malachova, A.; Jestoi, M.; Varga, E.; Michlmayr, H.; Adam, G.; Sievilainen, E.; Berthiller, F.; Peltonen, K. Simultaneous determination of major type A and B trichothecenes, zearalenone and certain modified metabolites in Finnish cereal grains with a novel liquid chromatography-tandem mass spectrometric method. *Anal. Bioanal. Chem.* **2015**, *407*, 4745–4755. [CrossRef]
93. Capriotti, A.L.; Cavaliere, C.; Foglia, P.; Samperi, R.; Stampachiacchiere, S.; Ventura, S.; Lagana, A. Multiclass analysis of mycotoxins in biscuits by high performance liquid chromatography-tandem mass spectrometry. Comparison of different extraction procedures. *J. Chromatogr. A* **2014**, *1343*, 69–78. [CrossRef]
94. Malachova, A.; Sulyok, M.; Beltran, E.; Berthiller, F.; Krska, R. Optimization and validation of a quantitative liquid chromatography-tandem mass spectrometric method covering 295 bacterial and fungal metabolites including all regulated mycotoxins in four model food matrices. *J. Chromatogr. A* **2014**, *1362*, 145–156. [CrossRef]
95. Tsiplakou, E.; Anagnostopoulos, C.; Liapis, K.; Haroutounian, S.A.; Zervas, G. Determination of mycotoxins in feedstuffs and ruminant's milk using an easy and simple LC-MS/MS multiresidue method. *Talanta* **2014**, *130*, 8–19. [CrossRef]
96. Yang, L.X.; Liu, Y.P.; Miao, H.; Dong, B.; Yang, N.J.; Chang, F.Q.; Yang, L.X.; Sun, J.B. Determination of aflatoxins in edible oil from markets in Hebei Province of China by liquid chromatography-tandem mass spectrometry. *Food Addit. Contam. Part B* **2011**, *4*, 244–247. [CrossRef] [PubMed]
97. Bao, L.; Trucksess, M.W.; White, K.D. Determination of aflatoxins B1, B2, G1, and G2 in olive oil, peanut oil, and sesame oil. *J. AOAC Int.* **2010**, *93*, 936–942. [CrossRef] [PubMed]
98. Junsai, T.; Poapolathep, S.; Sutjarit, S.; Giorgi, M.; Zhang, Z.; Logrieco, A.F.; Li, P.; Poapolathep, A. Determination of Multiple Mycotoxins and Their Natural Occurrence in Edible Vegetable Oils Using Liquid Chromatography-Tandem Mass Spectrometry. *Foods* **2021**, *10*, 2795. [CrossRef]
99. Deng, H.; Su, X.; Wang, H. Simultaneous Determination of Aflatoxin B1, Bisphenol A, and 4-Nonylphenol in Peanut Oils by Liquid-Liquid Extraction Combined with Solid-Phase Extraction and Ultra-High Performance Liquid Chromatography-Tandem Mass Spectrometry. *Food Anal. Methods* **2017**, *11*, 1303–1311. [CrossRef]
100. Hidalgo-Ruiz, J.L.; Romero-Gonzalez, R.; Martinez Vidal, J.L.; Garrido Frenich, A. A rapid method for the determination of mycotoxins in edible vegetable oils by ultra-high performance liquid chromatography-tandem mass spectrometry. *Food Chem.* **2019**, *288*, 22–28. [CrossRef]
101. Zhu, Z.; Feng, M.; Zuo, L.; Zhu, Z.; Wang, F.; Chen, L.; Li, J.; Shan, G.; Luo, S.Z. An aptamer based surface plasmon resonance biosensor for the detection of ochratoxin A in wine and peanut oil. *Biosens. Bioelectron.* **2015**, *65*, 320–326. [CrossRef]
102. Idris, Y.M.; Mariod, A.A.; Elnour, I.A.; Mohamed, A.A. Determination of aflatoxin levels in Sudanese edible oils. *Food Chem. Toxicol.* **2010**, *48*, 2539–2541. [CrossRef]
103. Majerus, P.; Graf, N.; Kramer, M. Rapid determination of zearalenone in edible oils by HPLC with fluorescence detection. *Mycotoxin Res.* **2009**, *25*, 117–121. [CrossRef]
104. Wang, H.; Zhao, L.; Yang, H.; Guo, Q.; Shi, H.; Pan, H.; Zhao, L.; Qian, C. Determination of benzo(a)pyrene and aflatoxins (B1, B2, G1, G2) in vegetable oil by GPC-HPLC-FLD. *Anal. Methods* **2014**, *6*, 1545–1549. [CrossRef]

105. Xie, J.; Peng, T.; He, J.L.; Shao, Y.; Fan, C.L.; Chen, Y.; Jiang, W.X.; Chen, M.; Wang, Q.; Pei, X.Y.; et al. Preparation and characterization of an immunoaffinity column for the selective extraction of aflatoxin B1 in 13 kinds of foodstuffs. *J. Chromatogr. B* **2015**, *998–999*, 50–56. [CrossRef]
106. Han, X.; Xu, W.; Zhang, J.; Xu, J.; Li, F. Co-Occurrence of Beauvericin and Enniatins in Edible Vegetable Oil Samples, China. *Toxins* **2019**, *11*, 100. [CrossRef]
107. Drzymala, S.S.; Weiz, S.; Heinze, J.; Marten, S.; Prinz, C.; Zimathies, A.; Garbe, L.A.; Koch, M. Automated solid-phase extraction coupled online with HPLC-FLD for the quantification of zearalenone in edible oil. *Anal. Bioanal. Chem.* **2015**, *407*, 3489–3497. [CrossRef] [PubMed]
108. Yu, X.; Yang, H. Pyrethroid residue determination in organic and conventional vegetables using liquid-solid extraction coupled with magnetic solid phase extraction based on polystyrene-coated magnetic nanoparticles. *Food Chem.* **2017**, *217*, 303–310. [CrossRef] [PubMed]
109. Sheng, J.; Zuo, J.; Liu, K.; Ma, L.; Li, C.; Li, Y.; Kong, D. Highly selective enrichment of aflatoxin B1 from edible oil using polydopamine-modified magnetic nanomaterials. *Food Sci. Technol.* **2021**, *41*, 321–327. [CrossRef]
110. Ferracane, R.; Tafuri, A.; Logieco, A.; Galvano, F.; Balzano, D.; Ritieni, A. Simultaneous determination of aflatoxin B1 and ochratoxin A and their natural occurrence in Mediterranean virgin olive oil. *Food Addit. Contam.* **2007**, *24*, 173–180. [CrossRef]
111. Giménez, I.; Herrera, M.; Escobar, J.; Ferruz, E.; Lorán, S.; Herrera, A.; Ariño, A. Distribution of deoxynivalenol and zearalenone in milled germ during wheat milling and analysis of toxin levels in wheat germ and wheat germ oil. *Food Control* **2013**, *34*, 268–273. [CrossRef]
112. Jiang, Y.; Li, Y.; Jiang, Y.; Li, J.; Pan, C. Determination of multiresidues in rapeseed, rapeseed oil, and rapeseed meal by acetonitrile extraction, low-temperature cleanup, and detection by liquid chromatography with tandem mass spectrometry. *J. Agric. Food Chem.* **2012**, *60*, 5089–5098. [CrossRef]
113. Payanan, T.; Leepipatpiboon, N.; Varanusupakul, P. Low-temperature cleanup with solid-phase extraction for the determination of polycyclic aromatic hydrocarbons in edible oils by reversed phase liquid chromatography with fluorescence detection. *Food Chem.* **2013**, *141*, 2720–2726. [CrossRef]
114. Urusov, A.E.; Petrakova, A.V.; Vozniak, M.V.; Zherdev, A.V.; Dzantiev, B.B. Rapid immunoenzyme assay of aflatoxin B1 using magnetic nanoparticles. *Sensors* **2014**, *14*, 21843–21857. [CrossRef]
115. Yu, X.; Li, Z.; Zhao, M.; Lau, S.C.S.; Ru Tan, H.; Teh, W.J.; Yang, H.; Zheng, C.; Zhang, Y. Quantification of aflatoxin B1 in vegetable oils using low temperature clean-up followed by immuno-magnetic solid phase extraction. *Food Chem.* **2019**, *275*, 390–396. [CrossRef]
116. Zhao, H.; Chen, X.; Shen, C.; Qu, B. Determination of 16 mycotoxins in vegetable oils using a QuEChERS method combined with high-performance liquid chromatography-tandem mass spectrometry. *Food Addit. Contam. Part A* **2017**, *34*, 255–264. [CrossRef] [PubMed]
117. Qian, M.; Zhang, H.; Wu, L.; Jin, N.; Wang, J.; Jiang, K. Simultaneous determination of zearalenone and its derivatives in edible vegetable oil by gel permeation chromatography and gas chromatography-triple quadrupole mass spectrometry. *Food Chem.* **2015**, *166*, 23–28. [CrossRef] [PubMed]
118. Alshannaq, A.; Yu, J.H. Occurrence, Toxicity, and Analysis of Major Mycotoxins in Food. *Int. J. Environ. Res. Public. Health* **2017**, *14*, 632. [CrossRef] [PubMed]
119. Santos, A.; Vaz, A.; Rodrigues, P.; Veloso, A.; Venâncio, A.; Peres, A. Thin Films Sensor Devices for Mycotoxins Detection in Foods: Applications and Challenges. *Chemosensors* **2019**, *7*, 3. [CrossRef]
120. Shephard, G.S. Current Status of Mycotoxin Analysis: A Critical Review. *J. AOAC Int.* **2016**, *99*, 842–848. [CrossRef] [PubMed]
121. Schwack, W.; Pellissier, E.; Morlock, G. Analysis of unauthorized Sudan dyes in food by high-performance thin-layer chromatography. *Anal. Bioanal. Chem.* **2018**, *410*, 5641–5651. [CrossRef]
122. Gauthier, M.S.; O'Brien, E.L.; Bigornia, S.; Mott, M.; Cacicedo, J.M.; Xu, X.J.; Gokce, N.; Apovian, C.; Ruderman, N. Decreased AMP-activated protein kinase activity is associated with increased inflammation in visceral adipose tissue and with whole-body insulin resistance in morbidly obese humans. *Biochem. Biophys. Res. Commun.* **2011**, *404*, 382–387. [CrossRef]
123. Sun, Y.; Wang, H.; Wang, W.; Hu, B.; Zhou, L.; Ye, H.; Zeng, X. Changes in molecular structure of chickpea starch during processing treatments: A thin layer chromatography study. *Food Chem.* **2018**, *243*, 186–191. [CrossRef]
124. Villani, T.S.; Reichert, W.; Ferruzzi, M.G.; Pasinetti, G.M.; Simon, J.E.; Wu, Q. Chemical investigation of commercial grape seed derived products to assess quality and detect adulteration. *Food Chem.* **2015**, *170*, 271–280. [CrossRef]
125. Wacoo, A.P.; Wendiro, D.; Vuzi, P.C.; Hawumba, J.F. Methods for Detection of Aflatoxins in Agricultural Food Crops. *J. Appl. Chem.* **2014**, *2014*, 706291. [CrossRef]
126. Cabezudo, I.; Salazar, M.O.; Ramallo, I.A.; Furlan, R.L.E. Effect-directed analysis in food by thin-layer chromatography assays. *Food Chem.* **2022**, *390*, 132937. [CrossRef] [PubMed]
127. Lin, L.; Zhang, J.; Wang, P.; Wang, Y.; Chen, J. Thin-layer chromatography of mycotoxins and comparison with other chromatographic methods. *J. Chromatogr. A* **1998**, *815*, 3–20. [CrossRef]
128. Sereshti, H.; Poursorkh, Z.; Aliakbarzadeh, G.; Zarre, S. Quality control of saffron and evaluation of potential adulteration by means of thin layer chromatography-image analysis and chemometrics methods. *Food Control* **2018**, *90*, 48–57. [CrossRef]

129. Li, Y.; Zhao, C.; Lu, C.; Zhou, S.; Tian, G.; He, L.; Bao, Y.; Fauconnier, M.L.; Xiao, H.; Zheng, J. Simultaneous determination of 14 bioactive citrus flavonoids using thin-layer chromatography combined with surface enhanced Raman spectroscopy. *Food Chem.* **2021**, *338*, 128115. [CrossRef] [PubMed]
130. Morlock, G.E.; Klingelhofer, I. Liquid chromatography-bioassay-mass spectrometry for profiling of physiologically active food. *Anal. Chem.* **2014**, *86*, 8289–8295. [CrossRef] [PubMed]
131. Tao, F.; Yao, H.; Hruska, Z.; Burger, L.W.; Rajasekaran, K.; Bhatnagar, D. Recent development of optical methods in rapid and non-destructive detection of aflatoxin and fungal contamination in agricultural products. *TrAC Trends Anal. Chem.* **2018**, *100*, 65–81. [CrossRef]
132. Wu, Q.; Xie, L.; Xu, H. Determination of toxigenic fungi and aflatoxins in nuts and dried fruits using imaging and spectroscopic techniques. *Food Chem.* **2018**, *252*, 228–242. [CrossRef]
133. Yadav, N.; Yadav, S.S.; Chhillar, A.K.; Rana, J.S. An overview of nanomaterial based biosensors for detection of Aflatoxin B1 toxicity in foods. *Food Chem. Toxicol.* **2021**, *152*, 112201. [CrossRef]
134. Karoui, R.; Blecker, C. Fluorescence Spectroscopy Measurement for Quality Assessment of Food Systems—A Review. *Food Bioprocess Technol.* **2010**, *4*, 364–386. [CrossRef]
135. Ma, H.; Ran, C.; Li, M.; Gao, J.; Wang, X.; Zhang, L.; Bian, J.; Li, J.; Jiang, Y. Graphene oxide-coated stir bar sorptive extraction of trace aflatoxins from soy milk followed by high performance liquid chromatography-laser-induced fluorescence detection. *Food Addit. Contam. Part A* **2018**, *35*, 772–781. [CrossRef]
136. Hruska, Z.; Yao, H.; Kincaid, R.; Brown, R.; Cleveland, T.; Bhatnagar, D. Fluorescence Excitation–Emission Features of Aflatoxin and Related Secondary Metabolites and Their Application for Rapid Detection of Mycotoxins. *Food Bioprocess Technol.* **2014**, *7*, 1195–1201. [CrossRef]
137. Chen, M.; He, X.; Pang, Y.; Shen, F.; Fang, Y.; Hu, Q. Laser induced fluorescence spectroscopy for detection of Aflatoxin B1 contamination in peanut oil. *J. Food Meas. Charact.* **2021**, *15*, 2231–2239. [CrossRef]
138. Qu, J.H.; Liu, D.; Cheng, J.H.; Sun, D.W.; Ma, J.; Pu, H.; Zeng, X.A. Applications of near-infrared spectroscopy in food safety evaluation and control: A review of recent research advances. *Crit. Rev. Food Sci. Nutr.* **2015**, *55*, 1939–1954. [CrossRef] [PubMed]
139. Rohman, A. Infrared spectroscopy for quantitative analysis and oil parameters of olive oil and virgin coconut oil: A review. *Int. J. Food Prop.* **2016**, *20*, 1447–1456. [CrossRef]
140. Nicolaï, B.M.; Beullens, K.; Bobelyn, E.; Peirs, A.; Saeys, W.; Theron, K.I.; Lammertyn, J. Nondestructive measurement of fruit and vegetable quality by means of NIR spectroscopy: A review. *Postharvest Biol. Technol.* **2007**, *46*, 99–118. [CrossRef]
141. Uddin, M.; Okazaki, E. Classification of Fresh and Frozen-thawed Fish by Near-infrared Spectroscopy. *J. Food Sci.* **2004**, *69*, C665–C668. [CrossRef]
142. Kurz, C.; Leitenberger, M.; Carle, R.; Schieber, A. Evaluation of fruit authenticity and determination of the fruit content of fruit products using FT-NIR spectroscopy of cell wall components. *Food Chem.* **2010**, *119*, 806–812. [CrossRef]
143. Kuligowski, J.; Carrión, D.; Quintás, G.; Garrigues, S.; de la Guardia, M. Direct determination of polymerised triacylglycerides in deep-frying vegetable oil by near infrared spectroscopy using Partial Least Squares regression. *Food Chem.* **2012**, *131*, 353–359. [CrossRef]
144. Balabin, R.M.; Smirnov, S.V. Melamine detection by mid- and near-infrared (MIR/NIR) spectroscopy: A quick and sensitive method for dairy products analysis including liquid milk, infant formula, and milk powder. *Talanta* **2011**, *85*, 562–568. [CrossRef]
145. Ravikanth, L.; Jayas, D.S.; White, N.D.G.; Fields, P.G.; Sun, D.-W. Extraction of Spectral Information from Hyperspectral Data and Application of Hyperspectral Imaging for Food and Agricultural Products. *Food Bioprocess Technol.* **2016**, *10*, 1–33. [CrossRef]
146. Morsy, N.; Sun, D.W. Robust linear and non-linear models of NIR spectroscopy for detection and quantification of adulterants in fresh and frozen-thawed minced beef. *Meat Sci.* **2013**, *93*, 292–302. [CrossRef] [PubMed]
147. Luo, J.; Liu, T.; Liu, Y. FT-NIR and Confocal Microscope Raman Spectroscopic Studies of Sesame Oil Adulteration. In Proceedings of the Computer and Computing Technologies in Agriculture V, Berlin, Heidelberg, 29 October 2011; pp. 24–31.
148. Mignani, A.G.; Ciaccheri, L.; Ottevaere, H.; Thienpont, H.; Conte, L.; Marega, M.; Cichelli, A.; Attilio, C.; Cimato, A. Visible and near-infrared absorption spectroscopy by an integrating sphere and optical fibers for quantifying and discriminating the adulteration of extra virgin olive oil from Tuscany. *Anal. Bioanal. Chem.* **2011**, *399*, 1315–1324. [CrossRef] [PubMed]
149. Sinelli, N.; Cerretani, L.; Egidio, V.D.; Bendini, A.; Casiraghi, E. Application of near (NIR) infrared and mid (MIR) infrared spectroscopy as a rapid tool to classify extra virgin olive oil on the basis of fruity attribute intensity. *Food Res. Int.* **2010**, *43*, 369–375. [CrossRef]
150. Luna, A.S.; da Silva, A.P.; Pinho, J.S.; Ferre, J.; Boque, R. Rapid characterization of transgenic and non-transgenic soybean oils by chemometric methods using NIR spectroscopy. *Spectrochim. Acta Part A* **2013**, *100*, 115–119. [CrossRef]
151. Yao, W.; Liu, R.; Xu, Z.; Zhang, Y.; Deng, Y.; Guo, H.; Cozzolino, D. Rapid Determination of Aflatoxin B1 Contamination in Peanut Oil by Fourier Transform Near-Infrared Spectroscopy. *J. Spectro.* **2022**, *2022*, 1–9. [CrossRef]
152. Song, H.; Li, F.; Guang, P.; Yang, X.; Pan, H.; Huang, F. Detection of Aflatoxin B1 in Peanut Oil Using Attenuated Total Reflection Fourier Transform Infrared Spectroscopy Combined with Partial Least Squares Discriminant Analysis and Support Vector Machine Models. *J. Food Prot.* **2021**, *84*, 1315–1320. [CrossRef] [PubMed]
153. Yang, Y.; Zhang, Y.; He, C.; Xie, M.; Luo, H.; Wang, Y.; Zhang, J. Rapid screen of aflatoxin-contaminated peanut oil using Fourier transform infrared spectroscopy combined with multivariate decision tree. *Int. J. Food Sci. Technol.* **2018**, *53*, 2386–2393. [CrossRef]

154. Rawson, A.; Sunil, C.K. Recent Advances in Terahertz Time-Domain Spectroscopy and Imaging Techniques for Automation in Agriculture and Food Sector. *Food Anal. Methods* **2021**, *15*, 498–526. [CrossRef]
155. Jan, A.; Pandiselvam, R.; Kothakota, A.; Sruthi, N.; Ramesh, S. Terahertz Spectroscopy Imaging Technique: Non-Destructive Tool For Evaluation Of Quality And Safety Of Food Products. In *Handbook of Research on Food Processing and Preservation Technologies*; Apple Academic Press: New York, NY, USA, 2021; pp. 141–157.
156. Feng, C.H.; Otani, C. Terahertz spectroscopy technology as an innovative technique for food: Current state-of-the-Art research advances. *Crit. Rev. Food Sci. Nutr.* **2021**, *61*, 2523–2543. [CrossRef]
157. Afsah-Hejri, L.; Hajeb, P.; Ara, P.; Ehsani, R.J. A Comprehensive Review on Food Applications of Terahertz Spectroscopy and Imaging. *Compr. Rev. Food Sci. Food Saf.* **2019**, *18*, 1563–1621. [CrossRef]
158. Kawano, Y.; Ishibashi, K. An on-chip near-field terahertz probe and detector. *Nat. Photonics* **2008**, *2*, 618–621. [CrossRef]
159. McIntosh, A.I.; Yang, B.; Goldup, S.M.; Watkinson, M.; Donnan, R.S. Terahertz spectroscopy: A powerful new tool for the chemical sciences? *Chem. Soc. Rev.* **2012**, *41*, 2072–2082. [CrossRef] [PubMed]
160. Wang, K.; Sun, D.-W.; Pu, H. Emerging non-destructive terahertz spectroscopic imaging technique: Principle and applications in the agri-food industry. *Trends Food Sci. Technol.* **2017**, *67*, 93–105. [CrossRef]
161. Chen, M.; Lijuan, X. A Preliminary Study of Aflatoxin B1 Detection in Peanut Oil by Terahertz Time-Domain Spectroscopy. *Trans. ASABE* **2014**, *57*, 1793–1799. [CrossRef]
162. Liu, W.; Zhao, P.; Wu, C.; Liu, C.; Yang, J.; Zheng, L. Rapid determination of aflatoxin B1 concentration in soybean oil using terahertz spectroscopy with chemometric methods. *Food Chem.* **2019**, *293*, 213–219. [CrossRef]
163. Wang, Q.; Xie, L.; Ying, Y. Overview of imaging methods based on terahertz time-domain spectroscopy. *Appl. Spectrosc. Rev.* **2021**, *57*, 249–264. [CrossRef]
164. Zhang, W.; Ma, J.; Sun, D.W. Raman spectroscopic techniques for detecting structure and quality of frozen foods: Principles and applications. *Crit. Rev. Food Sci. Nutr.* **2021**, *61*, 2623–2639. [CrossRef]
165. Wu, Z.; Pu, H.; Sun, D.-W. Fingerprinting and tagging detection of mycotoxins in agri-food products by surface-enhanced Raman spectroscopy: Principles and recent applications. *Trends Food Sci. Technol.* **2021**, *110*, 393–404. [CrossRef]
166. He, H.; Sun, D.W.; Pu, H.; Chen, L.; Lin, L. Applications of Raman spectroscopic techniques for quality and safety evaluation of milk: A review of recent developments. *Crit. Rev. Food Sci. Nutr.* **2019**, *59*, 770–793. [CrossRef]
167. Deng, J.; Zhang, X.; Li, M.; Jiang, H.; Chen, Q. Feasibility study on Raman spectra-based deep learning models for monitoring the contamination degree and level of aflatoxin B1 in edible oil. *Microchem. J.* **2022**, *180*, 107613. [CrossRef]
168. Zhu, C.; Jiang, H.; Chen, Q. High Precisive Prediction of Aflatoxin B1 in Pressing Peanut Oil Using Raman Spectra Combined with Multivariate Data Analysis. *Foods* **2022**, *11*, 1565. [CrossRef] [PubMed]
169. Yang, M.; Liu, G.; Mehedi, H.M.; Ouyang, Q.; Chen, Q. A universal SERS aptasensor based on DTNB labeled GNTs/Ag core-shell nanotriangle and CS-Fe3O4 magnetic-bead trace detection of Aflatoxin B1. *Anal. Chim. Acta.* **2017**, *986*, 122–130. [CrossRef] [PubMed]
170. Chen, Q.; Yang, M.; Yang, X.; Li, H.; Guo, Z.; Rahma, M. A large Raman scattering cross-section molecular embedded SERS aptasensor for ultrasensitive Aflatoxin B1 detection using CS-Fe3O4 for signal enrichment. *Spectrochim. Acta Part A* **2018**, *189*, 147–153. [CrossRef] [PubMed]
171. Guo, M.; Hou, Q.; Waterhouse, G.I.N.; Hou, J.; Ai, S.; Li, X. A simple aptamer-based fluorescent aflatoxin B1 sensor using humic acid as quencher. *Talanta* **2019**, *205*, 120131. [CrossRef]
172. Chen, Q.; Jiao, T.; Yang, M.; Li, H.; Ahmad, W.; Hassan, M.M.; Guo, Z.; Ali, S. Pre etched Ag nanocluster as SERS substrate for the rapid quantification of AFB1 in peanut oil via DFT coupled multivariate calibration. *Spectrochim. Acta Part A* **2020**, *239*, 118411. [CrossRef]
173. Li, J.; Wang, Q.; Xiong, C.; Deng, Q.; Zhang, X.; Wang, S.; Chen, M.M. An ultrasensitive CH3NH3PbBr3 quantum dots@SiO2-based electrochemiluminescence sensing platform using an organic electrolyte for aflatoxin B1 detection in corn oil. *Food Chem.* **2022**, *390*, 133200. [CrossRef]
174. He, H.; Sun, D.W.; Pu, H.; Huang, L. Bridging Fe3O4@Au nanoflowers and Au@Ag nanospheres with aptamer for ultrasensitive SERS detection of aflatoxin B1. *Food Chem.* **2020**, *324*, 126832. [CrossRef]
175. Jiang, Y.; Sun, D.-W.; Pu, H.; Wei, Q. A simple and sensitive aptasensor based on SERS for trace analysis of kanamycin in milk. *J. Food Meas. Charact.* **2020**, *14*, 3184–3193. [CrossRef]
176. Toh, S.Y.; Citartan, M.; Gopinath, S.C.; Tang, T.H. Aptamers as a replacement for antibodies in enzyme-linked immunosorbent assay. *Biosens. Bioelectron.* **2015**, *64*, 392–403. [CrossRef]
177. Jiang, Y.; Sun, D.W.; Pu, H.; Wei, Q. Ultrasensitive analysis of kanamycin residue in milk by SERS-based aptasensor. *Talanta* **2019**, *197*, 151–158. [CrossRef]
178. Azadniya, E.; Goldoni, L.; Bandiera, T.; Morlock, G.E. Same analytical method for both (bio)assay and zone isolation to identify/quantify bioactive compounds by quantitative nuclear magnetic resonance spectroscopy. *J. Chromatogr. A* **2020**, *1628*, 461434. [CrossRef] [PubMed]
179. Malu, S.P.; Donatus, R.B.; Ugye, J.T.; Imarenezor, E.P.K.; Leubem, A. Determination of Aflaxtoxin in Some Edible Oils Obtained from Makurdi Metropolis, North Central Nigeria. *Am. J. Chem. Appl.* **2017**, *4*, 36–40.
180. Zhang, Y.; Li, M.; Cui, Y.; Hong, X.; Du, D. Using of Tyramine Signal Amplification to Improve the Sensitivity of ELISA for Aflatoxin B1 in Edible Oil Samples. *Food Anal. Methods* **2018**, *11*, 2553–2560. [CrossRef]

181. Pundir, C.S.; Yadav, N.; Chhillar, A.K. Occurrence, synthesis, toxicity and detection methods for acrylamide determination in processed foods with special reference to biosensors: A review. *Trends Food Sci. Technol.* **2019**, *85*, 211–225. [CrossRef]
182. Wang, X.; Shan, Y.; Gong, M.; Jin, X.; Lv, l.; Jiang, M.; Xu, J. A novel electrochemical sensor for ochratoxin A based on the hairpin aptamer and double report DNA via multiple signal amplification strategy. *Sens. Actuators B* **2019**, *281*, 595–601. [CrossRef]
183. Wang, Z.; Ma, B.; Shen, C.; Cheong, L.Z. Direct, selective and ultrasensitive electrochemical biosensing of methyl parathion in vegetables using Burkholderia cepacia lipase@MOF nanofibers-based biosensor. *Talanta* **2019**, *197*, 356–362. [CrossRef]
184. Lin, T.; Shen, Y. Fabricating electrochemical aptasensors for detecting aflatoxin B1 via layer-by-layer self-assembly. *J. Electroanal. Chem.* **2020**, *870*, 114247. [CrossRef]
185. Xuan, Z.; Liu, H.; Ye, J.; Li, L.; Tian, W.; Wang, S. Reliable and disposable quantum dot-based electrochemical immunosensor for aflatoxin B1 simplified analysis with automated magneto-controlled pretreatment system. *Anal. Bioanal. Chem.* **2020**, *412*, 7615–7625. [CrossRef]
186. Ben Rejeb, I.; Arduini, F.; Arvinte, A.; Amine, A.; Gargouri, M.; Micheli, L.; Bala, C.; Moscone, D.; Palleschi, G. Development of a bio-electrochemical assay for AFB1 detection in olive oil. *Biosens. Bioelectron.* **2009**, *24*, 1962–1968. [CrossRef]
187. Kim, M.; Iezzi, R., Jr.; Shim, B.S.; Martin, D.C. Impedimetric Biosensors for Detecting Vascular Endothelial Growth Factor (VEGF) Based on Poly(3,4-ethylene dioxythiophene) (PEDOT)/Gold Nanoparticle (Au NP) Composites. *Front. Chem.* **2019**, *7*, 234. [CrossRef]
188. Lu, Y.; Zhang, B.; Tian, Y.; Guo, Q.; Yang, X.; Nie, G. An enhanced photoelectrochemical sensor for aflatoxin B1 detection based on organic-inorganic heterojunction nanomaterial: Poly(5-formylindole)/NiO. *Mikrochim. Acta* **2020**, *187*, 467. [CrossRef] [PubMed]
189. Sharma, P.; Panchal, A.; Yadav, N.; Narang, J. Analytical techniques for the detection of glycated haemoglobin underlining the sensors. *Int. J. Biol. Macromol.* **2020**, *155*, 685–696. [CrossRef] [PubMed]
190. Yu, L.; Zhang, Y.; Hu, C.; Wu, H.; Yang, Y.; Huang, C.; Jia, N. Highly sensitive electrochemical impedance spectroscopy immunosensor for the detection of AFB1 in olive oil. *Food Chem.* **2015**, *176*, 22–26. [CrossRef] [PubMed]
191. Perumal, V.; Hashim, U. Advances in biosensors: Principle, architecture and applications. *J. Appl. Biomed.* **2014**, *12*, 1–15. [CrossRef]
192. Wang, H.; Zhang, Y.; Chu, Y.; Ma, H.; Li, Y.; Wu, D.; Du, B.; Wei, Q. Disposable competitive-type immunoassay for determination of aflatoxin B1 via detection of copper ions released from Cu-apatite. *Talanta* **2016**, *147*, 556–560. [CrossRef]
193. Farka, Z.; Jurik, T.; Kovar, D.; Trnkova, L.; Skladal, P. Nanoparticle-Based Immunochemical Biosensors and Assays: Recent Advances and Challenges. *Chem. Rev.* **2017**, *117*, 9973–10042. [CrossRef]
194. Wang, X.; Wu, X.; Lu, Z.; Tao, X. Comparative Study of Time-Resolved Fluorescent Nanobeads, Quantum Dot Nanobeads and Quantum Dots as Labels in Fluorescence Immunochromatography for Detection of Aflatoxin B1 in Grains. *Biomolecules* **2020**, *10*, 575. [CrossRef]
195. Perez-Fernandez, B.; de la Escosura-Muniz, A. Electrochemical biosensors based on nanomaterials for aflatoxins detection: A review (2015–2021). *Anal. Chim. Acta.* **2022**, *1212*, 339658. [CrossRef]
196. Sun, C.; Li, H.; Koidis, A.; Chen, Q. Quantifying Aflatoxin B1 in peanut oil using fabricating fluorescence probes based on upconversion nanoparticles. *Spectrochim. Acta Part A* **2016**, *165*, 120–126. [CrossRef]
197. Xie, G.; Zhu, M.; Liu, Z.; Zhang, B.; Shi, M.; Wang, S. Development and evaluation of the magnetic particle-based chemiluminescence immunoassay for rapid and quantitative detection of Aflatoxin B1 in foodstuff. *Food Agric. Immunol.* **2017**, *29*, 564–576. [CrossRef]
198. Neelam; Chhillar, A.K.; Rana, J.S. Enzyme nanoparticles and their biosensing applications: A review. *Anal. Biochem.* **2019**, *581*, 113345. [CrossRef] [PubMed]
199. Ding, L.; Bond, A.M.; Zhai, J.; Zhang, J. Utilization of nanoparticle labels for signal amplification in ultrasensitive electrochemical affinity biosensors: A review. *Anal. Chim. Acta.* **2013**, *797*, 1–12. [CrossRef] [PubMed]
200. Peltomaa, R.; Benito-Pena, E.; Moreno-Bondi, M.C. Bioinspired recognition elements for mycotoxin sensors. *Anal. Bioanal. Chem.* **2018**, *410*, 747–771. [CrossRef] [PubMed]
201. Miklos, G.; Angeli, C.; Ambrus, A.; Nagy, A.; Kardos, V.; Zentai, A.; Kerekes, K.; Farkas, Z.; Jozwiak, A.; Bartok, T. Detection of Aflatoxins in Different Matrices and Food-Chain Positions. *Front. Microbiol.* **2020**, *11*, 1916. [CrossRef]
202. Wang, Q.; Yang, Q.; Wu, W. Progress on Structured Biosensors for Monitoring Aflatoxin B1 From Biofilms: A Review. *Front. Microbiol.* **2020**, *11*, 408. [CrossRef]
203. Langone, J.J.; Van Vunakis, H. Aflatoxin B1: Specific antibodies and their use in radioimmunoassay. *J. Natl. Cancer Inst.* **1976**, *56*, 591–595. [CrossRef] [PubMed]
204. Bazin, I.; Tria, S.A.; Hayat, A.; Marty, J.L. New biorecognition molecules in biosensors for the detection of toxins. *Biosens. Bioelectron.* **2017**, *87*, 285–298. [CrossRef]
205. Ma, H.; Sun, J.; Zhang, Y.; Bian, C.; Xia, S.; Zhen, T. Label-free immunosensor based on one-step electrodeposition of chitosan-gold nanoparticles biocompatible film on Au microelectrode for determination of aflatoxin B1 in maize. *Biosens. Bioelectron.* **2016**, *80*, 222–229. [CrossRef] [PubMed]
206. Zhao, F.; Tian, Y.; Shen, Q.; Liu, R.; Shi, R.; Wang, H.; Yang, Z. A novel nanobody and mimotope based immunoassay for rapid analysis of aflatoxin B1. *Talanta* **2019**, *195*, 55–61. [CrossRef]
207. Shi, L.; Wang, Z.; Yang, G.; Yang, H.; Zhao, F. A novel electrochemical immunosensor for aflatoxin B1 based on Au nanoparticles-poly 4-aminobenzoic acid supported graphene. *Appl. Surf. Sci.* **2020**, *527*, 146934. [CrossRef]

208. Pfeiffer, F.; Mayer, G. Selection and Biosensor Application of Aptamers for Small Molecules. *Front. Chem.* **2016**, *4*, 25. [CrossRef] [PubMed]
209. Kong, H.Y.; Byun, J. Nucleic Acid aptamers: New methods for selection, stabilization, and application in biomedical science. *Biomol. Ther.* **2013**, *21*, 423–434. [CrossRef] [PubMed]
210. Kleinjung, F.; Klussmann, S.; Erdmann, V.; Scheller, F.; Fürste, J.; Bier, F. High-affinity RNA as a recognition element in a biosensor. *Anal. Chem.* **1998**, *70*, 328–331. [CrossRef]
211. Ma, X.; Wang, W.; Chen, X.; Xia, Y.; Wu, S.; Duan, N.; Wang, Z. Selection, identification, and application of Aflatoxin B1 aptamer. *Eur. Food Res. Technol.* **2014**, *238*, 919–925. [CrossRef]
212. Vidal, J.C.; Bonel, L.; Ezquerra, A.; Hernandez, S.; Bertolin, J.R.; Cubel, C.; Castillo, J.R. Electrochemical affinity biosensors for detection of mycotoxins: A review. *Biosens. Bioelectron.* **2013**, *49*, 146–158. [CrossRef]
213. Rivas, L.; Mayorga-Martinez, C.C.; Quesada-Gonzalez, D.; Zamora-Galvez, A.; de la Escosura-Muniz, A.; Merkoci, A. Label-free impedimetric aptasensor for ochratoxin-A detection using iridium oxide nanoparticles. *Anal. Chem.* **2015**, *87*, 5167–5172. [CrossRef]
214. McKeague, M.; Bradley, C.R.; De Girolamo, A.; Visconti, A.; Miller, J.D.; Derosa, M.C. Screening and initial binding assessment of fumonisin b(1) aptamers. *Int. J. Mol. Sci.* **2010**, *11*, 4864–4881. [CrossRef]
215. Sabet, F.S.; Hosseini, M.; Khabbaz, H.; Dadmehr, M.; Ganjali, M.R. FRET-based aptamer biosensor for selective and sensitive detection of aflatoxin B1 in peanut and rice. *Food Chem.* **2017**, *220*, 527–532. [CrossRef]
216. Chen, X.; Bai, X.; Li, H.; Zhang, B. Aptamer-based microcantilever array biosensor for detection of fumonisin B-1. *RSC Adv.* **2015**, *5*, 35448–35452. [CrossRef]
217. Guo, X.; Wen, F.; Zheng, N.; Li, S.; Fauconnier, M.L.; Wang, J. A qPCR aptasensor for sensitive detection of aflatoxin M1. *Anal. Bioanal. Chem.* **2016**, *408*, 5577–5584. [CrossRef]
218. Samokhvalov, A.V.; Safenkova, I.V.; Eremin, S.A.; Zherdev, A.V.; Dzantiev, B.B. Use of anchor protein modules in fluorescence polarisation aptamer assay for ochratoxin A determination. *Anal. Chim. Acta* **2017**, *962*, 80–87. [CrossRef] [PubMed]
219. Li, Q.; Lu, Z.; Tan, X.; Xiao, X.; Wang, P.; Wu, L.; Shao, K.; Yin, W.; Han, H. Ultrasensitive detection of aflatoxin B1 by SERS aptasensor based on exonuclease-assisted recycling amplification. *Biosens. Bioelectron.* **2017**, *97*, 59–64. [CrossRef] [PubMed]
220. Abnous, K.; Danesh, N.M.; Alibolandi, M.; Ramezani, M.; Sarreshtehdar Emrani, A.; Zolfaghari, R.; Taghdisi, S.M. A new amplified pi-shape electrochemical aptasensor for ultrasensitive detection of aflatoxin B1. *Biosens. Bioelectron.* **2017**, *94*, 374–379. [CrossRef]
221. Wang, S.; Zhang, Y.; Pang, G.; Zhang, Y.; Guo, S. Tuning the Aggregation/Disaggregation Behavior of Graphene Quantum Dots by Structure-Switching Aptamer for High-Sensitivity Fluorescent Ochratoxin A Sensor. *Anal. Chem.* **2017**, *89*, 1704–1709. [CrossRef] [PubMed]
222. Zhang, S.; Shen, Y.; Shen, G.; Wang, S.; Shen, G.; Yu, R. Electrochemical immunosensor based on Pd-Au nanoparticles supported on functionalized PDDA-MWCNT nanocomposites for aflatoxin B1 detection. *Anal. Biochem.* **2016**, *494*, 10–15. [CrossRef]
223. Ren, S.; Li, Y.; Guo, Q.; Peng, Y.; Bai, J.; Ning, B.; Gao, Z. Turn-on fluorometric immunosensor for diethylstilbestrol based on the use of air-stable polydopamine-functionalized black phosphorus and upconversion nanoparticles. *Mikrochim. Acta* **2018**, *185*, 429. [CrossRef] [PubMed]
224. Wu, J.; Ali, S.; Ouyang, Q.; Wang, L.; Rong, Y.; Chen, Q. Highly specific and sensitive detection of aflatoxin B1 in food based on upconversion nanoparticles-black phosphorus nanosheets aptasensor. *Microchem. J.* **2021**, *171*, 106847. [CrossRef]
225. Xia, X.; Wang, H.; Yang, H.; Deng, S.; Deng, R.; Dong, Y.; He, Q. Dual-Terminal Stemmed Aptamer Beacon for Label-Free Detection of Aflatoxin B1 in Broad Bean Paste and Peanut Oil via Aggregation-Induced Emission. *J. Agric. Food Chem.* **2018**, *66*, 12431–12438. [CrossRef]
226. Yang, Y.; Yin, Y.; Li, X.; Wang, S.; Dong, Y. Development of a chimeric aptamer and an AuNPs aptasensor for highly sensitive and specific identification of Aflatoxin B1. *Sens. Actuators B* **2020**, *319*, 128250. [CrossRef]
227. Zhong, T.; Li, S.; Li, X.; JiYe, Y.; Mo, Y.; Chen, L.; Zhang, Z.; Wu, H.; Li, M.; Luo, Q. A label-free electrochemical aptasensor based on AuNPs-loaded zeolitic imidazolate framework-8 for sensitive determination of aflatoxin B1. *Food Chem.* **2022**, *384*, 132495. [CrossRef]
228. Wang, N.; Liu, Q.; Hu, X.; Wang, F.; Hu, M.; Yu, Q.; Zhang, G. Electrochemical immunosensor based on AuNPs/Zn/Ni-ZIF-8-800@graphene for rapid detection of aflatoxin B1 in peanut oil. *Anal. Biochem.* **2022**, *650*, 114710. [CrossRef] [PubMed]
229. Zhao, Y.; Chen, F.; Li, Q.; Wang, L.; Fan, C. Isothermal Amplification of Nucleic Acids. *Chem. Rev.* **2015**, *115*, 12491–12545. [CrossRef] [PubMed]
230. Hu, Y.; Cecconello, A.; Idili, A.; Ricci, F.; Willner, I. Triplex DNA Nanostructures: From Basic Properties to Applications. *Angew. Chem. Int. Ed. Engl.* **2017**, *56*, 15210–15233. [CrossRef] [PubMed]
231. Bai, Y.; Zhang, H.; Zhao, L.; Wang, Y.; Chen, X.; Zhai, H.; Tian, M.; Zhao, R.; Wang, T.; Xu, H.; et al. A novel aptasensor based on HCR and G-quadruplex DNAzyme for fluorescence detection of Carcinoembryonic Antigen. *Talanta* **2021**, *221*, 121451. [CrossRef]
232. Zhang, Q. Application of Hybridization Chain Reaction (HCR) in Electrochemical Analysis. *Int. J. Electrochem. Sci.* **2022**, *17*, 220227. [CrossRef]
233. Dirks, R.M.; Pierce, N.A. Triggered amplification by hybridization chain reaction. *Proc. Natl. Acad. Sci. USA* **2004**, *101*, 15275–15278. [CrossRef]
234. Evanko, D. Hybridization chain reaction. *Nat. Methods* **2004**, *1*, 186. [CrossRef]

235. Zhou, R.; Zeng, Z.; Sun, R.; Liu, W.; Zhu, Q.; Zhang, X.; Chen, C. Traditional and new applications of the HCR in biosensing and biomedicine. *Analyst* **2021**, *146*, 7087–7103. [CrossRef]
236. Wang, Q.; Zhao, F.; Yang, Q.; Wu, W. Graphene oxide quantum dots based nanotree illuminates AFB1: Dual signal amplified aptasensor detection AFB1. *Sens. Actuators B* **2021**, *345*, 130387. [CrossRef]
237. Zou, L.; Zhang, M.; Li, M.; Xiao, Z.; Ling, L. Hybridization chain reaction and DNAzyme-based dual signal amplification strategy for sensitive fluorescent sensing of aflatoxin B1 by using the pivot of triplex DNA. *Food Res. Int.* **2022**, *158*, 111538. [CrossRef]
238. Molaei, M.J. Principles, mechanisms, and application of carbon quantum dots in sensors: A review. *Anal. Methods* **2020**, *12*, 1266–1287. [CrossRef]
239. Lim, S.Y.; Shen, W.; Gao, Z. Carbon quantum dots and their applications. *Chem. Soc. Rev.* **2015**, *44*, 362–381. [CrossRef]
240. Tajik, S.; Dourandish, Z.; Zhang, K.; Beitollahi, H.; Le, Q.V.; Jang, H.W.; Shokouhimehr, M. Carbon and graphene quantum dots: A review on syntheses, characterization, biological and sensing applications for neurotransmitter determination. *RSC Adv.* **2020**, *10*, 15406–15429. [CrossRef] [PubMed]
241. Ye, J.; Zheng, M.; Ma, H.; Xuan, Z.; Tian, W.; Liu, H.; Wang, S.; Zhang, Y. Development and Validation of an Automated Magneto-Controlled Pretreatment for Chromatography-Free Detection of Aflatoxin B1 in Cereals and Oils through Atomic Absorption Spectroscopy. *Toxins* **2022**, *14*, 454. [CrossRef]
242. Lu, Z.; Chen, X.; Wang, Y.; Zheng, X.; Li, C.M. Aptamer based fluorescence recovery assay for aflatoxin B1 using a quencher system composed of quantum dots and graphene oxide. *Microchim. Acta* **2014**, *182*, 571–578. [CrossRef]
243. Hussain, A.; Sun, D.W.; Pu, H. Bimetallic core shelled nanoparticles (Au@AgNPs) for rapid detection of thiram and dicyandiamide contaminants in liquid milk using SERS. *Food Chem.* **2020**, *317*, 126429. [CrossRef]
244. Jiao, T.; Ahmad, W.; Zhu, J.; Hassan, M.M.; Wang, J.; Rong, Y.; Guo, Z.; Li, H.; Ding, Z.; Lv, C.; et al. Aggregation triggered aflatoxin B1 determination in foodstuff employing 5-aminotetramethylrhodamine decorated gold–silver core–shell nanoparticles in surface enhanced Raman scattering. *Sens. Actuators B* **2021**, *331*, 129424. [CrossRef]
245. Xiong, Z.; Wang, Q.; Xie, Y.; Li, N.; Yun, W.; Yang, L. Simultaneous detection of aflatoxin B1 and ochratoxin A in food samples by dual DNA tweezers nanomachine. *Food Chem.* **2021**, *338*, 128122. [CrossRef] [PubMed]
246. Wu, S.S.; Wei, M.; Wei, W.; Liu, Y.; Liu, S. Electrochemical aptasensor for aflatoxin B1 based on smart host-guest recognition of β-cyclodextrin polymer. *Biosens. Bioelectron.* **2019**, *129*, 58–63. [CrossRef] [PubMed]
247. Ahmad, O.S.; Bedwell, T.S.; Esen, C.; Garcia-Cruz, A.; Piletsky, S.A. Molecularly Imprinted Polymers in Electrochemical and Optical Sensors. *Trends Biotechnol.* **2019**, *37*, 294–309. [CrossRef]
248. Uzun, L.; Turner, A.P. Molecularly-imprinted polymer sensors: Realising their potential. *Biosens. Bioelectron.* **2016**, *76*, 131–144. [CrossRef] [PubMed]
249. BelBruno, J.J. Molecularly Imprinted Polymers. *Chem. Rev.* **2019**, *119*, 94–119. [CrossRef] [PubMed]
250. Li, S.; Ge, Y.; Piletsky, S.A.; Lunec, J. *Molecularly Imprinted Sensors: Overview and Applications*; Elsevier: Amsterdam, The Netherlands, 2012.
251. Haupt, K. *Molecular Imprinting*; Springer Science & Business Media: Berlin/Heidelberg, Germany, 2012; Volume 325.
252. Mirsky, V.M.; Yatsimirsky, A. *Artificial Receptors for Chemical Sensors*; John Wiley & Sons: Hoboken, NJ, USA, 2010.
253. Sergey, A.; Piletsky, M.J.W. *Designing Receptors for the Next Generation of Biosensors*, 1st ed.; Springer: Berlin/Heidelberg, Germany, 2013; Volume 12, p. 264.
254. Lee, S.-W.; Kunitake, T. *Handbook of Molecular Imprinting: Advanced Sensor Applications*; CRC Press: Boca Raton, FL, USA, 2012.
255. Wang, S.; Zhu, X.; Zhao, M.; Li, S.; Le, Y.; Li, H. Optical sensors based on molecularly imprinted nanomaterials. In *Smart Nanomaterials for Sensor Application*; Bentham: Oak Park, IL, USA, 2012; pp. 60–75.
256. Alvarez-Lorenzo, C. *Handbook of Molecularly Imprinted Polymers*; Smithers Rapra: Shawbury, UK, 2013.
257. Chen, L.; Wang, X.; Lu, W.; Wu, X.; Li, J. Molecular imprinting: Perspectives and applications. *Chem. Soc. Rev.* **2016**, *45*, 2137–2211. [CrossRef]
258. Sergeyeva, T.; Yarynka, D.; Piletska, E.; Linnik, R.; Zaporozhets, O.; Brovko, O.; Piletsky, S.; El'skaya, A. Development of a smartphone-based biomimetic sensor for aflatoxin B1 detection using molecularly imprinted polymer membranes. *Talanta* **2019**, *201*, 204–210. [CrossRef]
259. Gu, Y.; Wang, Y.; Wu, X.; Pan, M.; Hu, N.; Wang, J.; Wang, S. Quartz crystal microbalance sensor based on covalent organic framework composite and molecularly imprinted polymer of poly(o-aminothiophenol) with gold nanoparticles for the determination of aflatoxin B1. *Sens. Actuators B* **2019**, *291*, 293–297. [CrossRef]
260. Gui, R.; Jin, H.; Guo, H.; Wang, Z. Recent advances and future prospects in molecularly imprinted polymers-based electrochemical biosensors. *Biosens. Bioelectron.* **2018**, *100*, 56–70. [CrossRef] [PubMed]
261. Cavaliere, C.; Foglia, P.; Guarino, C.; Nazzari, M.; Samperi, R.; Lagana, A. Determination of aflatoxins in olive oil by liquid chromatography-tandem mass spectrometry. *Anal. Chim. Acta* **2007**, *596*, 141–148. [CrossRef]
262. Tan, H.; Ma, L.; Guo, T.; Zhou, H.; Chen, L.; Zhang, Y.; Dai, H.; Yu, Y. A novel fluorescence aptasensor based on mesoporous silica nanoparticles for selective and sensitive detection of aflatoxin B1. *Anal. Chim. Acta* **2019**, *1068*, 87–95. [CrossRef]
263. Xia, X.; Wang, Y.; Yang, H.; Dong, Y.; Zhang, K.; Lu, Y.; Deng, R.; He, Q. Enzyme-free amplified and ultrafast detection of aflatoxin B1 using dual-terminal proximity aptamer probes. *Food Chem.* **2019**, *283*, 32–38. [CrossRef]
264. Xiao, M.W.; Bai, X.L.; Liu, Y.M.; Yang, L.; Liao, X. Simultaneous determination of trace Aflatoxin B1 and Ochratoxin A by aptamer-based microchip capillary electrophoresis in food samples. *J. Chromatogr. A* **2018**, *1569*, 222–228. [CrossRef]

265. Li, R.; Wang, X.; Zhou, T.; Yang, D.; Wang, Q.; Zhou, Y. Occurrence of four mycotoxins in cereal and oil products in Yangtze Delta region of China and their food safety risks. *Food Control* **2014**, *35*, 117–122. [CrossRef]
266. Chen, Y.; Yang, Y.; Wang, Y.; Peng, Y.; Nie, J.; Gao, G.; Zhi, J. Development of an Escherichia coli-based electrochemical biosensor for mycotoxin toxicity detection. *Bioelectrochemistry* **2020**, *133*, 107453. [CrossRef] [PubMed]

molecules

Article

Analysis and Comparison of Aroma Compounds of Brown Sugar in Guangdong, Guangxi and Yunnan Using GC-O-MS

Erbao Chen [1], Shuna Zhao [1,2,*,†], Huanlu Song [1,*,†], Yu Zhang [1] and Wanyao Lu [2,3]

1 Beijing Engineering and Technology Research Center of Food Additives, School of Food and Health, Beijing Technology and Business University, Beijing 100048, China
2 Beijing Engineering Laboratory of Geriatric Nutrition & Foods, COFCO Nutrition and Health Research Institute Co., Ltd., Beijing 102209, China
3 COFCO Sugar Co., Ltd., Key Laboratory of Quality & Safety Control for Sugar Crops and Tomato, Ministry of Agriculture of the PRC, Changji 831100, China
* Correspondence: zhaoshuna@cofco.com (S.Z.); songhl@th.btbu.edu.cn (H.S.)
† These authors contributed equally to this work.

Abstract: Guangdong, Guangxi and Yunnan are the three provinces in China that yield the most brown sugar, a brown-red colored solid or powdered sugar product made from sugar cane. In the present study, the differences between odor compounds of brown sugar from Guangdong, Guangxi, and Yunnan provinces in China were compared and analyzed by gas chromatography-olfactometry-mass spectrometry (GC-O-MS). A total of 80 odor compounds, including 5 alcohols, 9 aldehydes, 8 phenols, 21 acids, 14 ketones, 5 esters, 12 pyrazines, and 6 other compounds, were detected. The fingerprint analysis of the brown sugar odor compounds showed 90% similarity, indicating a close relationship among the odor properties of brown sugar in each province. Moreover, the orthogonal partial least squares discriminant analysis (OPLS-DA) was performed to identify the compounds contributing to the volatile classification of the brown sugar from three provinces, which confirmed that OPLS-DA could be a potential tool to distinguish the brown sugar of three origins.

Keywords: non-centrifugal cane sugar (NCS); GC-O-MS; fingerprint; orthogonal partial least squares discriminant analysis (OPLS-DA)

Citation: Chen, E.; Zhao, S.; Song, H.; Zhang, Y.; Lu, W. Analysis and Comparison of Aroma Compounds of Brown Sugar in Guangdong, Guangxi and Yunnan Using GC-O-MS. *Molecules* **2022**, *27*, 5878. https://doi.org/10.3390/molecules27185878

Academic Editors: Weiying Lu and Yanping Chen

Received: 15 August 2022
Accepted: 1 September 2022
Published: 10 September 2022

1. Introduction

Brown sugar, a traditional sweetener with a distinctive flavor, is mainly made from sugarcane through extraction, clarification, and boiling [1]. It is also called non-centrifugal cane sugar (NCS), which does not separate molasses, so it retains the original flavor and nutrients of sugarcane. Brown sugar is rich in flavonoids and phenols that may act as antioxidants and, therefore, exert benefits on organisms [2–4]. Furthermore, it exerts immunomodulatory, cytoprotective, anti-carcinogenic, and anti-cancer properties [5].

A study on the physicochemical properties and storage stability of brown sugar revealed darker color, increased water content and water activity, but decreased glucose and fructose contents due to the Maillard reaction [6]. Similarly, a study on the odor components of brown sugar revealed that acetaldehyde, 2-methylbutyraldehyde, 3-methylbutyraldehyde, 2,6-dimethylpyrazine, nonanal, 2,6-diethylpyrazine, 2,3,5-trimethy lpyrazine, furfural, 2,3-dimethylpyrazine, decanal, and 2-acetylpyrrole were the primary components based on their relative concentration [7]. Juliana et al. [8] extracted a total of six odor compounds from brown sugar beverages through simultaneous steam distillation-solvent extraction using a mixture of diethyl ether-pentane (1:1, *w/w*) as the solvent. Of the six components, 2-methylpyrazine was the key aroma compound in this beverage. Our previous research has proved that heating of syrup was the primary production step affecting the brown sugar flavor because of the production of a large number of pyrazine compounds [9].

Brown sugar has a green and a strong caramel aroma. Some aroma compounds are inherent in sugarcane, while others are produced by microbial metabolism and Maillard reaction. Sugarcane varieties, growing regions, processing methods, storage conditions and other factors will affect the flavor of brown sugar [10]. The composition and concentration of odor compounds and nutrients in sugarcane from different producing areas are different, which leads to great differences in the flavor composition of brown sugar. However, it is difficult to distinguish the origin of brown sugars only by sensory evaluation. As an intuitive and reproducible method, GC-MS analysis has been effectively applied in origin differentiation studies [11]. Li et al. [12] and Zhao et al. [13] used GC-MS to analyze the volatile odor compounds of ham and rice, respectively, and the results proved that GC-MS played an important role in food odor analysis and origin identification.

Previous studies on brown sugar mostly focused on the identification of key aroma, and there is no study on the flavor differences of brown sugar in different regions. Guangdong, Guangxi and Yunnan are the three major producing areas of brown sugar in China. To the best of our knowledge, the discrimination of brown sugar according to origin has not been reported previously. Therefore, the purpose of this study is to (1) identify the odor compounds of the 18 brown sugar samples from Guangdong, Guangxi, and Yunnan using LLE/GC-O-MS; (2) determine the key odor compounds in brown sugar by calculating OAV; (3) establish the fingerprints of brown sugar from three different origins and (4) find out the compounds that cause the difference using OPLS, so as to provide the basis for selecting brown sugar from different regions when producing foods with different flavor characteristics.

2. Results and Discussion

2.1. Volatile Aroma Components Analysis

A total of 80 odor compounds, including 5 alcohols, 9 aldehydes, 8 phenols, 21 acids, 14 ketones, 5 esters, 12 pyrazines, and 6 other compounds, were detected in 18 samples from three different regions (Table 1). The brown sugar samples from Guangdong, Guangxi and Yunnan contained 72, 60 and 75 odor compounds, respectively. There are four kinds of alcohols in all three regions, but the types of acid compounds are quite different, with Guangdong and Yunnan containing 20 and 19 acid compounds, respectively, while Guangxi contained only 12 acid compounds. The types of pyrazines, aldehydes, ketones and phenols in the three regions are very close. By comparing the odor compounds in the three regions, it was found that the unique odor compounds of the brown sugar samples in Guangdong were 2-acetyl-5-methylpyrazine, 2-methylbutanoic acid and 3-phenylpropionic acid; the unique odor compound in Guangxi was propylene glycol; and the unique odor compounds in Yunnan were 1,3-dimethoxy-2-hydroxybenzene, 3-hydroxyl-2-methyl-4*H*-pyran-4-one, 3-methyl-1,2-cyclopentanedione, 4-methylpentanoic acid and γ-butyrolactone. These unique odor compounds are expected to be important indicators to distinguish the origin of brown sugar samples.

The average contents of odor compounds in brown sugar samples from the three regions are shown in Figure 1. It can be seen that the highest contents of acid compounds were found in all three regions with 25,595.06, 21,632.44 and 25,187.12 ng/g, followed by phenolic compounds with average contents of 111,69.29, 12,115.37 and 11,744.16 ng/g. In contrast, alcohols, esters, pyrazines and ketones had lower contents.

Table 1. Volatile aroma components of brown sugars from different producing areas.

No.	Compounds	Odor description	RI	Identification	Guangdong (ng/g)						Guangxi (ng/g)						Yunnan (ng/g)					
					Guang dong1	Guang dong2	Guang dong3	Guang dong4	Guang dong5	Guang dong6	Guang xi1	Guang xi2	Guang xi3	Guang xi4	Guang xi5	Guang xi6	Yun nan1	Yun nan2	Yun nan3	Yun nan4	Yun nan5	Yun nan6
1	2,3-butanediol	fruit, onion	1568	MS/RI/O	265.09 ± 24.94	184.68 ± 20.31	76.15 ± 9.36	209.60 ± 19.39	83.16 ± 6.00	179.33 ± 13.56	296.97 ± 15.27	-	114.44 ± 10.07	256.59 ± 21.35	81.45 ± 5.45	1019.88 ± 47.28	173.87 ± 12.84	256.31 ± 27.60	466.38 ± 31.83	410.50 ± 35.41	165.52 ± 4.61	50.48 ± 2.32
2	propylene glycol	sweet	1603	MS/RI/O	-	-	-	-	-	-	-	-	75.58 ± 12.71	-	-	-	-	-	-	-	-	-
3	furfuryl alcohol	burnt	1644	MS/RI	474.08 ± 15.80	480.63 ± 13.93	604.00 ± 40.01	130.47 ± 16.31	131.31 ± 6.74	535.66 ± 48.23	2006.99 ± 23.40	1502.75 ± 148.15	675.84 ± 18.94	1027.58 ± 93.40	205.87 ± 11.41	409.98 ± 34.42	881.75 ± 15.61	786.48 ± 36.72	681.50 ± 47.48	589.18 ± 31.74	84.99 ± 5.94	581.65 ± 39.11
4	5-methyl furfuryl alcohol	sweet, caramellic	1705	MS/RI	-	-	-	-	-	295.98 ± 33.01	779.98 ± 180.99	147.60 ± 7.97	-	303.36 ± 121.91	87.16 ± 26.45	325.62 ± 12.36	-	-	-	-	-	189.50 ± 59.19
5	benzyl alcohol	sweet, flower	1865	MS/RI	118.05 ± 16.92	153.98 ± 11.55	299.03 ± 22.63	-	-	17.22 ± 1.07	-	-	-	-	-	-	-	169.70 ± 13.38	-	380.32 ± 22.02	100.30 ± 7.22	-
	Content of total alcohols				857.22	819.29	979.18	340.07	214.47	1028.19	3083.94	1650.35	865.86	1587.53	374.48	1755.48	1055.62	1212.49	1147.88	1380	350.81	821.63
6	hexanal	grass, tallow, fat	1075	MS/RI/O	399.76 ± 21.53	279.98 ± 19.85	313.08 ± 16.32	308.33 ± 6.52	150.05 ± 4.55	260.23 ± 19.05	230.54 ± 19.85	57.51 ± 4.66	130.21 ± 11.55	294.62 ± 15.77	209.07 ± 19.25	182.99 ± 15.58	207.86 ± 7.53	36.23 ± 2.97	231.15 ± 20.79	241.62 ± 17.27	155.67 ± 9.51	182.12 ± 16.88
7	furfural	bread, almond, sweet	1455	MS/RI	-	-	-	50.05 ± 3.69	-	-	97.89 ± 3.70	111.40 ± 5.84	95.82 ± 6.77	-	108.26 ± 5.01	161.29 ± 9.46	-	-	-	-	-	80.62 ± 4.36
8	(*E*)-2-nonenal	cucumber, fat, green	1507	MS/RI/O	56.20 ± 2.12	95.35 ± 2.05	68.21 ± 4.20	71.00 ± 3.80	36.96 ± 3.69	85.51 ± 2.13	-	59.88 ± 0.21	34.18 ± 3.20	60.66 ± 2.67	73.78 ± 5.50	79.64 ± 8.90	44.27 ± 2.08	70.81 ± 6.62	109.32 ± 12.26	80.28 ± 8.59	-	-
9	benzaldehyde	almond, caramel	1514	MS/RI/O	57.98 ± 6.17	109.82 ± 6.24	50.92 ± 3.04	-	-	168.28 ± 3.28	-	-	46.19 ± 5.23	102.28 ± 7.28	52.62 ± 7.13	94.52 ± 2.68	-	49.57 ± 4.48	69.14 ± 6.60	58.67 ± 4.52	54.31 ± 1.43	-
10	5-methyl furfural	almond, caramel	1560	MS/RI	65.10 ± 5.32	60.46 ± 3.14	76.48 ± 2.04	-	-	-	121.80 ± 7.66	58.28 ± 2.64	71.86 ± 6.37	48.57 ± 3.84	27.43 ± 1.07	84.62 ± 4.31	-	74.75 ± 5.59	89.67 ± 8.14	65.65 ± 4.47	-	32.39 ± 2.98
11	2-hydroxy methyl-5-furfural	cardboard	2512	MS/RI	527.94 ± 1.84	543.97 ± 14.99	840.13 ± 26.60	204.68 ± 16.87	-	-	-	835.97 ± 5.84	965.27 ± 12.27	-	387.49 ± 6.35	559.31 ± 6.19	-	182.63 ± 3.23	826.44 ± 9.41	600.61 ± 18.03	-	806.60 ± 45.30
12	4-hydroxy-3-methoxyben zaldehyde	vanilla	2520	MS/RI	1163.20 ± 28.06	938.21 ± 12.55	659.00 ± 12.82	-	930.42 ± 38.08	1330.74 ± 17.04	1123.39 ± 38.79	1146.19 ± 38.02	1256.95 ± 36.57	796.62 ± 26.25	707.06 ± 24.88	813.25 ± 12.95	282.85 ± 23.12	337.09 ± 7.33	1009.19 ± 14.03	694.32 ± 19.89	562.95 ± 26.69	851.26 ± 25.56
13	3,5-dimeth oxy-4-hydroxy benzaldehyde	sweet, cocoa, nutty	2905	MS/RI/O	4115.06 ± 183.57	3970.11 ± 230.98	3258.76 ± 103.02	4530.25 ± 115.07	1861.44 ± 137.60	1393.28 ± 152.55	5558.03 ± 199.50	-	-	-	-	-	4106.21 ± 220.69	2356.17 ± 61.68	4986.34 ± 98.27	1882.87 ± 82.12	1255.94 ± 14.65	3220.19 ± 164.10
14	4-hydroxy benzaldehyde	creamy, musty	2908	MS/RI	-	-	-	-	-	1990.86 ± 53.37	-	6374.49 ± 82.32	6982.29 ± 125.16	-	3014.51 ± 138.30	-	-	-	-	-	-	-

Table 1. *Cont.*

No.	Compounds	Odor description	RI	Identification	Guangdong (ng/g)						Guangxi (ng/g)						Yunnan (ng/g)					
					Guangdong1	Guangdong2	Guangdong3	Guangdong4	Guangdong5	Guangdong6	Guangxi1	Guangxi2	Guangxi3	Guangxi4	Guangxi5	Guangxi6	Yunnan1	Yunnan2	Yunnan3	Yunnan4	Yunnan5	Yunnan6
	Content of total aldehydes				6385.24	5997.9	5266.58	5164.31	2978.87	5228.9	7131.65	8643.72	9582.77	1302.75	4580.22	1975.62	4641.19	3107.25	7321.25	3624.02	2028.87	5173.18
15	2,6-di-tert-butyl-4-methyl phenol	camphor	1904	MS/RI/O	6272.71 ± 42.00	6702.74 ± 109.26	6540.27 ± 49.48	5238.74 ± 73.22	4705.80 ± 38.67	5559.88 ± 55.06	6468.03 ± 79.39	7106.41 ± 74.89	6733.20 ± 53.34	5149.51 ± 72.96	5177.12 ± 46.93	7033.33 ± 25.56	6293.40 ± 47.17	5659.68 ± 67.47	7638.60 ± 69.28	7938.76 ± 88.45	5115.72 ± 52.86	3851.64 ± 64.59
16	4-ethenyl-2-methoxy phenol	clove, curry	2168	MS/RI	3571.11 ± 211.98	3749.85 ± 126.32	3296.40 ± 110.64	3640.61 ± 173.25	2405.55 ± 54.02	3822.00 ± 212.87	3438.47 ± 193.67	1672.14 ± 135.59	4249.96 ± 176.21	4588.34 ± 344.50	1419.61 ± 117.81	2766.58 ± 282.45	3400.23 ± 161.64	3112.76 ± 217.96	4595.74 ± 345.42	3222.71 ± 211.17	2317.34 ± 178.28	1808.79 ± 133.86
17	2,4-di-tert-butylphenol	phenolic	2292	MS/RI	644.13 ± 80.76	456.81 ± 31.92	1052.57 ± 17.01	765.57 ± 39.53	332.60 ± 19.04	947.69 ± 13.41	1292.16 ± 26.55	-	1359.89 ± 78.31	467.47 ± 30.26	323.04 ± 21.63	1207.83 ± 109.14	446.63 ± 20.55	684.35 ± 35.98	675.77 ± 60.09	705.89 ± 49.64	498.28 ± 14.81	558.50 ± 25.05
18	2-methoxy-4-propenylphenol	flower	2250	MS/RI	-	-	-	-	149.00 ± 15.25	-	-	-	-	-	-	-	-	-	-	-	-	391.82 ± 19.80
19	(*E*)-4-propenyl-2-methoxy phenol	flower	2315	MS/RI	630.45 ± 15.19	622.41 ± 11.80	689.57 ± 19.91	-	-	372.06 ± 21.68	605.14 ± 18.49	-	-	-	-	-	619.56 ± 48.92	563.39 ± 38.89	907.11 ± 87.87	662.54 ± 23.60	434.79 ± 35.66	-
20	4-allyl-2,6-dimethoxy phenol	sweet, flower	2510	MS/RI/O	-	-	-	-	-	72.86 ± 9.71	440.73 ± 32.29	-	243.43 ± 15.50	-	-	-	129.33 ± 13.43	-	-	269.60 ± 18.11	132.49 ± 14.66	-
21	4-ethenyl-2,6-dimethoxy-phenol	animal leather	2541	MS/RI	893.16 ± 23.65	799.51 ± 17.03	916.48 ± 36.36	655.80 ± 23.22	321.18 ± 28.46	835.89 ± 38.06	1183.04 ± 31.14	803.61 ± 72.23	1461.19 ± 29.31	1365.67 ± 76.92	746.52 ± 42.58	634.98 ± 13.65	572.43 ± 51.73	807.85 ± 70.45	1249.40 ± 36.45	1069.77 ± 73.55	718.19 ± 34.31	1237.55 ± 79.52
22	2-methoxy-4-acetylphenol	vanilla	2640	MS/RI/O	-	-	-	-	352.34 ± 24.92	-	693.73 ± 55.31	808.51 ± 34.53	638.98 ± 11.43	1094.03 ± 62.69	935.46 ± 64.81	584.08 ± 42.12	855.15 ± 69.12	-	-	-	669.33 ± 59.70	649.86 ± 47.29
	Content of total phenols				12,011.56	12,331.32	12,495.29	10,300.72	8266.47	11,610.38	14,121.3	10,390.67	14,686.65	12,665.02	8601.75	12,226.8	12,316.73	10,828.03	15,066.62	13,869.27	9886.14	8498.16
23	acetic acid	sour	1415	MS/RI/O	10,184.79 ± 464.53	10,757.27 ± 305.67	2778.13 ± 46.39	6350.81 ± 222.21	6432.09 ± 175.47	5900.64 ± 582.42	6650.69 ± 245.63	4210.80 ± 330.79	2977.92 ± 249.84	10,999.92 ± 407.67	5047.46 ± 287.21	11,141.62 ± 393.70	4237.56 ± 172.97	1068.38 ± 53.52	12,748.83 ± 163.89	9209.62 ± 249.19	5570.25 ± 363.18	10,334.03 ± 261.79
24	formic acid	acetic, astringent, fruity	1489	MS/RI/O	249.89 ± 11.00	123.64 ± 7.91	219.04 ± 9.56	195.29 ± 11.26	84.47 ± 6.03	96.96 ± 8.18	-	-	-	-	-	-	91.75 ± 13.52	-	389.25 ± 20.61	426.15 ± 37.81	-	286.39 ± 23.78
25	propanoic acid	pungent, rancid, soy	1526	MS/RI/O	821.24 ± 27.91	619.01 ± 37.09	644.68 ± 16.61	396.79 ± 26.64	303.37 ± 5.15	260.37 ± 16.25	332.73 ± 19.43	147.19 ± 14.68	329.19 ± 2.45	457.22 ± 9.98	280.06 ± 14.46	653.58 ± 28.53	425.26 ± 19.38	299.24 ± 25.79	1143.61 ± 18.50	994.93 ± 49.00	515.40 ± 13.38	344.61 ± 26.73

Table 1. *Cont.*

No.	Compounds	Odor description	RI	Identification	Guangdong (ng/g)						Guangxi (ng/g)						Yunnan (ng/g)					
					Guang dong1	Guang dong2	Guang dong3	Guang dong4	Guang dong5	Guang dong6	Guang xi1	Guang xi2	Guang xi3	Guang xi4	Guang xi5	Guang xi6	Yun nan1	Yun nan2	Yun nan3	Yun nan4	Yun nan5	Yun nan6
26	2-methylpropionic acid	rancid, butter, cheese	1563	MS/RI	519.47 ± 21.24	325.82 ± 19.73	264.85 ± 9.33	120.82 ± 9.90	166.03 ± 2.88	4.00 ± 0.24	-	-	88.45 ± 8.92	278.92 ± 25.60	272.22 ± 18.48	322.66 ± 11.67	-	153.86 ± 29.30	621.60 ± 24.32	768.20 ± 37.42	342.05 ± 10.18	-
27	butanoic acid	rancid, cheese, sweat	1607	MS/RI/O	2660.81 ± 89.68	2644.03 ± 91.30	614.21 ± 9.49	2513.84 ± 26.28	1227.91 ± 39.85	1421.39 ± 43.73	810.96 ± 56.86	192.43 ± 16.79	1269.77 ± 80.47	1178.36 ± 29.86	168.86 ± 8.63	440.72 ± 17.86	1266.55 ± 13.63	837.56 ± 53.38	4204.64 ± 71.97	2118.54 ± 37.34	1154.60 ± 33.74	-
28	3-methyl butanoic acid	sweat, acid, rancid	1665	MS/RI/O	1633.63 ± 125.32	1076.75 ± 113.20	645.70 ± 71.56	306.69 ± 15.93	-	12.90 ± 0.34	271.65 ± 16.23	116.04 ± 14.73	218.33 ± 20.09	884.46 ± 15.52	810.07 ± 53.66	710.73 ± 29.15	1123.49 ± 87.06	566.92 ± 16.79	1390.95 ± 38.95	1609.20 ± 36.46	-	-
29	2-methyl butanoic acid	cheese, sweat	1651	MS/RI/O	-	-	-	-	-	373.25 ± 28.50	-	-	-	-	-	-	-	-	-	-	-	-
30	2-methyl pentanoic acid	buttery, creamy	1728	MS/RI/O	-	-	-	183.26 ± 5.52	-	37.89 ± 3.86	-	-	-	-	-	-	-	-	-	-	302.40 ± 19.20	-
31	4-methyl pentanoic acid	pungent cheese	1820	MS/RI/O	-	-	-	-	-	-	-	-	-	-	-	-	-	-	-	-	119.80 ± 6.83	56.28 ± 4.95
32	hexanoic acid	sweat	1826	MS/RI/O	300.51 ± 7.89	597.73 ± 14.47	487.13 ± 17.43	280.13 ± 27.04	283.56 ± 12.23	435.09 ± 35.20	280.31 ± 18.91	-	155.31 ± 12.61	453.98 ± 23.33	119.62 ± 11.23	112.67 ± 9.39	580.84 ± 13.70	441.59 ± 28.07	727.74 ± 38.34	728.69 ± 15.97	383.25 ± 13.93	78.31 ± 3.80
33	octanoic acid	sweat, cheese	2083	MS/RI/O	228.77 ± 6.47	267.60 ± 21.47	-	-	-	-	-	-	-	-	-	-	-	262.06 ± 13.93	252.12 ± 24.62	487.06 ± 38.82	-	-
34	nonanoic acid	green, fat	2147	MS/RI/O	-	-	-	-	-	413.10 ± 28.63	-	-	-	-	-	-	-	-	-	-	-	227.94 ± 12.15
35	levulinic acid	acidic, sweet, creamy	2312	MS/RI	587.58 ± 22.24	584.31 ± 18.74	589.85 ± 16.89	-	-	-	-	-	306.63 ± 20.02	-	338.16 ± 17.34	530.63 ± 35.66	-	-	653.12 ± 14.73	424.09 ± 35.11	-	-
36	benzoic acid	urine	2392	MS/RI/O	3080.65 ± 130.04	3564.46 ± 241.54	4062.78 ± 303.13	2709.59 ± 234.74	1609.97 ± 79.53	1795.59 ± 129.38	3194.93 ± 194.09	1982.81 ± 136.80	2736.59 ± 235.52	3573.53 ± 145.40	1480.40 ± 134.07	2837.50 ± 175.53	2965.81 ± 197.60	1857.28 ± 192.93	4446.03 ± 192.35	4751.21 ± 272.41	1872.63 ± 65.71	2262.00 ± 138.28
37	dodecanoic acid	metal	2517	MS/RI	555.78 ± 15.49	708.53 ± 18.03	575.78 ± 16.57	215.12 ± 19.98	-	-	-	-	-	-	-	-	-	438.26 ± 33.22	750.95 ± 39.27	504.14 ± 42.56	222.18 ± 21.86	-
38	phenylacetic acid	honey, flower	2551	MS/RI/O	1468.96 ± 56.27	1173.77 ± 75.04	1187.97 ± 92.84	880.89 ± 63.45	730.30 ± 47.81	47.37 ± 3.63	1267.41 ± 64.32	-	945.30 ± 61.58	1180.95 ± 82.76	741.94 ± 40.05	1353.01 ± 89.13	526.12 ± 26.32	903.11 ± 25.21	1487.08 ± 19.13	1976.49 ± 49.24	747.98 ± 54.78	606.57 ± 35.96
39	3-phenylpropionic acid	balsamic	2650	MS/RI/O	-	-	-	156.03 ± 11.33	-	10.79 ± 0.95	-	-	-	-	-	-	-	-	-	-	-	-

Table 1. *Cont.*

No.	Compounds	Odor description	RI	Identification	Guangdong (ng/g)						Guangxi (ng/g)						Yunnan (ng/g)					
					Guangdong1	Guangdong2	Guangdong3	Guangdong4	Guangdong5	Guangdong6	Guangxi1	Guangxi2	Guangxi3	Guangxi4	Guangxi5	Guangxi6	Yunnan1	Yunnan2	Yunnan3	Yunnan4	Yunnan5	Yunnan6
40	tetradecanoic acid	sweet, spicy, carnation	2674	MS/RI	821.53 ± 67.39	1273.66 ± 98.64	457.82 ± 19.58	-	-	173.22 ± 16.83	-	-	-	-	-	-	-	498.30 ± 17.36	975.39 ± 21.75	631.33 ± 10.41	252.25 ± 17.50	-
41	pentadecanoic acid	waxy	2784	MS/RI/O	904.37 ± 37.76	1205.19 ± 44.08	867.02 ± 48.07	423.32 ± 20.01	-	806.02 ± 71.41	586.50 ± 17.13	-	-	-	-	-	-	733.11 ± 17.63	1093.45 ± 76.47	886.36 ± 59.46	381.02 ± 26.80	-
42	3-phenyl-2-propenoic acid	balsam, sweet, storax	2815	MS/RI	-	502.70 ± 18.02	-	-	329.63 ± 27.74	555.52 ± 31.40	693.22 ± 19.57	-	-	1366.71 ± 46.48	904.32 ± 31.86	502.43 ± 39.39	-	-	871.20 ± 27.75	491.13 ± 25.65	416.66 ± 18.74	-
43	n-hexadecanoic acid	fatty	2903	MS/RI	11,549.66 ± 283.92	12,914.61 ± 160.72	8039.99 ± 159.00	6586.75 ± 270.87	4432.49 ± 192.77	8965.43 ± 247.84	6422.49 ± 176.26	13,536.74 ± 152.68	3748.63 ± 233.13	17,908.48 ± 240.92	5811.71 ± 96.26	3458.70 ± 197.36	9174.60 ± 92.17	6027.10 ± 231.08	15,086.68 ± 176.07	9036.63 ± 202.89	3926.92 ± 280.51	4354.01 ± 214.68
	Content of total carboxylic acids				35,567.64	38,339.08	21,434.95	21,319.33	15,599.82	21,309.53	20,510.89	20,186.01	12,776.12	38,282.53	15,974.82	22,064.25	20,391.98	14,086.77	46,842.64	35,043.77	16,207.39	18,550.14
44	2-methyl-4,5-dihydro-3(2*H*)-furanone	nutty, creamy	1253	MS/RI/O	341.22 ± 28.42	1177.87 ± 50.20	513.32 ± 23.30	276.67 ± 4.52	70.39 ± 4.46	327.17 ± 28.81	202.04 ± 17.85	291.29 ± 28.07	93.57 ± 6.69	712.73 ± 89.42	141.35 ± 12.32	173.07 ± 3.56	701.59 ± 29.98	1141.42 ± 76.00	391.82 ± 12.71	578.35 ± 18.67	726.04 ± 22.75	1389.22 ± 47.00
45	3-hydroxy-2-butanone	butter, cream	1272	MS/RI	91.84 ± 7.28	126.67 ± 7.48	124.74 ± 9.98	103.62 ± 3.31	-	163.28 ± 2.03	253.98 ± 9.70	145.05 ± 8.21	47.80 ± 1.02	123.45 ± 4.66	32.46 ± 4.69	68.52 ± 1.15	97.36 ± 7.88	74.92 ± 5.14	79.07 ± 3.53	80.41 ± 3.66	56.30 ± 2.47	105.69 ± 8.71
46	1-hydroxy-2-propanone	sweet	1287	MS/RI/O	457.94 ± 39.12	582.30 ± 18.77	605.26 ± 9.93	438.74 ± 12.52	206.31 ± 12.44	652.59 ± 47.59	841.13 ± 59.74	715.56 ± 23.75	503.38 ± 2.41	505.10 ± 33.50	220.33 ± 5.58	211.55 ± 20.44	458.82 ± 7.71	360.04 ± 29.17	382.45 ± 17.01	461.70 ± 11.83	355.89 ± 14.98	438.74 ± 30.24
47	1-hydroxy-2-butanone	oily, alcoholic	1375	MS/RI/O	39.22 ± 2.39	111.06 ± 5.97	41.37 ± 3.74	43.91 ± 3.57	-	16.70 ± 0.68	-	76.40 ± 7.75	82.88 ± 4.40	62.58 ± 4.71	-	24.26 ± 1.31	56.34 ± 6.77	46.29 ± 6.89	63.27 ± 6.83	56.43 ± 3.28	-	-
48	1-acetoxy-2-propanone	fruity, nutty	1451	MS/RI/O	118.22 ± 4.77	230.25 ± 15.43	187.71 ± 12.77	168.77 ± 15.67	-	66.52 ± 6.06	218.30 ± 12.12	159.99 ± 17.64	115.45 ± 11.12	245.41 ± 19.90	139.08 ± 6.91	235.11 ± 24.82	176.14 ± 11.14	191.49 ± 11.15	132.61 ± 7.14	177.62 ± 9.60	125.30 ± 4.14	131.45 ± 4.25
49	4,5-dihydro-5-methyl-2(3*H*)-furanone	sweet, cocoa, woody	1590	MS/RI	65.34 ± 7.38	113.07 ± 5.93	89.27 ± 2.10	-	-	-	-	-	-	-	-	-	-	-	-	90.38 ± 5.27	-	-
50	2(5*H*)-furanone	buttery	1727	MS/RI/O	475.08 ± 10.84	581.86 ± 9.05	479.55 ± 13.75	-	-	145.49 ± 10.40	394.39 ± 11.56	265.72 ± 18.69	286.88 ± 27.60	221.84 ± 19.51	203.23 ± 15.77	363.54 ± 28.70	-	232.91 ± 14.77	712.75 ± 37.46	266.08 ± 24.57	-	154.22 ± 8.91
51	3-methyl-1,2-cyclopentanedione	sweet, maple, bready	1781	MS/RI/O	-	-	-	-	-	-	-	-	-	-	-	-	-	-	-	-	-	361.67 ± 24.53

Table 1. *Cont.*

No.	Compounds	Odor description	RI	Identification	Guangdong (ng/g)						Guangxi (ng/g)						Yunnan (ng/g)					
					Guang dong1	Guang dong2	Guang dong3	Guang dong4	Guang dong5	Guang dong6	Guang xi1	Guang xi2	Guang xi3	Guang xi4	Guang xi5	Guang xi6	Yun nan1	Yun nan2	Yun nan3	Yun nan4	Yun nan5	Yun nan6
52	2-hydroxy-3-methyl-2-cyclopenten-1-one	caramellic	1807	MS/RI/O	468.60 ± 24.21	482.48 ± 34.60	707.42 ± 68.00	283.22 ± 26.02	182.61 ± 11.36	588.56 ± 57.85	879.99 ± 59.16	726.67 ± 69.62	745.18 ± 66.72	519.02 ± 24.12	186.46 ± 22.90	476.39 ± 46.95	681.71 ± 49.56	637.34 ± 38.58	716.33 ± 14.99	684.63 ± 28.94	248.07 ± 2.51	-
53	3-hydroxyl-2-methyl-4H-pyran-4-one	caramel	1931	MS/RI	-	-	-	-	-	-	-	-	-	-	-	-	-	-	-	-	201.59 ± 23.74	94.06 ± 7.11
54	2(3*H*)-furanone	cotton candy	2002	MS/RI/O	974.60 ± 62.46	721.47 ± 64.38	1069.93 ± 92.52	351.84 ± 39.07	292.08 ± 13.56	470.52 ± 30.39	631.95 ± 22.11	615.99 ± 36.85	655.84 ± 55.61	705.84 ± 16.73	496.09 ± 31.24	977.50 ± 56.20	560.98 ± 8.63	596.89 ± 59.04	1123.58 ± 46.28	1324.65 ± 17.01	397.11 ± 16.87	323.38 ± 29.23
55	2,5-dimethyl-4-hydroxy-3(2*H*)-furanone	caramel	2012	MS/RI/O	271.76 ± 26.35	326.77 ± 19.05	454.38 ± 18.37	230.03 ± 13.34	123.31 ± 4.96	731.60 ± 15.54	2168.59 ± 55.50	1158.64 ± 13.07	298.58 ± 23.33	471.80 ± 39.52	69.64 ± 3.26	144.05 ± 13.78	415.29 ± 14.57	861.94 ± 90.58	625.51 ± 26.54	529.53 ± 18.15	-	141.69 ± 16.00
56	4-hydroxy-5-methyl-3-(2*H*)-furanone	caramel	2113	MS/RI/O	-	-	-	181.54 ± 21.75	-	19.69 ± 1.35	612.25 ± 48.80	-	-	-	-	-	-	-	-	-	-	99.22 ± 7.93
57	4-hydroxy acetophenone	sweet	2958	MS/RI/O	1461.87 ± 17.59	1892.62 ± 25.11	1469.22 ± 34.48	-	-	-	1546.24 ± 57.40	-	-	-	-	-	-	1656.02 ± 57.98	2234.53 ± 34.42	1370.88 ± 77.57	1132.23 ± 64.94	-
	Content of total ketones				4765.69	6346.42	5742.17	2078.34	874.7	3182.12	7748.86	4155.31	2829.56	3567.77	1488.64	2673.99	3148.23	5799.26	6461.92	5620.66	3242.53	3239.34
58	dimethyl butanedioate	sweet, fruity, green	1558	MS/RI	-	-	-	-	-	-	-	-	-	590.16 ± 16.86	-	-	120.03 ± 12.32	-	-	-	-	-
59	γ-butyro lactone	caramel, sweet	1647	MS/RI	-	-	-	-	-	-	-	-	-	-	-	-	-	-	-	-	-	193.04 ± 34.38
60	dimethyl glutarate	floral	1687	MS/RI	-	-	-	-	-	-	-	-	-	2030.19 ± 99.86	-	-	1039.76 ± 40.69	-	-	-	-	-
61	benzyl benzoate	balsamic, oil, herb	2592	MS/RI	-	-	-	205.81 ± 17.70	-	503.82 ± 28.75	621.78 ± 52.33	-	503.64 ± 41.22	-	431.15 ± 37.87	482.89 ± 31.13	510.18 ± 36.70	381.49 ± 27.23	682.59 ± 34.84	514.18 ± 27.27	-	-
62	dibutyl phthalate	faint odor	2705	MS/RI	1264.25 ± 33.84	1113.96 ± 65.74	810.31 ± 68.93	1061.36 ± 22.53	979.13 ± 65.98	373.31 ± 25.22	1089.36 ± 16.24	-	1202.60 ± 47.72	977.92 ± 29.04	735.97 ± 19.95	-	2530.75 ± 52.67	-	1333.62 ± 38.03	2536.41 ± 162.39	955.08 ± 69.59	1694.09 ± 93.37
	Content of total esters				1264.25	1113.96	810.31	1267.17	979.13	877.13	1711.14	0	1706.24	3598.27	1167.12	482.89	4200.72	381.49	2016.21	3050.59	955.08	1887.13

Table 1. *Cont.*

No.	Compounds	Odor description	RI	Identification	Guangdong (ng/g)						Guangxi (ng/g)						Yunnan (ng/g)					
					Guangdong1	Guangdong2	Guangdong3	Guangdong4	Guangdong5	Guangdong6	Guangxi1	Guangxi2	Guangxi3	Guangxi4	Guangxi5	Guangxi6	Yunnan1	Yunnan2	Yunnan3	Yunnan4	Yunnan5	Yunnan6
63	2-methylpyrazine	popcorn	1259	MS/RI	366.92 ± 30.48	662.33 ± 40.09	159.11 ± 11.85	402.41 ± 7.65	112.41 ± 5.11	854.10 ± 77.77	651.18 ± 27.24	500.27 ± 17.30	235.82 ± 8.38	561.03 ± 42.71	76.46 ± 1.56	46.44 ± 5.46	289.74 ± 9.95	196.65 ± 14.01	172.90 ± 15.26	203.79 ± 21.19	88.40 ± 1.38	-
64	2,5-dimethylpyrazine	cocoa, nutty, roast beef	1321	MS/RI/O	924.49 ± 34.20	1938.05 ± 101.57	111.73 ± 10.50	963.30 ± 10.96	201.81 ± 14.82	1827.76 ± 35.72	1343.97 ± 86.27	989.09 ± 53.50	1309.65 ± 15.59	1748.65 ± 121.67	88.92 ± 6.07	70.87 ± 2.23	1114.25 ± 44.64	703.42 ± 34.33	419.79 ± 28.03	430.64 ± 21.63	375.15 ± 3.73	93.56 ± 8.17
65	2,6-dimethylpyrazine	nutty, cocoa, roast beef	1326	MS/RI/O	682.76 ± 22.04	1035.34 ± 80.61	144.76 ± 15.74	983.96 ± 18.81	72.49 ± 6.85	1094.56 ± 26.51	1444.15 ± 48.60	1045.87 ± 46.22	800.00 ± 42.14	946.87 ± 86.05	68.92 ± 14.85	85.72 ± 11.44	693.95 ± 41.35	273.74 ± 19.98	230.77 ± 22.95	289.35 ± 16.10	139.77 ± 2.60	208.04 ± 17.25
66	2,3-dimethylpyrazine	nutty, cocoa, meat	1343	MS/RI	125.46 ± 9.27	231.83 ± 13.14	56.13 ± 2.20	-	38.84 ± 3.35	325.66 ± 36.06	253.09 ± 14.24	142.65 ± 9.66	72.47 ± 5.15	216.09 ± 18.81	10.89 ± 1.38	-	152.68 ± 11.65	73.56 ± 4.48	75.96 ± 6.30	70.09 ± 5.01	42.31 ± 1.02	-
67	2-ethyl-6-methylpyrazine	roasted hazelnut	1382	MS/RI	69.27 ± 2.41	157.44 ± 5.95	-	92.61 ± 1.87	13.81 ± 2.78	88.18 ± 9.04	165.64 ± 5.86	114.68 ± 7.43	58.06 ± 6.05	167.14 ± 12.11	14.91 ± 1.33	19.91 ± 1.53	153.89 ± 9.44	66.03 ± 4.67	52.17 ± 4.62	46.08 ± 2.79	21.31 ± 0.76	5.15 ± 0.38
68	2-ethyl-5-methylpyrazine	fruit, sweet	1376	MS/RI	55.63 ± 6.94	227.21 ± 18.30	-	-	-	-	151.24 ± 10.05	139.24 ± 7.77	123.72 ± 3.04	178.42 ± 6.03	-	21.87 ± 0.65	88.57 ± 8.00	105.86 ± 13.64	129.60 ± 16.42	77.52 ± 8.13	-	-
69	2,3,5-trimethylpyrazine	roast, potato, must	1405	MS/RI/O	-	-	-	180.76 ± 15.90	35.78 ± 3.42	306.68 ± 21.34	249.04 ± 14.64	250.84 ± 14.85	40.86 ± 3.21	-	46.57 ± 3.30	-	-	125.18 ± 7.11	-	-	-	33.86 ± 3.99
70	2,5-dimethyl-3-ethylpyrazine	potato, roast	1445	MS/RI/O	151.21 ± 12.91	315.82 ± 12.23	-	157.83 ± 2.51	-	338.49 ± 32.43	-	-	-	-	-	-	109.14 ± 7.37	201.75 ± 16.35	138.27 ± 14.70	119.00 ± 1.38	-	-
71	2,6-dimethyl-3-ethylpyrazine	potato	1455	MS/RI/O	-	-	-	-	79.78 ± 4.61	186.16 ± 12.73	412.18 ± 36.51	272.34 ± 16.48	203.53 ± 6.05	286.15 ± 18.85	-	-	-	-	-	-	82.88 ± 8.05	-
72	2-methyl-6-vinylpyrazine	hazelnut	1487	MS/RI	-	-	-	-	-	177.67 ± 24.80	688.17 ± 39.07	665.08 ± 41.61	311.55 ± 14.08	-	181.63 ± 8.49	550.08 ± 34.15	-	133.21 ± 8.77	-	-	-	-
73	2-acetyl-5-methylpyrazine	popcorn	1664	MS/RI/O	-	-	-	156.34 ± 13.27	-	218.85 ± 16.63	-	-	-	-	-	-	-	-	-	-	-	-
74	2-acetyl-6-methylpyrazine	coffee, cocoa, popcorn	1673	MS/RI/O	251.42 ± 21.79	419.45 ± 28.68	-	141.33 ± 9.04	-	245.73 ± 14.16	183.55 ± 23.20	144.23 ± 7.83	121.93 ± 11.22	218.58 ± 16.46	-	-	-	408.61 ± 9.22	-	113.67 ± 8.19	96.94 ± 7.01	-
	Content of total pyrazines				2627.16	4987.47	471.73	3078.54	554.92	5663.84	5542.21	4264.29	3277.59	4322.93	488.3	794.89	2602.22	2288.01	1219.46	1350.14	846.76	340.61
75	2-acetyl pyrrole	nutty, walnut, bread	1947	MS/RI	679.69 ± 54.14	856.53 ± 39.09	1676.37 ± 66.60	531.22 ± 38.42	111.84 ± 5.12	791.31 ± 37.88	2227.41 ± 88.13	2964.65 ± 58.72	2703.33 ± 101.11	747.17 ± 33.20	461.32 ± 47.83	1263.31 ± 41.84	855.32 ± 80.10	1610.00 ± 150.67	942.57 ± 99.06	1761.17 ± 84.12	1298.42 ± 59.86	971.33 ± 70.56

Table 1. Cont.

No.	Compounds	Odor description	RI	Identification	Guangdong (ng/g)						Guangxi (ng/g)						Yunnan (ng/g)					
					Guang dong1	Guang dong2	Guang dong3	Guang dong4	Guang dong5	Guang dong6	Guang xi1	Guang xi2	Guang xi3	Guang xi4	Guang xi5	Guang xi6	Yun nan1	Yun nan2	Yun nan3	Yun nan4	Yun nan5	Yun nan6
76	2-acetyl furan	balsamic	1490	MS/RI/O	-	-	-	-	-	5.27 ± 0.85	-	231.24 ± 17.91	103.74 ± 7.62	-	37.99 ± 2.05	72.73 ± 1.11	-	-	-	-	-	54.65 ± 3.48
77	(+)-limonene	citrus, mint	1201	MS/RI	-	-	-	-	48.67 ± 3.12	-	210.31 ± 21.01	-	-	-	71.37 ± 2.85	99.25 ± 2.63	69.37 ± 7.74	69.29 ± 5.81	89.42 ± 6.55	51.53 ± 3.80	-	-
78	phenylethylene	balsamic, gasoline	1247	MS/RI/O	651.29 ± 54.83	613.81 ± 14.74	573.68 ± 15.42	451.99 ± 11.84	160.07 ± 12.09	457.61 ± 37.97	546.32 ± 24.83	469.44 ± 28.01	511.72 ± 22.34	984.77 ± 64.37	984.42 ± 73.21	1025.24 ± 14.21	500.79 ± 9.91	459.13 ± 34.72	564.51 ± 15.57	601.32 ± 18.81	471.01 ± 19.63	162.20 ± 23.00
79	methyl sulfoxide	garlic	1576	MS/RI	151.90 ± 19.28	176.80 ± 8.92	63.28 ± 7.19	89.59 ± 3.82	49.16 ± 4.14	2.89 ± 1.67	-	-	-	-	-	-	-	178.52 ± 18.72	237.91 ± 15.70	207.83 ± 13.45	151.60 ± 18.19	42.30 ± 4.61
80	1,3-dimethoxy-2-hydroxybenzene	medicine, phenol, smoke	2296	MS/RI	-	-	-	-	-	-	-	-	-	-	-	-	1138.32 ± 54.20	-	-	-	-	-
	Content of other compounds				1482.88	1647.14	2313.33	1072.8	369.74	1257.08	2984.04	3665.33	3318.79	1731.94	1555.1	2460.53	2563.8	2316.94	1834.41	2621.85	1921.03	1230.48
	Total identified/detected				64,961.64	71,582.58	49,513.54	44,621.28	29,838.12	50,157.17	62,834.03	52,955.68	49,043.58	67,058.74	34,230.43	44,434.45	50,920.49	40,020.24	81,910.39	66,560.3	35,438.61	39,740.67

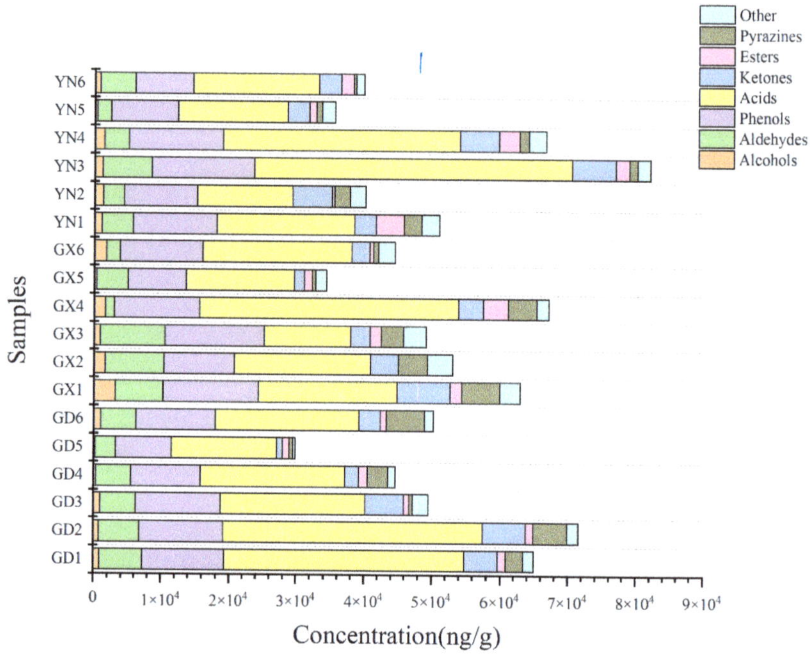

Figure 1. The average content of different kinds of compounds in three regions.

2.2. Analysis of Key Aroma Compounds in Brown Sugar Samples

A total of 46 aroma-active compounds were identified in 18 brown sugar samples by olfactometry, including 4 alcohols, 4 aldehydes, 3 phenols, 15 acids, 11 ketones, 7 pyrazines, and 2 other compounds. According to the odor properties of the aroma active compounds, these compounds can be classified into nine types: sweet/caramel, fruity, green/grassy, sour, sweaty/cheese, nutty, roasted, fatty and potato, which indicated that the aroma profile of brown sugar was the result of the synergistic effect of various odors.

In fact, it is the OAV of the aroma compound, and not its amount, that determines the contribution of the aroma compound. Aroma activity is generally defined as compounds with OAVs greater than 1 [14]. Therefore, the calculation of OAV was carried out for aroma compounds that can be sniffed (Table 2). Among the 18 brown sugar samples, 26 compounds with OAV >1 were considered as the key aroma active compounds of the brown sugar samples in this study and contributed to the overall flavor.

Alcohols: Among the four alcohols that can be sniffed, only furfuryl alcohol had OAV >1 and was only found in Guangxi and Yunnan. The content of furfuryl alcohol in Guangxi and Yunnan was 971.50 and 392.70 ng/g, respectively, and it contributed sweet, toast and caramel aroma to brown sugar. Sugar and amino acids react readily at elevated temperatures to form this compound [15]. The furfuryl alcohol contained in soy sauce has been considered to be one of the main components responsible for its odor, exhibiting a caramel scent, which contributes to the overall flavor of the sample [16].

Table 2. OAV of key odor compounds in brown sugar.

No.	Compounds [a]	OT (ng/g) [b]	GD1	GD2	GD3	GD4	GD5	GD6	GX1	GX2	GX3	GX4	GX5	GX6	YN1	YN2	YN3	YN4	YN5	YN6
1	pentadecanoic acid	500	2	2	2	1	-	2	1	-	-	-	-	-	-	1	2	2	1	-
2	2-methylbutanoic acid	20	-	-	-	-	-	19	-	-	-	-	-	-	-	-	-	-	-	-
3	3-methylbutanoic acid	1.8	908	598	359	170	-	7	151	64	121	491	450	395	624	315	773	894	-	-
4	4-methylpentanoic acid	1.9	-	-	-	-	-	-	-	-	-	-	-	-	-	-	-	-	63	30
5	acetic acid	13	783	827	214	489	495	454	512	324	229	846	388	857	326	82	981	708	428	795
6	benzoic acid	1000	3	4	4	3	2	2	3	2	3	4	1	3	3	2	4	5	2	2
7	butanoic acid	20	133	132	31	126	61	71	41	10	63	59	8	22	63	42	210	106	58	-
8	hexanoic acid	4.8	63	125	101	58	59	91	58	-	32	95	25	23	121	92	152	152	80	16
9	nonanoic acid	1.6	-	-	-	-	-	258	-	-	-	-	-	-	-	-	-	-	-	142
10	octanoic acid	22	10	12	-	-	-	-	-	-	-	-	-	-	-	12	11	22	-	-
11	phenylacetic acid	17	86	69	70	52	43	3	75	-	56	69	44	80	31	53	87	116	44	36
12	hexanal	1.4	286	200	224	220	107	186	165	41	93	210	149	131	148	26	165	173	111	130
13	(*E*)-2-nonenal	0.19	296	502	359	374	195	450	-	315	180	319	388	419	233	373	575	423	-	-
14	3,5-dimethoxy-4-hydroxybenzaldehyde	1900	2	2	2	2	1	1	3	-	-	-	-	-	2	1	3	1	1	2
15	benzaldehyde	60	1	2	1	-	-	3	-	-	1	2	1	2	-	1	1	1	1	-
16	3-methyl-1,2-cyclopentanedione	26	-	-	-	-	-	-	-	-	-	-	-	-	-	-	-	-	-	14
17	2-hydroxy-3-methyl-2-cyclopenten-1-one	10	47	48	71	28	18	59	88	73	75	52	19	48	68	64	72	68	25	-
18	2,5-dimethyl-4-hydroxy-3(2*H*)-furanone	1.6	170	204	284	144	77	457	1355	724	187	295	44	90	260	539	391	331	-	89
19	4-hydroxy-5-methyl-3-(2*H*)-furanone	500	-	-	-	<1	-	<1	1	-	-	-	-	-	-	-	-	-	-	<1
20	phenylethylene	37	18	17	16	12	4	12	15	13	14	27	27	28	14	12	15	16	13	4
21	furfuryl alcohol	1415	<1	<1	<1	<1	<1	<1	1	1	<1	1	<1	<1	1	1	<1	<1	<1	<1
22	2,3,5-trimethylpyrazine	23	-	-	-	8	2	13	11	11	2	-	2	-	-	5	-	-	-	1
23	2,5-dimethylpyrazine	80	12	24	1	12	3	23	17	12	16	22	1	1	14	9	5	5	5	1
24	2,6-dimethyl-3-ethylpyrazine	0.04	-	-	-	-	1995	4654	10,305	6809	5088	7154	-	-	-	-	-	-	2072	-
25	2,6-dimethylpyrazine	250	3	4	1	4	<1	4	6	4	3	4	<1	<1	3	1	1	1	1	1
26	2-acetyl-6-methylpyrazine	300	1	1	-	<1	-	1	1	<1	<1	1	-	-	-	1	-	<1	<1	-

[a] Volatile compounds that can be smelled at sniffer port. [b] Odor thresholds were referenced in a book, named: *odor thresholds compilations of odor threshold values in air, water and other media.*

Aldehydes: Among the aldehydes, there are four aldehydes with OAV >1, namely hexanal, (*E*)-2-nonenal, 3,5-dimethoxy-4-hydroxybenzaldehyde and benzaldehyde. (*E*)-2-nonenal and hexanal are probably oxidation products of polyunsaturated fatty acids [17], with high OAV due to their higher concentration and lower odor threshold, and are key aroma compounds among aldehydes, contributing to the green odor of brown sugar. The average content of benzaldehyde in Guangdong was higher than that in Guangxi and Yunnan, and it may be the degradation product of phenylalanine [14], contributing nutty and caramel aromas to the brown sugar. 3,5-dimethoxy-4-hydroxybenzaldehyde showed close OAV in Guangdong and Yunnan, and was higher than that in Guangxi, contributing sweet and nutty aroma to brown sugars. According to Chen, Song, Li, Chen, Wang, Che, Zhang and Zhao [9], 3,5-dimethoxy-4-hydroxybenzaldehyde is formed during brown sugar production, and the difference in content might be related to the raw materials and processing technology.

Ketones: Four ketones with OAV >1 were found in brown sugar samples, including 3-methyl-1,2-cyclopentanedione, 2-hydroxy-3-methyl-2-cyclopenten-1-one, 2,5-dimethyl-4-hydroxy-3(2*H*)-furanone, and 4-hydroxy-5-methyl-3(2*H*)-furanone. 2,5-Dimethyl-4-hydroxy-3(2*H*)-furanone has the highest OAV and contributes a strong caramel flavor to brown sugar, which is most likely formed by the Maillard reaction through deoxy sugars and is most abundant in strawberries [18,19]. 2-Hydroxy-3-methyl-2-cyclopenten-1-one has a strong caramel aroma and is one of the key odor compounds that contribute to the caramel odor in black tea, soy sauce and molasses [20–22]. 3-Methyl-1,2-cyclopentanedione was detected only in Yunnan brown sugar with OAV=14, which contributed sweet and bready aroma to Yunnan brown sugar. 2,5-Dimethyl-4-hydroxy-3(2*H*)-furanone was detected in all the three regions' samples, but the OAV was greater than 1 only in Guangxi brown sugar, which was caused by its high concentration in Guangxi brown sugar.

Pyrazines: Many products possess a distinctive aroma resulting from pyrazines, which are special Maillard reaction compounds [23,24]. Pyrazine is formed by condensing two α-aminocarbonyl compounds and forming a dihydropyrazine, which oxidizes spontaneously to form the pyrazine [23,25,26]. Among the twelve pyrazines detected in the eighteen samples, there are five kinds of pyrazines with OAV greater than 1, namely 2,3,5-trimethylpyrazine, 2,5-dimethylpyrazine, 2,6-dimethyyl-3-ethylpyrazine, 2,6-dimethyl-3-ethylpyrazine and 2-acetyl-6-methylpyrazine. 2,6-Dimethyl-3-ethylpyrazine exhibited the highest OVA due to its low threshold (OT=0.04 ng/g), contributing a strong roasted potato flavor to brown sugar. 2,5-Dimethylpyrazine and 2,6-dimethylpyrazine were previously reported to be key odor compounds in coffee, exhibiting strong roasted and nutty aroma [27].

Acids: A total of 21 kinds of acid compounds were detected in 18 brown sugars, among which the OAV of 11 kinds of acid compounds was greater than 1. Acetic acid, one of the most abundant compounds in brown sugar, had the highest OAV and contributed sour aroma to the samples. 2-Methylbutanoic acid and 3-methylbutanoic acid exhibited a sour aroma and had been reported to be the key aroma components in Japanese sweet rice wine, which played an important role in the overall flavor of sweet rice wine [28]. Benzoic acid, however, has an unpleasant urine-like odor, which may be caused by phenylalanine under the action of phenylalanine ammonia-lyase in plants [29].

2.3. *Fingerprint Analysis of Sugar Products from Three Different Regions*

A food fingerprint can be defined as molecular markers that indicate a characteristic state or condition of food, thus enabling more accurate product identification [30]. Each sample is regarded as a multidimensional space vector. If two samples are more similar, their space will be closer, and the angle between the two samples' space vectors will be smaller, which leads the cosine of the angle between the two vectors to move closer to 1. Therefore, the similarity of samples can be expressed by the cosine of the included angle. On the contrary, if the difference between the two samples is greater, the cosine of the included angle becomes smaller. In this study, the samples were determined by GC-O-MS, and the odor-active compounds were selected for fingerprint and similarity evaluation.

It is worth mentioning that the similarity of samples becomes higher when the similarity or the cosine of the angle is above 90%. As depicted in Table 3 and Figure 2, of the six samples in Guangdong, except for Guangdong3, the similarity and cosine of the included angle of the other five samples were above 90%. This indicated that the odor properties of Guangdong3 were quite different than the other five samples, which might have happened due to different processing technology.

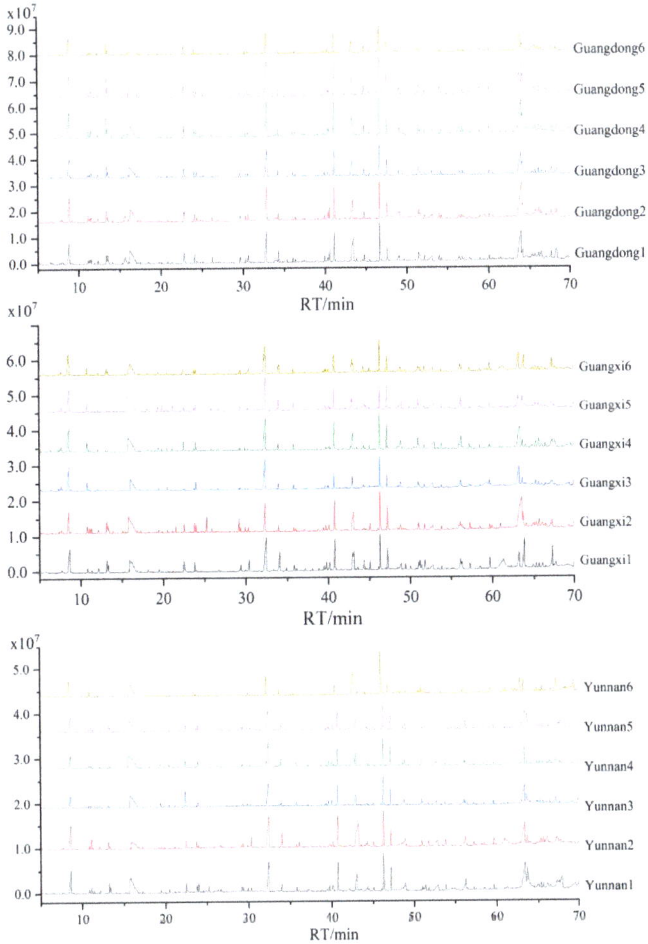

Figure 2. Fingerprint of brown sugar from Guangdong, Guangxi and Yunnan.

The cosine of the included angle of six samples in Guangxi was above 90%, and the similarity of Guangxi3 was just less than 90% (89.80%). This result indicated that the odor properties of these six samples in Guangxi were similar, without much difference

Of the six samples in Yunnan, only Yunnan2 had similarity and cosine of included angle lower than 90%, while the other five samples had similarity and cosine of included angle higher than 90%. This result indicated that the odor attributes of the other five samples were similar, but Yunnan2 had significant differences with them.

Table 3. Fingerprint results of brown sugar from each producing area.

No.	Compounds	Guangdong						Guangxi						Yunnan					
		Guang dong1	Guang dong2	Guang dong3	Guang dong4	Guang dong5	Guang dong6	Guang xi1	Guang xi2	Guang xi3	Guang xi4	Guang xi5	Guang xi6	Yun nan1	Yun nan2	Yun nan3	Yun nan4	Yun nan5	Yun nan6
1	2,3-butanediol	265.09	184.68	76.15	209.6	83.16	179.33	296.97	0	114.44	256.59	81.45	1019.88	173.87	256.31	466.38	410.5	165.52	50.48
2	propylene glycol	0	0	0	0	0	0	0	0	75.58	0	0	0	0	0	0	0	0	0
3	hexanal	399.76	279.98	313.08	308.33	150.05	260.23	230.54	57.51	130.21	294.62	209.07	182.99	207.86	36.23	231.15	241.62	155.67	182.12
4	(*E*)-2-nonenal	56.2	95.35	68.21	71	36.96	85.51	0	59.88	34.18	60.66	73.78	79.64	44.27	70.81	109.32	80.28	0	0
5	benzaldehyde	57.98	109.82	50.92	0	0	168.28	0	0	46.19	102.28	52.62	94.52	0	49.57	69.14	58.67	54.31	0
6	3,5-dimethoxy-4-hydroxybenzaldehyde	4115.06	3970.11	3258.76	4530.25	1861.44	1393.28	5558.03	0	0	0	0	0	4106.21	2356.17	4986.34	1882.87	1255.94	3220.19
7	2,6-di-tert-butyl-4-methylphenol	6272.71	6702.74	6540.27	5238.74	4705.8	5559.88	6468.03	7106.41	6733.2	5149.51	5177.12	7033.33	6293.4	5659.68	7638.6	7938.76	5115.72	3851.64
8	4-allyl-2,6-dimethoxyphenol	0	0	0	0	0	72.86	440.73	0	243.43	0	0	0	129.33	0	0	269.6	132.49	0
9	2-methoxy-4-acetylphenol	0	0	0	0	352.34	0	693.73	808.51	638.98	1094.03	935.46	584.08	855.15	0	0	0	669.33	649.86
10	acetic acid	10,184.79	10,757.27	2778.13	6350.81	6432.09	5900.64	6650.69	4210.8	2977.92	10,999.92	5047.46	11,141.62	4237.56	1068.38	12,748.83	9209.62	5570.25	10,334.03
11	methanoic acid	249.89	123.64	219.04	195.29	84.47	96.96	0	0	0	0	0	0	91.75	0	389.25	426.15	0	286.39
12	propanoic acid	821.24	619.01	644.68	396.79	303.37	260.37	332.73	147.19	329.19	457.22	280.06	653.58	425.26	299.24	1143.61	994.93	515.4	344.61
13	butanoic acid	2660.81	2644.03	614.21	2513.84	1227.91	1421.39	810.96	192.43	1269.77	1178.36	168.86	440.72	1266.55	837.56	4204.64	2118.54	1154.6	0
14	3-methylbutanoic acid	1633.63	1076.75	645.7	306.69	0	12.9	271.65	116.04	218.33	884.46	810.07	710.73	1123.49	566.92	1390.95	1609.2	0	0
15	2-methylbutanoic acid	0	0	0	0	0	373.25	0	0	0	0	0	0	0	0	0	0	0	0
16	2-methylpentanoic acid	0	0	0	183.26	0	37.89	0	0	0	0	0	0	0	0	0	0	302.4	0
17	4-methylpentanoic acid	0	0	0	0	0	0	0	0	0	0	0	0	0	0	0	0	119.8	56.28
18	hexanoic acid	300.51	597.73	487.13	280.13	283.56	435.09	280.31	0	155.31	453.98	119.62	112.67	580.84	441.59	727.74	728.69	383.25	78.31
19	octanoic acid	228.77	267.6	0	0	0	0	0	0	0	0	0	0	0	262.06	252.12	487.06	0	0
20	nonanoic acid	0	0	0	0	0	413.1	0	0	0	0	0	0	0	0	0	0	0	227.94
21	benzoic acid	3080.65	3564.46	4062.78	2709.59	1609.97	1795.59	3194.93	1982.81	2736.59	3573.53	1480.4	2837.5	2965.81	1857.28	4446.03	4751.21	1872.63	2262
22	phenylacetic acid	1468.96	1173.77	1187.97	880.89	730.3	47.37	1267.41	0	945.3	1180.95	741.94	1353.01	526.12	903.11	1487.08	1976.49	747.98	606.57
23	3-phenylpropionic acid	0	0	0	156.03	0	10.79	0	0	0	0	0	0	0	0	0	0	0	0
24	pentadecanoic acid	904.37	1205.19	867.02	423.32	0	806.02	586.5	0	0	0	0	0	0	733.11	1093.45	886.36	381.02	0
25	2-methyl-4,5-dihydro-3(2*H*)-furanone	341.22	1177.87	513.32	276.67	70.39	327.17	202.04	291.29	93.57	712.73	141.35	173.07	701.59	1141.42	391.82	578.35	726.04	1389.22
26	1-hydroxy-2-propanone	457.94	582.3	605.26	438.74	206.31	652.59	841.13	715.56	503.38	505.1	220.33	211.55	458.82	360.04	382.45	461.7	355.89	438.74
27	1-hydroxy-2-butanone	39.22	111.06	41.37	43.91	0	16.7	0	76.4	82.88	62.58	0	24.26	56.34	46.29	63.27	56.43	0	0
28	1-acetoxy-2-propanone	118.22	230.25	187.71	168.77	0	66.52	218.3	159.99	115.45	245.41	139.08	235.11	176.14	191.49	132.61	177.62	125.3	131.45
29	2(5*H*)-furanone	475.08	581.86	479.55	0	0	145.49	394.39	265.72	286.88	221.84	203.23	363.54	0	232.91	712.75	266.08	0	154.22
30	3-methyl-1,2-cyclopentanedione	0	0	0	0	0	0	0	0	0	0	0	0	0	0	0	0	0	361.67
31	2-hydroxy-3-methyl-2-cyclopenten-1-one	468.6	482.48	707.42	283.22	182.61	588.56	879.99	726.67	745.18	519.02	186.46	476.39	681.71	637.34	716.33	684.63	248.07	0
32	2(3*H*)-furanone	974.6	721.47	1069.93	351.84	292.08	470.52	631.95	615.99	655.84	705.84	496.09	977.5	560.98	596.89	1123.58	1324.65	397.11	323.38
33	2,5-dimethyl-4-hydroxy-3(2*H*)-furanone	271.76	326.77	454.38	230.03	123.31	731.6	2168.59	1158.64	298.58	471.8	69.64	144.05	415.29	861.94	625.51	529.53	0	141.69

Table 3. *Cont.*

No.	Compounds	Guangdong						Guangxi						Yunnan					
		Guang dong1	Guang dong2	Guang dong3	Guang dong4	Guang dong5	Guang dong6	Guang xi1	Guang xi2	Guang xi3	Guang xi4	Guang xi5	Guang xi6	Yun nan1	Yun nan2	Yun nan3	Yun nan4	Yun nan5	Yun nan6
34	4-hydroxy-5-methyl-3-(2*H*)-furanone	0	0	0	181.54	0	19.69	612.25	0	0	0	0	0	0	0	0	0	0	99.22
35	4-hydroxyace tophenone	1461.87	1892.62	1469.22	0	0	0	1546.24	0	0	0	0	0	0	1656.02	2234.53	1370.88	1132.23	0
36	2,5-dimethylpyrazine	924.49	1938.05	111.73	963.3	201.81	1827.76	1343.97	989.09	1309.65	1748.65	88.92	70.87	1114.25	703.42	419.79	430.64	375.15	93.56
37	2,6-dimethylpyrazine	682.76	1035.34	144.76	983.96	72.49	1094.56	1444.15	1045.87	800	946.87	68.92	85.72	693.95	273.74	230.77	289.35	139.77	208.04
38	2,3,5-trimethylpyrazine	0	0	0	180.76	35.78	306.68	249.04	250.84	40.86	0	46.57	0	0	125.18	0	0	0	33.86
39	2,5-dimethyl-3-ethylpyrazine	151.21	315.82	0	157.83	0	338.49	0	0	0	0	0	0	109.14	201.75	138.27	119	0	0
40	2,6-dimethyl-3-ethylpyrazine	0	0	0	0	79.78	186.16	412.18	272.34	203.53	286.15	0	0	0	0	0	0	32.88	0
41	2-acetyl-5-methylpyrazine	0	0	0	156.34	0	218.85	0	0	0	0	0	0	0	0	0	0	0	0
42	2-acetyl-6-methylpyrazine	251.42	419.45	0	141.33	0	245.73	183.55	144.23	121.93	218.58	0	0	0	408.61	0	113.67	96.94	0
43	2-acetylfuran	0	0	0	0	0	5.27	0	231.24	103.74	0	37.99	72.73	0	0	0	0	0	54.65
44	phenylethylene	651.29	613.81	573.68	451.99	160.07	457.61	546.32	469.44	511.72	984.77	984.42	1025.24	500.79	459.13	564.51	601.32	471.01	162.2
Cosine of included angle		0.9879	0.9888	0.8855	0.9800	0.9750	0.9643	0.9031	0.9463	0.9152	0.9527	0.9776	0.9671	0.9439	0.8155	0.9815	0.9839	0.9822	0.9189
Similarity		0.9850	0.9859	0.8562	0.9752	0.9781	0.9553	0.8824	0.9373	0.8980	0.9445	0.9762	0.9664	0.9300	0.7655	0.9773	0.9799	0.9787	0.9138

2.4. Verification of Fingerprint

In order to verify whether the fingerprint method is suitable for the analysis of brown sugar, the verification was carried out. Fingerprint verification includes three parts: stability experiment, precision experiment, and repeatability experiment. Following the sample preparation described in Section 2.4, a brown sugar sample was selected and analyzed by GC-MS after 0, 2, 4, 8, 16, and 24 h. Furthermore, the relative standard deviations (RSD) of the relative retention times (RT) and relative peak areas of the odor-active compounds were calculated. The results showed that the RSD of the relative RT of the odor-active compounds was less than 0.3%, and the RSD of the relative peak areas was less than 5%, indicating that the samples were stable within 24 h and met the requirements of the fingerprint method.

A brown sugar sample was extracted and concentrated with the organic solvent, and then the concentration was injected six times consecutively to calculate the RSD of relative RT and relative peak area of the odor-active compounds. These results showed that the RSD of the relative RT of the odor active compounds was less than 0.5%, and the RSD of the relative peak area was less than 6%, indicating that the precision of the instrument was good and met the requirements of the fingerprint method.

Five brown sugar samples were extracted and analyzed for their odor compounds, followed by the RSD of relative RT and relative peak area of the odor active compounds analysis. The results showed that the RSD of relative RT was less than 0.3%, and the RSD of the relative peak area was less than 7%, indicating that they had good repeatability and met the requirements of the fingerprint method.

2.5. Orthogonal Partial Least Squares Discriminant Analysis (OPLS-DA)

The fingerprinting analysis of samples from the three origins of Guangdong, Guangxi, and Yunnan revealed that the majority of samples within each province had similar odor types. In addition, a supervised OPLS-DA multivariate statistical analysis method was used to establish a statistical model in order to distinguish odor compounds between Guangdong and Guangxi, Guangdong and Yunnan, and Guangxi and Yunnan.

By conducting OPLS-DA analysis on the brown sugar, a variable importance of projection diagram (VIP) of the model was obtained. A VIP is a vector that summarizes the contribution of a variable to the explanation of the model. Variables with a VIP >1 are generally considered to contribute to the explanation of the model [31,32]. The samples were assessed as independent variables, and the OPLS-DA model was fitted automatically.

The OPLS-DA and VIP results (Figure 3) indicate that the brown sugars from Guangdong and Guangxi were well separated. The brown sugar from Guangdong and Guangxi showed the greatest degree of separation and low intra-group differences, facilitating an accurate exploration of the differences in composition. VIP diagram elucidated that 4-hydroxybenzaldehyde, 3,5-dimethoxy-4-hydroxybenzaldehyde, n-hexadecanoic acid, butanoic acid, acetic acid, 2-methoxy-4-acetylphenol, 2-acetylpyrrole, pentadecanoic acid, furfuryl alcohol, 4-hydroxyacetophenone, etc., were the main contributors to the distinction between Guangdong and Guangxi samples. These compounds were basically aldehydes, acids, ketones, and phenols. Among these, 3,5-dimethoxy-4-hydroxybenzaldehyde and 4-hydroxybenzaldehyde played an important role in classifying Guangdong and Guangxi. 4-Hydroxybenzaldehyde and 3,5-dimethoxy-4-hydroxybenzaldehyde presented a pleasant nutty and creamy odor. Previously, 4-hydroxybenzaldehyde and 3,5-dimethoxy-4-hydroxybenzaldehyde were identified as the major volatile constituents in brown sugars [33]. Acetic acid is also one of the key compounds that can distinguish brown sugar from two provinces. Acetate is a well-known product of the thermal degradation of saccharides, and it is primarily formed during the early stage of the Maillard reaction, under neutral and alkaline conditions. Acetic acid is formed exclusively by hydrolytic cleavage of β-dicarbonyl in hexose-based systems [34].

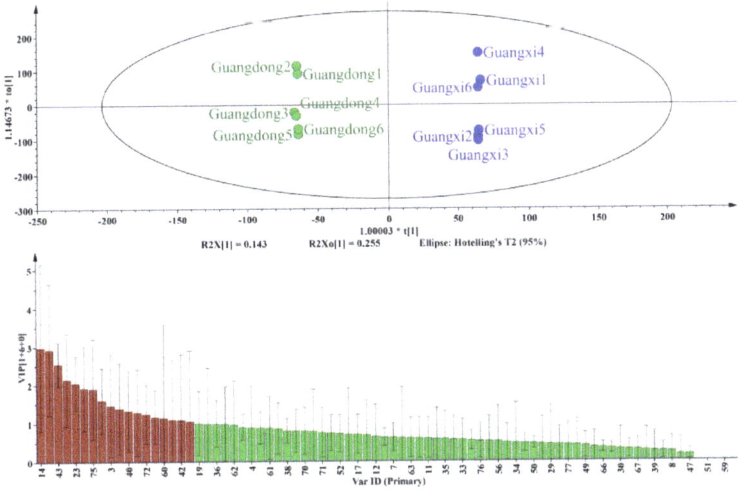

Figure 3. OPLS-DA analysis and VIP diagram of brown sugar in Guangdong and Guangxi.

As shown in Figure 4, OPLS-DA analysis and VIP results indicate that the brown sugars from Guangdong and Yunnan are distinguishable. The principal compounds contributing to this distinction include n-hexadecanoic acid, acetic acid, dibutylphthalate, 2-acetylpyrrole, 2,5-dimethylpyrazine, and 2-methylpyrazine. Of the compounds with VIP greater than 1, pyrazine compounds appeared, which indicated that pyrazine compounds played a significant role in distinguishing brown sugar between Guangdong and Yunnan. The average content of pyrazines in Guangdong and Yunnan was 2897.28 ng/g and 1441.20 ng/g, respectively, and the pyrazine contents in Guangdong samples were higher than in Yunnan. These compounds could impart a popcorn, nutty, and roasted aroma to brown sugar.

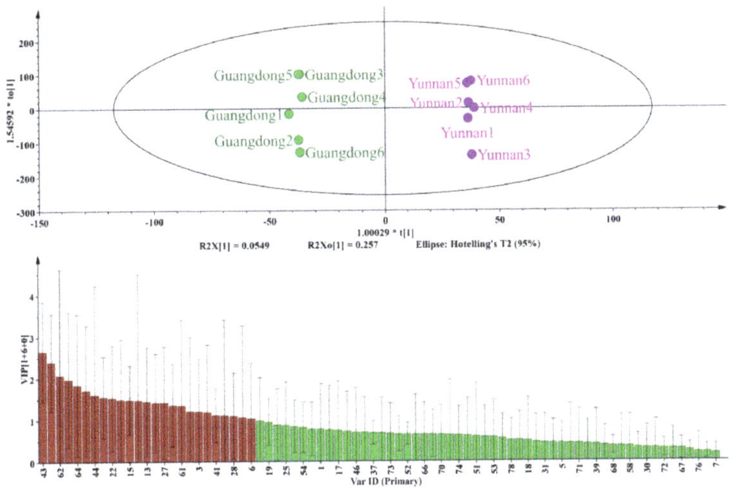

Figure 4. OPLS-DA analysis and VIP diagram of brown sugar in Guangdong and Yunnan.

Based on the VIP diagram and OPLS-DA analysis of brown sugar between Guangxi and Yunnan (Figure 5), they were well separated. A number of compounds contributed to the differentiation between the two provinces, including 4-hydroxybenzaldehyde, 3,5-

dimethoxy-4-hydroxybenzaldehyde, n-hexadecanoic acid, acetic acid, butanoic acid, and 4-hydroxyacetophenone. Of these volatile compounds, the contribution of 4-hydroxybenzalde hyde was the greatest. The average content of 4-hydroxybenzaldehyde in Guangxi was 2728.55 ng/g, while the samples from Guangxi had no odor compounds. The average contents of 3,5-dimethoxy-4-hydroxybenzaldehyde in Guangxi and Yunnan were 926.34 ng/g and 2967.95 ng/g and the contents in Yunnan were significantly higher than in Guangxi. Perhaps these compounds play an important role in distinguishing the sugars from Guangxi and Yunnan.

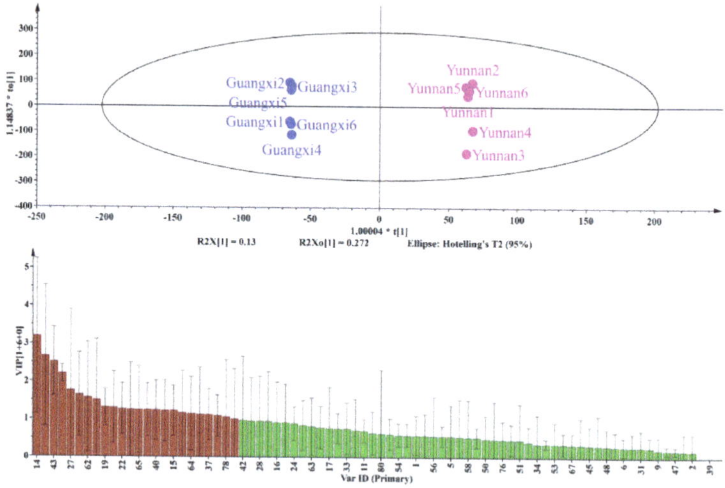

Figure 5. OPLS-DA analysis and VIP diagram of brown sugar in Guangxi and Yunnan.

3. Materials and Methods

3.1. Materials

Eighteen brown sugar samples from Guangdong, Guangxi and Yunnan were provided by COFCO. These samples were stored in a refrigerator at −80 °C before analysis.

3.2. Standards and Reagents

Ether (purity > 99%), dichloromethane (purity > 99%), anhydrous sodium sulfate, 2-methyl-3-heptanone (purity > 99%) and n-alkane (C_7-C_{30}) were purchased from Sigma-Aldrich (St. Louis, MO, USA), and carrier gas (helium) was purchased from Beijing AP Baif Gases Industry Co., Ltd. (Beijing, China).

3.3. Extraction of Odor Compounds from Sugars

The odor compounds in brown sugar were extracted by a liquid–liquid extraction (LLE) method according to Chen et al. [33]. In brief, 50.00 g of brown sugar was placed in a triangular flask, 50 mL of distilled water was added to dissolve the brown sugar, then, 50 mL of ether, 50 mL of dichloromethane and 5 μL of internal standard 2-methyl-3-heptanone (81.6 mg/mL) were added, and the mixture was magnetically stirred at 1000 rpm for 10 min. After centrifugation (Hitachi, Japan) for 30 min at 10,000 rpm, the extract containing the volatile aroma compounds was separated by a funnel. Subsequently, 150.0 g anhydrous sodium sulfate was added to the extract and put into a refrigerator at 4 °C to remove water for 12 h, and filtered with a filter paper. A gentle nitrogen stream was used to concentrate the volume into 100 μL, and the odor compounds were extracted and stored at −80 °C for further analysis.

3.4. GC-O-MS

Three well-trained panelists conducted a GC-O analysis of the concentrated distillate. The panelists were recruited from Beijing Technology and Business University's Molecular Sensory Laboratory. To identify and describe the aroma characteristics of the reference compounds, they smelled several concentrations of reference compounds in model solutions 2 h per day before analysis. The training lasted for one month. For the GC-O analysis, wet gas was delivered to the nose using a blank capillary column to improve the sensitivity of the panelists. The aroma perceptions, intensity, and RT were recorded by the panelists. If two or more panelists detected the aroma, an aroma-active compound was identified [35].

To determine the volatile aroma profile of sugars, an Agilent 7890A gas chromatograph (GC) coupled with an Agilent 5977B mass spectrometer (MS) and a sniffing port (Gerstel, Germany) was used. The aroma extract (1 μL) was injected into a DB-Wax column (60 m × 0.25 mm i.d., film thickness 0.25 μm, Agilent J&W) through splitless mode, and the flow rate of the helium carrier gas was maintained at 1.7 mL/min. The oven temperature was initially programmed at 40 °C, further raised to 100 °C at a rate of 4 °C/min, following a gradual increase up to 200 °C at a rate of 3 °C/min for 5 min, and after achieving an ultimate temperature of 230 °C at a rate of 3 °C/min, it was maintained for 10 min. The interface and ion source were set at 250 °C and 230 °C, respectively, while the electron-impact ionization was set at 70 eV, the acquisition range (m/z) at 35–350 amu, and the scan rate at 1.77 scans/s. The transmission line temperature of the olfactory detection port (ODP) was maintained at 235 °C.

3.5. Qualitative Analysis

The ionization of a molecule in a vacuum produces a characteristic group of ions of different masses. The plot of relative abundance versus mass of these ions constitutes a mass spectrum. The spectrum can be used to identify the molecule. The unknowns were identified by comparing the fragments with the National Institute of Standards and Technology (NIST) MS Spectral Library (Version 2020), by comparing the odor percepts with the database (http://www.thegoodscentscompany.com) and by calculating the linear retention indices (LRIs) using a homologous series of n-alkanes (C_7-C_{30}). The use of multiple methods can increase the accuracy of qualitative results. Using the internal standard area, the resulting peaks were calibrated, and the aroma compound contents were expressed as nanograms per gram of sample [10].

3.6. Odor Activity Value (OAV)

In order to evaluate the contribution of each odorant to the overall aroma of brown sugar, the OAV (ratio of concentration to its odor threshold) was calculated [36]. These threshold values were derived from the literature in water [37].

3.7. Statistical Analysis

All experiments in this study were conducted in triplicates, and the data were expressed as mean ± standard deviation. The bar graph was drawn by OriginPro 2022 (OriginLab Corp., Northampton, MA, USA), the OPLS-DA analysis was conducted by SIMCA 14.1 (MKS Instruments, Andover, MA, USA), and the tables were organized by Microsoft Excel 2021 (Microsoft Corp., Redmond, WA, USA).

4. Conclusions

In summary, a total of 80 odor compounds, including 5 alcohols, 9 aldehydes, 8 phenols, 21 acids, 14 ketones, 5 esters, 12 pyrazines, and 6 other compounds, were detected in 18 brown sugar samples from three different provinces. The fingerprint analysis showed 90% similarity, indicating a close relationship among the odor components of brown sugars from each province without much difference. Further, the stability, accuracy, and repeatability of the fingerprint method were verified, and speculated that the method could meet the requirements of the fingerprint. In the future, fingerprint might have wider applications due to its characteristic of distinguishing geographical origin and food adulteration. Additionally, the OPLS-DA was employed to identify the tracing of brown sugar and to identify the compounds contributing to brown sugars' volatile classification. The results demonstrated that 4-hydroxybenzaldehyde, 3,5-dimethoxy-4-hydroxybenzaldehyde, n-hexadecanoic acid, and acetic acid were the essential components in distinguishing the sugars from Guangdong, Guangxi, and Yunnan, validating the efficiency of OPLS-DA.

Author Contributions: Methodology, H.S.; Software, Y.Z. and W.L.; Writing—original draft preparation, E.C.; Writing—review and editing, H.S. and S.Z.; Supervision, H.S. and Y.Z. All authors have read and agreed to the published version of the manuscript.

Funding: COFCO Nutrition and Health Research Institute Co., Ltd.: No number.

Institutional Review Board Statement: Not applicable.

Informed Consent Statement: Not applicable.

Data Availability Statement: Not applicable.

Acknowledgments: The present work was supported by China Oil and Food Import and Export Corporation (COFCO). No project number.

Conflicts of Interest: The authors declare no conflict of interest.

Sample Availability: Not available.

Abbreviations

GC-O-MS	gas chromatography-olfactometry-mass spectrometry
LLE	liquid–liquid extraction
NCS	non-centrifugal cane sugar
COFCO	China Oil and Food Import and Export Corporation
GC	gas chromatography
MS	mass spectrometer
ODP	olfactory detection port
RSD	relative standard deviation
RT	retention time
RI	retention index
OPLS-DA	orthogonal partial least squares discriminant analysis
VIP	variable importance of projection
OT	odor threshold

References

1. Jader, R.; Fabian, V.; John, E.; Sebastian, E.; Oscar, M. Thermal performance evaluation of production technologies for non-centrifuged sugar for improvement in energy utilization. *Energy* **2018**, *152*, 858–865. [CrossRef]
2. Feng, S.; Luo, Z.; Zhang, Y.; Zhong, Z.; Lu, B. Phytochemical contents and antioxidant capacities of different parts of two sugarcane (Saccharum officinarum L.) cultivars. *Food Chem.* **2014**, *151*, 452–458. [CrossRef] [PubMed]
3. Lee, J.S.; Ramalingam, S.; Jo, I.G.; Kwon, Y.S.; Bahuguna, A.; Oh, Y.S.; Kwon, O.J.; Kim, M. Comparative study of the physico-chemical, nutritional, and antioxidant properties of some commercial refined and non-centrifugal sugars. *Food Res. Int.* **2018**, *109*, 614–625. [CrossRef] [PubMed]
4. Scalbert, A.; Johnson, I.T.; Saltmarsh, M. Polyphenols: Antioxidants and beyond. *Am. J. Clin. Nutr.* **2005**, *81*, 215S–217S. [CrossRef]
5. EFSA Panel on Dietetic Products, N. Allergies, Scientific Opinion on the substantiation of health claims related to various food (s)/food constituent (s) and protection of cells from premature aging, antioxidant activity, antioxidant content and antioxidant properties, and protection of DNA, proteins and lipids from oxidative damage pursuant to Article 13 (1) of Regulation (EC) No 1924/2006. *EFSA J.* **2010**, *8*, 1489.
6. Asikin, Y.; Kamiya, A.; Mizu, M.; Takara, K.; Tamaki, H.; Wada, K. Changes in the physicochemical characteristics, including flavour components and Maillard reaction products, of non-centrifugal cane brown sugar during storage. *Food Chem.* **2014**, *149*, 170–177. [CrossRef]
7. Huang, S.R.; Hang, F.X.; Wei, C.B.; Xie, C.F.; Li, K. Analysis of volatile aroma components in brown sugar. *China Condiment* **2019**, *44*, 146–151.

8. Juliana, M.G.; Paulo, C.N.; Francisco, J H.; Alvaro, O.; Coralia, O. Physicochemical and sensory (aroma and colour) characterisation of a non-centrifugal cane sugar ("panela") beverage. *Food Chem.* **2017**, *228*, 7–13.
9. Chen, E.; Song, H.; Li, Y.; Chen, H.; Wang, B.; Che, X.; Zhang, Y.; Zhao, S. Analysis of aroma components from sugarcane to non-centrifugal cane sugar using GC-O-MS. *RSC Adv.* **2020**, *10*, 32276–32289. [CrossRef]
10. Asikin, Y.; Takahara, W.; Takahashi, M.; Hirose, N.; Ito, S.; Wada, K. Compositional and Electronic Discrimination Analyses of Taste and Aroma Profiles of Non-Centrifugal Cane Brown Sugars. *Food Anal. Methods* **2017**, *10*, 1844–1856. [CrossRef]
11. Cheng, H.; Wu, W.; Chen, J.; Haibo, P.; Xu, E.; Chen, S.; Ye, X.; Chen, J. Establishment of anthocyanin fingerprint in black wolfberry fruit for quality and geographical origin identification. *LWT* **2022**, *157*, 113080. [CrossRef]
12. Li, W.; Chen, Y.P.; Blank, I.; Li, F.; Li, C.; Liu, Y. GC × GC-ToF-MS and GC-IMS based volatile profile characterization of the Chinese dry-cured hams from different regions. *Food Res. Int.* **2021**, *142*, 110222. [CrossRef]
13. Zhao, Q.; Xi, J.; Xu, D.; Jin, Y.; Wu, F.; Tong, Q.; Yin, Y.; Xu, X. A comparative HS-SPME/GC-MS-based metabolomics approach for discriminating selected japonica rice varieties from different regions of China in raw and cooked form. *Food Chem.* **2022**, *385*, 132701. [CrossRef]
14. Hofmann, T.; Schieberle, P. Formation of aroma-active Strecker-aldehydes by a direct oxidative degradation of Amadori compounds. *J. Agric. Food Chem.* **2000**, *48*, 4301–4305. [CrossRef]
15. Moon, J.K.; Shibamoto, T. Role of roasting conditions in the profile of volatile flavor chemicals formed from coffee beans. *J. Agric. Food Chem.* **2009**, *57*, 5823–5831. [CrossRef]
16. Wang, X.; Guo, M.; Song, H.; Meng, Q. Characterization of key aroma compounds in traditional Chinese soy sauce through the molecular sensory science technique. *Lwt* **2020**, *128*, 109413. [CrossRef]
17. Liu, Y.; He, C.; Song, H. Comparison of SPME versus SAFE processes for the analysis of flavor compounds in watermelon juice. *Food Anal. Methods* **2018**, *11*, 1677–1689. [CrossRef]
18. Raab, T.; López-Ráez, J.A.; Klein, D.; Caballero, J.L.; Moyano, E.; Schwab, W.; Muñoz-Blanco, J. FaQR, Required for the Biosynthesis of the Strawberry Flavor Compound 4-Hydroxy-2,5-Dimethyl-3-(2*H*)-Furanone, Encodes an Enone Oxidoreductase. *Plant. Cell* **2006**, *18*, 1023–1037. [CrossRef]
19. Roscher, R.; Herderich, M.; Steffen, J.-P.; Schreier, P.; Schwab, W. 2,5-Dimethyl-4-hydroxy-3-[2*H*]-furanone 6′O-malonyl-β-d-glucopyranoside in strawberry fruits. *Phytochemistry* **1996**, *43*, 155–159. [CrossRef]
20. Kumazawa, K.; Masuda, H.; Nishimura, O.; Kato, T. Identification of potent aroma components in brewed black tea. *J. Jpn. Soc. Food Sci. Technol. Jpn.* **1998**, *45*, 728–734. [CrossRef]
21. Lee, K.E.; Lee, S.M.; Choi, Y.H.; Hurh, B.S.; Kim, Y.S. Comparative volatile profiles in soy sauce according to inoculated microorganisms. *Biosci. Biotechnol. Biochem.* **2013**, *77*, 2192–2200. [CrossRef]
22. Kallio, H. In Comparison and characteristics of aroma compounds from maple and birch syrups. In *Proceedings of the 5th International Flavor Conference, Porto Carras, Greece, 1987*; J-GLOBAL: Tokyo, Japan, 1988; pp. 241–248. Available online: https://jglobal.jst.go.jp/en/detail?JGLOBAL_ID=200902006786988250 (accessed on 2 August 2022).
23. Adams, A.; Polizzi, V.; Van Boekel, M.; De Kimpe, N. Formation of Pyrazines and a Novel Pyrrole in Maillard Model Systems of 1,3-Dihydroxyacetone and 2-Oxopropanal. *J. Agric. Food Chem.* **2008**, *56*, 2147–2153. [CrossRef]
24. Scalone, G.L.L.; Cucu, T.; De Kimpe, N.; De Meulenaer, B. Influence of Free Amino Acids, Oligopeptides, and Polypeptides on the Formation of Pyrazines in Maillard Model Systems. *J. Agric. Food Chem.* **2015**, *63*, 5364–5372. [CrossRef]
25. Cha, J.; Debnath, T.; Lee, K.G. Analysis of α-dicarbonyl compounds and volatiles formed in Maillard reaction model systems. *Sci. Rep.* **2019**, *9*, 5325. [CrossRef]
26. Vernin, G.; Parkanyi, C. *Occurrence and Formation of Heterocyclic Compounds*; Ellis Horwood: New York, NY, USA, 1982; pp. 192–198.
27. Pickard, S.; Becker, I.; Merz, K.H.; Richling, E. Determination of the alkylpyrazine composition of coffee using stable isotope dilution–gas chromatography–mass spectrometry (SIDA-GC-MS). *J. Agric. Food Chem.* **2013**, *61*, 6274–6281. [CrossRef]
28. Kaneko, S.; Kumazawa, K. Aroma compounds in Japanese sweet rice wine (Mirin) screened by aroma extract dilution analysis (AEDA). *Biosci. Biotechnol. Biochem.* **2015**, *79*, 484–487. [CrossRef]
29. Gonda, I.; Davidovich-Rikanati, R.; Bar, E.; Lev, S.; Jhirad, P.; Meshulam, Y.; Wissotsky, G.; Portnoy, V.; Burger, J.; Schaffer, A.A.; et al. Differential metabolism of L–phenylalanine in the formation of aromatic volatiles in melon (*Cucumis melo* L.) fruit. *Phytochemistry* **2018**, *148*, 122–131. [CrossRef]
30. Medina, S.; Pereira, J.A.; Silva, P.; Perestrelo, R.; Câmara, J.S. Food fingerprints-A valuable tool to monitor food authenticity and safety. *Food Chem.* **2019**, *278*, 144–162. [CrossRef]
31. Beatriz, G.P.; Eriksson, L.; Trygg, J. Variable influence on projection (VIP) for orthogonal projections to latent structures (OPLS). *J. Chemom.* **2014**, *28*, 623–632.
32. Huang, B.; Chen, T.; Xiao, S.; Zha, Q.; Luo, P.; Wang, Y.; Cui, X.; Liu, L.; Zhou, H. A new approach for authentication of four ginseng herbs and their related products based on the simultaneous quantification of 19 ginseng saponins by UHPLC-TOF/MS coupled with OPLS-DA. *RSC Adv.* **2017**, *7*, 46839–46851. [CrossRef]
33. Chen, E.; Song, H.; Zhao, S.; Liu, C.; Tang, L.; Zhang, Y. Comparison of odor compounds of brown sugar, muscovado sugar, and brown granulated sugar using GC-O-MS. *LWT* **2021**, *142*, 111002. [CrossRef]
34. Davidek, T.; Gouézec, E.; Devaud, S.; Blank, I. Origin and yields of acetic acid in pentose-based Maillard reaction systems. *Ann. N. Y. Acad Sci* **2008**, *1126*, 241–243. [CrossRef] [PubMed]

35. Yang, F.; Liu, Y.; Wang, B.; Song, H.; Zou, T. Screening of the volatile compounds in fresh and thermally treated watermelon juice via headspace-gas chromatography-ion mobility spectrometry and comprehensive two-dimensional gas chromatography-olfactory-mass spectrometry analysis. *Lwt* **2021**, *137*, 110478. [CrossRef]
36. Zhai, X.; Granvogl, M. Characterization of the key aroma compounds in two differently dried Toona sinensis (A. Juss.) Roem. by means of the molecular sensory science concept. *J. Agric. Food Chem.* **2019**, *67*, 9885–9894. [CrossRef]
37. Van Gemert, L.J. *Compilations of Odour Threshold Values in Air, Water and Other Media*, 2nd ed.; Oliemans, Punter & Partners BV: Utrecht, The Netherlands, 2011; p. 485. Available online: https://agris.fao.org/agris-search/search.do?recordID=US201300091148 (accessed on 2 August 2022).

molecules MDPI

Review

Raman Method in Identification of Species and Varieties, Assessment of Plant Maturity and Crop Quality—A Review

Aneta Saletnik, Bogdan Saletnik * and Czesław Puchalski

Department of Bioenergetics, Food Analysis and Microbiology, Institute of Food Technology and Nutrition, College of Natural Science, Rzeszow University, Ćwiklińskiej 2D, 35-601 Rzeszow, Poland; asaletnik@ur.edu.pl (A.S.); cpuchalski@ur.edu.pl (C.P.)
* Correspondence: bsaletnik@ur.edu.pl

Abstract: The present review covers reports discussing potential applications of the specificity of Raman techniques in the advancement of digital farming, in line with an assumption of yield maximisation with minimum environmental impact of agriculture. Raman is an optical spectroscopy method which can be used to perform immediate, label-free detection and quantification of key compounds without destroying the sample. The authors particularly focused on the reports discussing the use of Raman spectroscopy in monitoring the physiological status of plants, assessing crop maturity and quality, plant pathology and ripening, and identifying plant species and their varieties. In recent years, research reports have presented evidence confirming the effectiveness of Raman spectroscopy in identifying biotic and abiotic stresses in plants as well as in phenotyping and digital selection of plants in farming. Raman techniques used in precision agriculture can significantly improve capacities for farming management, crop quality assessment, as well as biological and chemical contaminant detection, thereby contributing to food safety as well as the productivity and profitability of agriculture. This review aims to increase the awareness of the growing potential of Raman spectroscopy in agriculture among plant breeders, geneticists, farmers and engineers.

Keywords: Raman spectroscopy; digital farming; harvest maturity assessment; fruit and seeds quality diagnostics; non-invasive phenotyping

Citation: Saletnik, A.; Saletnik, B.; Puchalski, C. Raman Method in Identification of Species and Varieties, Assessment of Plant Maturity and Crop Quality—A Review. *Molecules* **2022**, *27*, 4454. https://doi.org/10.3390/molecules27144454

Academic Editors: Weiying Lu and Yanping Chen

Received: 21 June 2022
Accepted: 11 July 2022
Published: 12 July 2022

1. Introduction

Population growth and consequently the increasing demand for food as well as the decreasing availability of fertile land have led to reduced agricultural expansion and higher costs of agriculture and food. All these factors, combined with the long history of conventional agriculture, have forced food producers to seek new solutions. Owing to advancements in technology, these problems can be solved by introducing innovative methods in agriculture. The transformation of farming processes with the use of smart technologies, referred to as digital agriculture, aims to develop innovative technological methods to be used to maximise yield and minimise the environmental impact of agriculture [1–4]. Digital agriculture requires advanced methodologies for cultivation and selection of plants [5,6], detection and identification of biotic and abiotic stresses in plants, and for the acquisition of information about the health of plants and growth stages directly on the plantation. These methods are highly needed by crop producers [7].

In the related literature, there is more and more evidence showing the effectiveness of Raman spectroscopy in identifying the physiological condition of plants, crop quality, plant species or varieties, in assessing the maturity of plants and in pre-symptomatic diagnostics, or in detecting abiotic and biotic stresses in plants [1,8–17]. Raman spectroscopy also enables phenotyping and digital selection of plants in breeding. Owing to the immediate access to information about plant health, it is possible to detect and identify bacterial infections, secondary diseases, insect infestations, fungal infections, and other pathogens and the diseases they transmit in greenhouses and fields [8,18,19]. The information acquired this way is used to perform precise and site-specific chemical treatments which may prevent the spread of biotic stresses and save up to 30% of the crop. Rapid detection of physiological drought or nutrient deficiencies allows for supply of the nutrients promptly and accurately. By applying fertilisers to a strictly defined area, it is possible to reduce contamination of the crops and the soil [20,21].

Moreover, a tool effectively assessing the ripeness of the fruit and the quality of the crop helps the farmer to make sure that the crop is harvested at the right time. Researchers have demonstrated in a number of studies that Raman spectroscopy is a perfect method for enabling monitoring of field and greenhouse crops because of its high selectivity and specificity [1]. Recent research findings show that Raman spectroscopy (RS) can effectively be used in diagnostics of biotic and abiotic stresses [22–25]. RS is a label-free laser technique that does not require chemicals to perform analysis of plant material, so the farmer does not incur any costs by purchasing reagents. Furthermore, it takes only one second to perform analysis of a plant in order to detect pathogens or to identify the origin of abiotic stresses [26].

Agriculture and crop production are relatively new areas in research involving Raman spectroscopy [8]. This review focuses particularly on RS application for evaluating the physiological status of plants as well as the stage in the ripening process, for assessing quality of crops (fruit and seeds), and for identifying plant species, varieties, and their origin. This review aims to increase the awareness of the potential of Raman spectroscopy in agriculture among plant breeders, geneticists, farmers, and engineers.

2. Principle, Instrumentation

Raman spectroscopy is used to examine the structure, dynamics of changes, and function of biomolecules. It is a vibrational technique providing insight into the structure of tissues and their components at the molecular level [27–30].

In principle, Raman spectroscopy measures the frequency shift in the inelastic scattering of light when a photon of incident light strikes a particle and produces a scattered photon [31–36]. In the light scattered by the test medium there is, for the most part, a component of the same frequency as in the incident light (Rayleigh scattering, elastic scattering) [36], while in a minority of cases there are variable frequency components associated with the change in photon energy (inelastic scattering, Raman scattering). The outgoing scattered light can be a photon with a frequency lower than the incident photon and in such cases we call it Stokes Raman scattering, or it is of a frequency that is higher, and then it is known as anti-Stokes Raman scattering [29]. The Stokes band forms when the molecule, after interacting with the excitation radiation, shifts to a higher oscillatory level and the scattered photon has energy lower than the energy difference between the levels of vibrational energy. On the other hand, the anti-Stokes band may appear if the molecule was at the excited oscillatory level before the impact of the excitation radiation – that way there is a high probability that it returns to the basic oscillatory level. The scattered photon has an energy greater than the difference in energy of the oscillating energy levels [36]. Since the Stokes bands are of higher intensity than the anti-Stokes bands, in Raman spectroscopy, the measurement most often concerns only the Stokes part of the Raman spectrum, presented in the range from 4000 to 0 Δ cm^{-1} (the so-called Raman shift) [29].

The phenomenon of Raman scattering of light by particles was predicted by Smekal in 1923 and observed experimentally for the first time by Sir Chandrasekhar Venkat Raman and his student Kariamanickam Srinivas Krishnan in 1928 [37,38].

In honour of the discoverer, the phenomenon of inelastic scattering of light is called Raman scattering [39]. An advantage presented by Raman spectroscopy lies in the fact that when it is used to examine biological materials, the spectra containing large amounts of information can be acquired from intact tissues [40,41], without interfering in their structure. This way it is possible to perform detailed chemical analysis of the biological material despite its high complexity. Organic compounds and functional groups have characteristic spectral patterns, the so-called "fingerprints", enabling their identification, while the intensity of the bands can be used to calculate the concentration in the sample analysed [42]. The Raman spectrum can be used as a fingerprinting tool for various compounds [43]. Table 1 shows the assignment of bands in the Raman spectra of cell wall polysaccharides based on the literature, and Figure 1 shows the Raman spectra of the pure cell wall components pectin (A), xyloglucan (B), cellulose (C), and the Raman spectrum of the tomato cell wall (D) [44].

Table 1. Assignment of bands in the Raman spectra of cell wall polysaccharides based on the literature [8]. Copyright Front. Plant Sci. 2021.

Band (cm^{-1})	Vibration Mode	Assigment
480	C–C–O and C–C–C Deformations; Related to glycosidic ring skeletal deformations δ(C–C–C) + τ(C–O) Scissoring of C–C–C and out-of-plane bending of C–O	Carbohydrates
520	ν(C–O–C) Glycosidic	Cellulose
747	γ(C–O–H) of COOH	Pectin
849–853	(C_6–C_5–O_5–C_1–O_1)	Pectin
917	ν(C–O–C) In plane, symmetric	Cellulose and phenylpropanoids
964–969	δ(CH_2)	Aliphatics
1000–1005	In plane CH_3 rocking of polyene aromatic ring of phenylalanine	Carotenoids and protein
1048	ν(C–O) + ν(C–C) + δ(C–O–H)	Cellulose and phenylpropanoids
1080	ν(C–O) + ν(C–C) + δ(C–O–H)	Carbohydrates
1115–1119	Sym ν(C–O–C); C–O–H bending	Cellulose
1155	C–C Stretching; ν(C–O–C), ν(C–C) in glycosidic linkages, asymmetric ring breathing	Carotenoids and carbohydrates
1185	ν(C–O–H) Next to aromatic ring + δ(CH)	Carotenoids
1218	δ(C–C–H)	Carotenoids, xylan
1265	Guaiacyl ring breathing, C–O stretching (aromatic); –C=C–	Phenylpropanoids, unsaturated fatty acids
1286	δ(C–C–H)	Aliphatics
1301	δ(C–C–H) + δ(O–C–H) + δ(C–O–H)	Carbohydrates
1327	δCH_2 Bending	Aliphatics, cellulose, and phenylpropanoids
1339	ν(C–O); δ(C–O–H)	Carbohydrates
1387	δCH_2 Bending	Aliphatics
1443–1446	δ(CH_2) + δ(CH_3)	Aliphatics
1515–1535	–C=C– (in plane)	Carotenoids
1606–1632	ν(C–C) Aromatic ring + δ(CH)	Phenylpropanoids
1654–1660	–C=C–, C=O Stretching, amide I	Unsaturated fatty acids
1682	COOH	Carboxylic acids
1748	C=O Stretching	Esters, aldehydes, carboxylic acids and ketones

The obtained analyte spectrum can be treated as a qualitative analysis of unknown samples or mixtures of components [45]. Importantly, Raman spectroscopy is sensitive even to small structural changes, which is useful in comparative studies [41]. Raman scattering in tissues provides a wealth of information about the vibrational structure of their constituent proteins, GAGs, lipids, and DNA. Another important advantage of Raman spectroscopy is the low intensity of water bands, which in other spectroscopic methods makes the analysis of biological materials difficult [39]. Raman spectroscopy extracts spatial information from complex biological samples, and as a result, it is an extremely accurate tool for examining a variety of plant materials [46], such as pollen [47,48], fruits [49,50], roots [51], and wood of various origins [52–57]. Its mechanism makes it possible to perform analysis of both the molecular composition and molecular structure of cell walls [58–60].

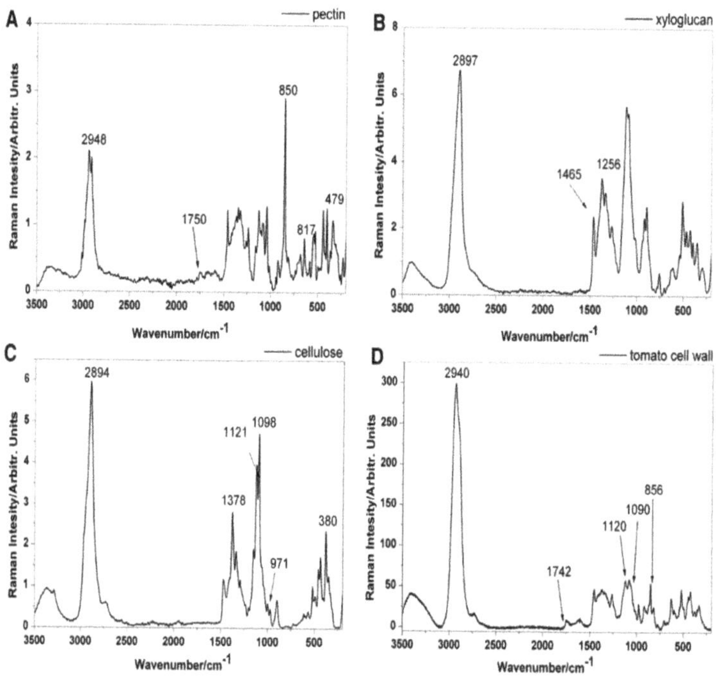

Figure 1. Raman spectra of the pure cell wall components (laser with green light at λ = 532 nm, with a power of 10 mW, the spectra were recorded within the range of 3500–150 cm−1): pectin (**A**), xyloglucan (**B**), cellulose (**C**), and the Raman spectrum of the tomato cell wall (**D**). Copyright Plant Methods 2014.

A Raman spectrometer consists of a laser source which generates a stream of photons. The light is then directed through a beam splitter and focused through a lens onto the test sample. This leads to scattering of light, which is then usually collected using the same optics and directed into the spectrometer. Elastically scattered photons are cut off by long-pass filters before entering the spectrometer. After the inelastically scattered photons (Raman photons) are scattered on the spectrometer gratings according to their energies, they are captured by the CCD [1,61]. An exemplary diagram of the Raman spectrometer operation is shown in Figure 2 (other settings are possible).

Raman spectroscopy was for a long time used exclusively as a laboratory technique, but there are now several commercially available handheld spectrometers. These instruments typically have laser excitation in the green (λ = 532 nm), red (λ = 785 or 830 nm), or infrared (λ = 1064 nm) parts of the electromagnetic spectrum [22–24,62–64]. The beam diameter or laser spot size on portable Raman spectrometers ranges from a few dozen microns to a few millimetres. Excitation wavelength is one of the most important parameters to pay attention to in spectroscopic examinations of plants. Results acquired by researchers show that by using radiation in the blue and green parts of the electromagnetic spectrum it is possible, in particular, to visualise carotenoid signals. This results from the strong absorption of carotenoids in this part of the electromagnetic spectrum [65]. Lasers with wavelengths above 561 nm and below 700 nm are not suitable for structural analysis of living plants due to the extremely strong fluorescence of chlorophyll. Chlorophyll fluorescence decreases exponentially at wavelengths above 700 nm. Therefore, laser excitations of 785–830 nm provide sufficient signal spectra for plant leaf noise [66]. Handheld Raman spectrometers provide accurate access to physiological responses in plants under field conditions [66], both for plant producers and researchers [67].

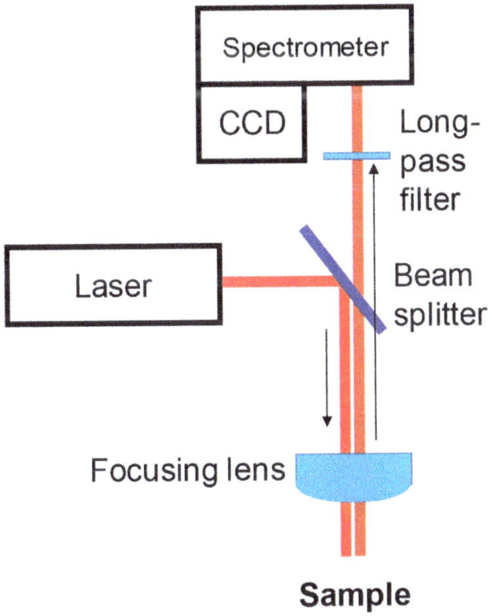

Figure 2. Schematic representation of a Raman spectrometer. Copyright Plant Methods 2021.

The Raman method is an optical spectroscopic technique which can effectively be used to detect and quantify key compounds without destroying the sample. As a result, it is a powerful tool for monitoring plant physiological status and assessing crop quality, pathology, and plant maturation [67–72].

Raman spectroscopy is not yet used widely in agriculture and in response to food analysis but has been used recently [73]. Ramaskie techniques do not require a labor-intensive access service, and they can be tested directly in the environment with pits [74]. The Raman method is fast and non-invasive; there is a text, a spectrometric manual, that prevents spectral registration and fruit testing on the plant or during sorting. An additional advantage of Raman spectroscopy is a small queue for water or its steps, thanks to which you can analyse the attendance and dried fruit. The Raman methodology also enables food safety to be controlled by glass or polymer [75,76]. Raman techniques used in precision agriculture can significantly improve capacities for farming management, crop quality assessment, as well as biological and chemical contaminant detection, thereby contributing to food safety as well as the productivity and profitability of agriculture [77].

3. Raman Spectroscopy in Assessment of Changes during Growth as Well as Harvest Maturity of Plants

Analysis of the spectra collected from various crops at specific vegetation stages can be successfully used to determine the optimal harvest time [78,79]. As an example, Piot and colleagues applied confocal Raman microscopy to examine wheat grain and follow the evolution of protein content and structure during the growth of wheat grains of different cultivars. The cultivars differed in the level of grain hardness and the aptitude to separation of peripheral layers during milling process [80]. The study showed that RS is a suitable tool for acquiring information on grain structure and composition and, most importantly, can detect molecules present at low concentrations, such as α-helical protein. Researchers have noted that concentration of this protein increases as the grains harden during the maturation process [1].

Chylińska and her team applied confocal spectroscopy to assess the ripening stage of tomatoes. Fruit was harvested at the mature green and red ripe stages. The samples were examined using a Raman spectrometer equipped with a green light laser (λ = 532 nm). The researchers investigated the effect of biochemical parameters (such as cell wall polysaccharide content, phenolic compounds, ascorbic acid, and pectinolytic enzyme activity) on cell wall microstructure and changes in polysaccharide distribution during the process of physiological development of tomato fruit (Solanum lycopersicum cv Cerise). The study showed that confocal Raman spectroscopy makes it possible to visualise changes in the spatial distribution of polysaccharides in the plant cell wall (including the central lamella region). In mature green tomato, pectin concentrations were visible particularly in cell corners, whereas ripe red tomato were found with a homogeneous distribution of pectin in the cell wall (Figure 3). The study by Chylińska et al. demonstrates that the Raman technique can visualise the changes occurring in the cell wall (mainly degradation of pectin polysaccharides) during tomato ripening [81].

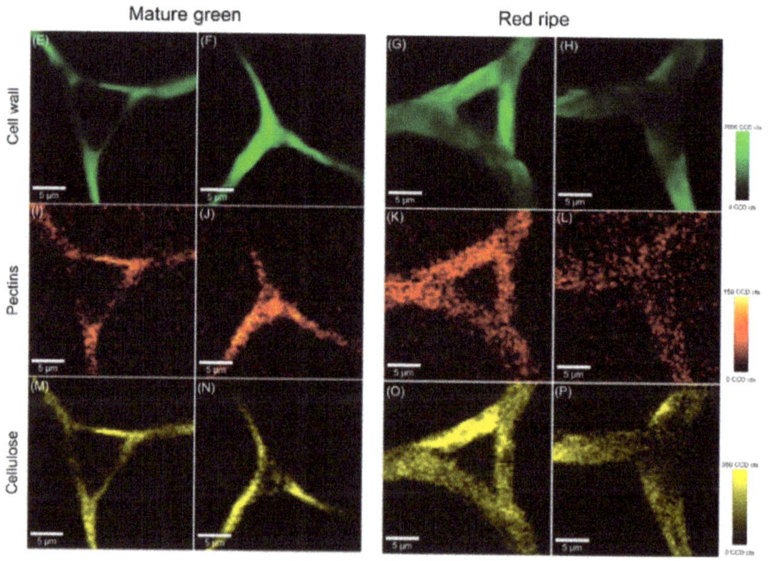

Figure 3. Raman images of cross sections of tomato cell wall from mesocarp at laser with green light (λ = 532 nm); a power of 25 mW and integration time at 0.1 s was chosen. Raman images of all primary cell wall polysaccharides at 2936 cm^{-1}, γ(CH) (**E–H**). Raman images of pectin at 854 cm^{-1}, the (COC) skeletal (**I–L**). Raman images of cellulose at 1096 cm^{-1} and 1115 cm^{-1} (glycosidic bond) (**M–P**). Copyright Plant Physiology and Biochemistry 2017.

Lopez-Sanchez et al. [82] used the potential of Raman spectroscopy to assess changes taking place in the process of olive fruit growth and ripening. For this purpose, the researchers measured the spectra of different parts of olive fruit (skin, flesh, and stone) at different stages of development. They observed an increase in carotenoids and phenolic compounds during olive growth and a decrease during the ripening phase. The evolution of the different spectral bands was linked to the content of the olive constituents, such as triglycerides, water, carotenoids, and phenolic compounds. The study demonstrated the ability of RS to track oil accumulation in olive fruit. Increased intensity of the peaks at 1440 cm^{-1} correlates well with oil content in the fruit measured using the standard Soxhlet extraction method. The researchers showed that increase in carotenoid and phenolic contents during olive growth and decrease in these values during the ripening stage can effectively be monitored using vibration bands at 1525 and 1605 cm^{-1}.

Szymańska-Chargot et al. [83] applied confocal Raman microscopy to assess changes in the distribution of polysaccharides in the cell wall of apple flesh in ripening fruit and during storage after harvest. The apples were collected on three dates: at 1 month and at 2 weeks before the optimum harvest date and on the optimum harvest date. Apples collected on the optimum harvest date were kept in storage for 3 months. The researchers acquired Raman maps for each harvest date and after 1, 2, and 3 months of storage. Raman images of apple cell walls show significant changes in the quantity and distribution of the main cell wall polysaccharides. Analysis of the Raman maps showed degradation of pectins distributed in the middle lamella and primary cell wall. These findings were consistent with the results of chemical assays. During the process of apple ripening and ageing, the analyses based on Raman spectroscopy showed changes in the distribution of pectins, which in young fruit were dispersed along the cell walls, whereas in ripe fruit and those kept in storage they were concentrated in the cell wall corners. Analysis of apples after 3 months in storage showed a significant decrease in pectin content. The findings reported by this research team show that Raman imaging can be a very useful tool for early identification of changes in plant tissue composition during development.

Qin et al. [84] report that it is possible to use Raman spectroscopy to visualise the content of lycopene, the main carotenoid in tomatoes. According to those researchers, imaging of changes in lycopene content during fruit ripening is a good method for monitoring the stage of tomato maturity. During the study, the research team developed a laboratory-based point-source Raman chemical imaging system to detect and visualise the internal distribution of lycopene during the process of fruit ripening and post-harvest. The researchers analysed lycopene content in tomato fruit samples representing different stages of ripeness (i.e., green, breaker, turning, pink, light red, and red). Qin et al. applied the spatially offset Raman spectroscopy (SORS) technique for subsurface detection of a Teflon slab placed under samples of the outer pericarp from green and pink tomatoes. The findings showed that the Teflon spectrum acquired this way can be extracted from SORS measurements of tomato pericarp placed above Teflon. These results suggest a potential for the development of the SORS method in imaging lycopene concentration as a marker of tomato fruit ripeness.

Martin et al. [85] developed a tomato ripening model based on vibrational bands of carotenoids in Raman spectra. Tomato fruit during the growth phase and during the post-harvest ripening stage were analysed using a laboratory Raman spectrometer equipped with a 532 nm laser. The researchers observed increase in the carotenoid signal at the start of the turning stage of the fruit ripening. The acquired data were used by the team to build a model describing the stages of the tomato ripening process and helping to accurately assess post-harvest fruit quality.

A similar study using handheld Raman spectrometers for hot pepper was first carried out by Langer et al. [86] Scientists in their reports described the Raman signals of carotenoids typical of hot pepper fruits and followed their evolution during maturation (Figure 4). Researchers proposed and compared a multivariate chemometric model and a simple one-dimensional model, resulting in a four-point scale for grading the ripeness of hot pepper fruit. The authors suggest that the model proposed is appropriate for assessing the ripening stage of fruit containing carotenoid, and thus for determining the maturity on site or during the sorting process in an automated manner.

Bands in the range of approximately 800 to 1600 cm^{-1} have been assigned to carotenoids. The band at about 860 cm $^{-1}$ can be attributed to the asymmetric stretching of the C–O–C glycosidic bond in acid pectins. A faint band at 1327 cm^{-1} is attributed to chlorophyll a, which is known to have the highest intensities in the pure chlorophyll spectrum. The most intense bands in the spectrum were assigned to the carotenoids, observed at 1150–1170 and 1500–1550 cm^{-1}. The band is formed as a result of vibrations in the C = C phase and C–C stretching of the polyene chain. In the range of 1000–1020 cm^{-1}, methyl groups attached to the polyene chain were recognized, showing moderate intensity [86].

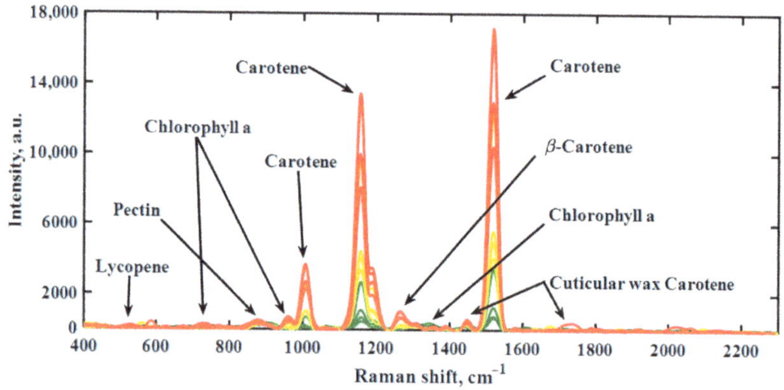

Figure 4. Raman spectra of hot peppers during the maturation process obtained with handheld Raman spectrometer at 785 nm laser wavelength with a power of 100 mW. Unripe pepper (green), ripening pepper (yellow), and fully ripe fruit (red) are each represented by four deliberately selected spectra. Copyright MDPI 2021.

Cabrales and team applied CRM to examine cross-sections of growing cotton fibres during the main five developmental stages. In course of the study, the researchers analysed vibrational bands at 383 cm^{-1}, attributed to cellulose. The intensity of the Raman spectra for cellulose was significantly lower for fibres harvested at 21 days post-anthesis (dpa) compared to 56 dpa. The sub-micron resolution of the CRM provides insight into the deposition of cellulose in the secondary cell wall. The findings showed that it is possible to obtain information on the chemical composition and structure of deposited cellulose in developing cotton fibres. CRM can be used as a tool to differentiate between cotton fibres at different stages of growth [87].

4. Raman Spectroscopy for Fruit and Seeds Quality Control

RS can also be applied to perform non-invasive assessment of the nutritional value and quality of plants, fruits, and seeds, and as a result, it is a perfect tool to be used in digital agronomy [88,89]. Nekvapil et al. [41] investigated the relationship between the freshness of selected citrus fruit varieties and their Raman spectra. Citrus fruit freshness is associated with the appearance and colour and consequently the net carotenoid content of the peel. In that study the research team assessed freshness of commercially available citrus fruits (clementines and different varieties of mandarins) using a handheld Raman spectrometer. Evaluation of fruit freshness can be performed using Raman instruments quickly, objectively, and without destroying the material. The samples were excited with 532 nm and 785 nm waves. Nekvapil et al. evaluated the fruit for carotenoid contents in the peel during the time span between fresh fruit delivery and their physical degradation. The analyses found a strong correlation between carotenoid content of the peel and the intensity of the Raman signal. The findings showed that the intensity of the Raman signal for carotenoids is a good indicator of fruit freshness. Based on this, the researchers introduced a Raman coefficient of freshness (CFresh), which decreases linearly in time with a different slope for different citrus groups. After 7 days of the experiment, a decrease in signal intensity was observed in the case of all the samples, but this change was varied. It was observed that the decrease in Raman signal intensity was smaller in the case of fruit stored in daylight compared to fruit kept in dark storage. The Raman coefficient of freshness (CFresh) calculated for specific fruit appears to be a user-friendly, fast, and sensitive method for assessing citrus fruit freshness with a portable Raman spectrometer. The portable system shows great potential for extensive use in the evaluation of citrus freshness, both in ripening crops and in fruit supplied to the market or to the food industry.

Similarly, Nikbakht and colleagues [90] applied RS to determine tomato fruit quality. The researchers demonstrated that RS can be used to measure important tomato quality parameters such as soluble solids content (SSC), acidity (pH), and colour. They showed that RS can be very effective in assessing the quality of both the external and internal properties of tomatoes.

Zhu et al. [89] applied Raman spectroscopy to examine the mechanisms underlying fruit lignification. Knowledge of the fruit lignification process would facilitate operations aimed to optimise storage and preservation strategies and to reduce post-harvest deterioration. By investigating lignin deposition in fruit at cellular level it may be possible to work out the mechanisms underlying fruit lignification. The study mainly aimed to establish a procedure for applying Raman microspectroscopy to visualise cellular-level lignification in loquat fruit. Fruit lignification leads to increased fruit firmness and is important from the viewpoint of optimum post-harvest handling of fruit to minimise deterioration. The findings of the study showed that Raman spectroscopy can effectively be used to assess fruit lignification in order to determine fruit maturity.

The group of Morey and Kurouski et al. [91] used RS to assess nutrient content of potato tubers. The researchers showed that the intensity of the 479 cm^{-1} band (correlating with starch) increases linearly with growing starch content in the test samples. These findings suggest that RS can effectively be used to measure starch content in intact potato tubers. Using such calibration curves, the researchers were able to accurately determine the absolute starch concentration in potatoes. The findings showed differences in the spectra collected from samples with different starch contents; the Raman spectra acquired from a sample containing 12% starch (6 g starch) were statistically different from the spectra collected from samples with starch at a content of 9% (4.5 g starch) and 15% (7.5 g starch). Similarly, the spectra acquired from samples with 15% starch content differed significantly from the spectra obtained from samples with 12 and 18% starch content. The results of the analyses and the standard deviations obtained suggest that this method can be used to identify starch content within 3% accuracy. It can be expected that more careful standardisation may change the prediction accuracy to 1% and below.

Krimmer et al. [92] used a Raman spectrometer to assess the nutrient content of maize grains. Maize is popular worldwide as an ingredient in human food and livestock feed as well as a raw material in industry and a biofuel. The researchers found that Raman spectroscopy can identify carbohydrates, fibre, carotenoids, and proteins in maize kernels [63]. Abreu and colleagues applied RS to monitor coffee quality. They collected spectra from coffee beans stored under different conditions for 0, 3, 6, 9, 12, and 18 months. The researchers observed that changes in kahweol visualised by means of vibrational bands can be helpful in predicting coffee quality and changes taking place in beans kept in storage.

5. Raman Spectroscopy for Determination of Species and Origin of Plants, Fruit, and Seeds

Related publications have reported successful attempts to build RS-based models enabling identification of fruit varieties. Feng and colleagues applied RS to study eight different citrus fruits. The team succeeded in building a model to distinguish between citrus varieties. This work shows that RS can be used to accurately, quickly, and efficiently identify varieties and assess citrus fruit quality [93].

Kurouski et al. [91] focused on using SR to distinguish potato varieties. Owing to their high starch content, simple cultivation process, and high yield, potatoes are one of the basic ingredients in the diet of people across the world. Potato tubers consist of approximately 83% water and 12% carbohydrates, whereas proteins, vitamins, and other trace elements account for the remaining 4%. The precise composition of potatoes varies relative to the type of potato and the place where they are grown. The researchers successfully used Raman spectroscopy to identify nine different potato varieties and to determine the origin of the crop. Using spatially offset Raman spectroscopy (SORS), the researchers observed that the peak intensity varied between potato varieties at 479 and 1125 cm^{-1} for starch,

1600 and 1630 cm^{-1} for phenylpropanoid, 1527 cm^{-1} for carotenoid content, and 1660 cm^{-1} for protein content. Based on the data obtained, Kurouski's team was able to identify the potato variety and determine the location of the potato crop with an accuracy of 81 to 100%.

Farber and colleagues [18] showed that by acquiring Raman spectroscopy spectra of peanut leaves it is possible to identify different peanut varieties and genotypes. This method can also be applied to determine plant resistance to nematodes and to measure the oleic/linoleic oil (O/L) ratio. Furthermore, analysis of peanut seeds based on Raman spectroscopy can be applied to accurately identify the genotype as well as carbohydrates, proteins, fibre, and other nutrients. In course of their experiment, the researchers grew plants of 10 different peanut genotypes and analysed their spectra. All the genotypes were found with similar profiles, with vibrational bands characteristic of carbohydrates, cellulose, pectins, carotenoids, phenylpropanoids, protein, and carboxylic acid. The findings acquired using the specially developed PLS-DA model showed that the Raman method makes it possible to identify peanut varieties with 80% accuracy. The results of the study suggest that the resistance of peanut plants to nematodes is related to changes in carotenoid- and phenylpropanoid-specific peaks. In the food industry, peanuts with a high oleic ratio are preferred because of their longer shelf life and consequently reduced rancidity. It has also been shown that high-oleic peanuts beneficially affect (i.e., decrease) serum cholesterol levels and reduce the risk of cardiovascular disease. RS showed that plants of this specific genotype had lower phenylpropanoid contents, while all other peaks remained almost identical. Farber and the team reported 82% accuracy of the Raman method in identifying peanuts with high versus normal oleic ratios. To compare the accuracy of the method, the researchers performed Raman analyses of peanut seeds. The findings show 95% accuracy of Raman spectroscopy in identification of peanut seeds, compared to 82% in the case of leaves.

Krimmer et al. [63] investigated the feasibility of Raman spectroscopy in identification of maize varieties. Using PLS-DA, Krimmer and colleagues showed that RS can be used to identify six different maize varieties based on their unique spectra. In the course of the experiment, Krimmer and colleagues collected over 600 spectra from six different maize varieties. All six varieties had similar spectral profiles, except for the scanned darker kernels, which had a lower intensity. This is due to the different light absorption and scattering properties of these maize kernels, which affect the scanning. This problem can be solved by normalisation, especially at the 1458 cm^{-1} peak displayed by all spectra display. The authors analysed the intensity of the bands at 479 cm^{-1} (starch), 1530 cm^{-1} (carotenoids), 1600/1632 cm^{-1} (both fibres), and 1640–1670 cm^{-1} (protein region) to quantify the carbohydrate, carotenoid, fibre, and protein content of maize. Krimmer and colleagues showed that RS in combination with chemometric methods can be used for highly accurate typing of maize varieties.

Abreu and colleagues in their study showed that RS can be used to discriminate Arabica coffee genotypes very accurately. [92]. Using Raman spectroscopy and principal component analysis, Figueiredo and co-workers were able to identify four Arabica coffee genotypes: one Mundo Novo line and three Bourbon lines, with an accuracy of ~ 80% [94]. Keidel and co-workers [95] showed that spectroscopic measurements of both ground and whole beans can be used to predict the geographical origin of coffee beans.

6. Conclusions

Raman spectroscopy has been used in analysis of plant biomass for nearly 30 years [96,97]. In plant science, the Raman spectroscope was first used by Atella and Agarwal in the 1980s [52,96,97]. Owing to recent technological advancements, today Raman spectroscopy and the related instruments are recognised as precise tools for studying plant tissues [97,98]. Currently, there are over 25 different types of Raman spectroscopy techniques. For example, a Fourier transformation (FT) Raman spectrometer using a near-infrared (NIR) laser solves the problem of fluorescence interference [99]. The surface-enhanced Raman spectroscopy technique enhances the Raman scattering signal [100]. Raman confocal microscopy provides three-dimensional images of the structure and composition of the material with micrometric resolution and clear image quality [101]. Coherent anti-Stokes Raman scattering–(CARS) provides spectral information with excellent sensitivity and low laser power [102]. Raman resonance scattering (RRS) allows the study of a spectrum of materials in the range of the photon energy itself [103]. Raman optical activity (ROA) monitors a small difference in the scattering of right- and left-circularly polarized light. ROA spectra are sensitive to chirality (e.g., to the enantiomeric excess) and fine variations in molecular geometry including conformational states [104].

Raman microscopy is widely used to examine the composition of the plant cell wall [105]. Numerous studies have reported that Raman imaging techniques can successfully be applied to investigate differences between tissues and to monitor changes in plant cells [35,80,82,106–112]. Raman spectroscopy can be used to acquire marker-free spectral maps of biological tissues and cells for the needs of chemical, structural, and environmental analyses. Raman images present information about molecular structure, composition, and interactions in micrometres or even nanometres [54,55]. An important advantage presented by the Raman techniques is the fact that Raman signal processing can be used to obtain analysis of several samples or plant fragments of one type [60].

Agriculture and plant growing, as well as plant pathology, are relatively new areas in RS-based research. The related literature published in the recent years shows that Raman spectroscopy can successfully be used in diagnostics of biotic as well as abiotic stresses in plants. Rapid assessment of plant phenotype allows farmers to intervene immediately and precisely to mitigate biotic and abiotic stresses [11,17,22–24,26]. Since they are highly sensitive to small changes in plant biochemistry, Raman spectrometers make it possible to identify plant species and plant varieties and allow farmers to select and grow plants with advantageous genotypes [18,63].

This review shows the potential of RS for the digital agriculture of the future. Portable and rapid Raman spectrometry analysis allows for quick detection of biotic and abiotic stresses in plants. Moreover, Raman techniques can be used as an advanced method for plant breeding and selection as they are both non-invasive and non-destructive. RS can also be used for plant phenotyping and nutrient analysis. Advantages presented by RS will certainly become more obvious to others, and the use of Raman spectrometry in digital agriculture will become more widespread. The relatively high cost of the related equipment is a significant drawback for farmers, adversely affecting widespread use of Raman spectroscopy in crop monitoring. It can be expected that continued advancements in the technology will bring the cost of these devices down in the near future. Furthermore, application of Raman spectrometry in agriculture is likely to be implemented in the near future as a service providing farmers with information on the condition of the field along with GPS coordinates of the assessed locations.

The above review takes a closer look at the potential of Raman spectrometers to be used for the advancement of digital agriculture. The related gains outlined here are gradually becoming more obvious for farmers and investors, and the use of Raman spectroscopy in digital agriculture will become more widespread [8].

Author Contributions: Conceptualization, A.S.; formal analysis, A.S.; investigation, A.S.; resources, A.S.; data curation, B.S.; writing—original draft preparation, A.S.; writing—review and editing, A.S. and B.S.; visualization, B.S.; supervision, B.S and C.P.; project administration, B.S.; funding acquisition, C.P. All authors have read and agreed to the published version of the manuscript.

Funding: The project is financed by the program of the Ministry of Science and Higher Education "Regional Initiative of Excellence" in the years 2019–2022, Project No.026/RID/2018/19, the amount of financing totaling PLN 9 542 500.00.

Institutional Review Board Statement: Not applicable.

Informed Consent Statement: Not applicable.

Data Availability Statement: Not applicable.

Conflicts of Interest: The authors declare no conflict of interest.

Abbreviations

CARS	Coherent anti-Stokes Raman scattering
CRM	Confocal Raman spectroscopy
DNA	Deoxyribonucleic acid
FT	Fourier transformation
GAGs	Glycosaminoglycans
NIR	Raman near-infrared
ROA	Raman optical activity
RS	Raman spectroscopy
RRS	Raman resonance scattering
SORS	Spatially offset Raman spectroscopy
PLS-DA	Partial least-squares discriminant analysis

References

1. Payne, W.Z.; Kurouski, D. Raman spectroscopy enables phenotyping and assessment of nutrition values of plants: A review. *Plant Methods* **2021**, *17*, 78. [CrossRef] [PubMed]
2. Brummer, E.C.; Barber, W.T.; Collier, S.; Cox, T.S.; Johnson, R.; Murray, S.C.; Olsen, R.T.; Pratt, R.C.; Thro, A.M. Plant breeding for harmony between agriculture and the environment. *Front. Ecol. Environ.* **2011**, *9*, 561–568. [CrossRef]
3. Teng, P.S.; Shane, W.W.; MacKenzie, D.R. Crop losses due to plant pathogens. *Crit. Rev. Plant Sci.* **1984**, *2*, 21–47. [CrossRef]
4. Savary, S.; Ficke, A.; Aubertot, J.-N.; Hollier, C. Crop losses due to diseases and their implications for global food production losses and food security. *Food Secur.* **2012**, *4*, 519–537. [CrossRef]
5. He, Y.; Borrego, E.J.; Gorman, Z.; Huang, P.C.; Kolomiets, M.V. Relative contribution of LOX10, green leaf volatiles and JA to wound-induced local and systemic oxylipin and hormone signature in *Zea mays* (maize). *Phytochemistry* **2020**, *174*, 112334. [CrossRef]
6. Wang, K.D.; Borrego, E.J.; Kenerley, C.M.; Kolomiets, M.V. Oxylipins other than Jasmonic acid are xylem-resident signals regulating systemic resistance induced by *Trichoderma virens* in maize. *Plant Cell* **2020**, *32*, 166–185. [CrossRef]
7. Mihaljev, Ž.; Jakšić, S.; Prica, N.B.; Ćupić, Ž.N.; Baloš, M.Ž. Comparison of the Kjeldahl method, Dumas method and NIR method for total nitrogen determination in meat and meat products. *J. Agroliment. Process. Technol.* **2015**, *21*, 365–370.
8. Payne, W.Z.; Kurouski, D. Raman-Based Diagnostics of Biotic and Abiotic Stresses in Plants. A Review. *Front. Plant Sci.* **2021**, *11*, 616672. [CrossRef]
9. Sanchez, L.; Ermolenkov, A.; Biswas, S.; Septiningshih, E.M.; Kurouski, D. Raman spectroscopy enables non-invasive and confirmatory diagnostics of salinity stresses, nitrogen, phosphorus, and potassium deficiencies in rice. *Front. Plant Sci.* **2020**, *11*, 573321. [CrossRef]
10. Altangerel, N.; Ariunbold, G.O.; Gorman, C.; Alkahtani, M.H.; Borrego, E.J.; Bohlmeyer, D.; Hemmer, P.; Kolomiets, M.V.; Yuan, J.S.; Scully, M.O. In vivo diagnostics of early abiotic plant stress response via Raman spectroscopy. *Proc. Natl. Acad. Sci. USA* **2017**, *114*, 3393–3396. [CrossRef]
11. Yeturu, S.; Vargas Jentzsch, P.; Ciobotă, V.; Guerrero, R.; Garrido, P.; Ramos, L.A. Handheld Raman spectroscopy for the early detection of plant diseases: Abutilon mosaic virus infecting Abutilon sp. *Anal. Methods* **2016**, *8*, 3450–3457. [CrossRef]
12. Farber, C.; Bryan, R.; Paetzold, L.; Rush, C.; Kurouski, D. Non-invasive characterization of single-, double- and triple-viral diseases of wheat with a hand-held Raman spectrometer. *Front. Plant Sci.* **2020**, *11*, 01300. [CrossRef]
13. Farber, C.; Sanchez, L.; Kurouski, D. Confirmatory non-invasive and non-destructive identification of poison ivy using a hand-held Raman spectrometer. *RCS Adv.* **2020**, *10*, 21530–21534. [CrossRef]
14. Farber, C.; Shires, M.; Ong, K.; Byrne, D.; Kurouski, D. Raman spectroscopy as an early detection tool for rose rosette infection. *Planta* **2019**, *250*, 1247–1254. [CrossRef]
15. Sanchez, L.; Ermolenkov, A.; Tang, X.T.; Tamborindeguy, C.; Kurouski, D. Non-invasive diagnostics of Liberibacter disease on tomatoes using a hand-held Raman spectrometer. *Planta* **2020**, *251*, 64. [CrossRef]
16. Sanchez, L.; Pant, S.; Mandadi, K.; Kurouski, D. Raman spectroscopy vs quantitative polymerase chain reaction in early stage Huanglongbing diagnostics. *Sci. Rep.* **2020**, *10*, 10101. [CrossRef]
17. Mandrile, L.; Rotunno, S.; Miozzi, L.; Vaira, A.M.; Giovannozzi, A.M.; Rossi, A.M.; Noris, E. Nondestructive Raman spectroscopy as a tool for early detection and discrimination of the infection of tomato plants by two economically important viruses. *Anal. Chem.* **2019**, *91*, 9025–9031. [CrossRef] [PubMed]
18. Farber, C.; Sanchez, L.; Rizevsky, S.; Ermolenkov, A.; McCutchen, B.; Cason, J.; Simpson, C.; Burow, M.; Kurouski, D. Raman spectroscopy enables non-invasive identification of Peanut genotypes and value-added traits. *Sci. Rep.* **2020**, *10*, 7730. [CrossRef]
19. Gan, Q.; Wang, X.; Wang, Y.; Xie, Z.; Tian, Y.; Lu, Y. Culture-free detection of crop pathogens at the single-cell level by micro-Raman spectroscopy. *Adv. Sci.* **2017**, *4*, 1700127. [CrossRef]

20. Mantri, N.; Patade, V.; Penna, S.; Ford, R.; Pang, E. Abiotic stress responses in plants: Present and future. In *Abiotic Stress Responses in Plants*; Parvaiz, A., Prasad, M.N.V., Eds.; Springer: New York, NY, USA, 2012; pp. 1–19.
21. Waqas, M.A.; Kaya, C.; Riaz, A.; Farooq, M.; Nawaz, I.; Wilkes, A.; Li, Y. Potential mechanisms of abiotic stress tolerance in crop plants induced by thiourea. *Front. Plant Sci.* **2019**, *10*, 1336. [CrossRef]
22. Egging, V.; Nguyen, J.; Kurouski, D. Detection and identification of fungal infections in intact wheat and sorghum grain using a hand-held Raman spectrometer. *Anal. Chem.* **2018**, *90*, 8616–8621. [CrossRef] [PubMed]
23. Farber, C.; Kurouski, D. Detection and identification of plant pathogens on maize kernels with a hand-held Raman spectrometer. *Anal. Chem.* **2018**, *90*, 3009–3012. [CrossRef] [PubMed]
24. Sanchez, L.; Farber, C.; Lei, J.; Zhu-Salzman, K.; Kurouski, D. Noninvasive and nondestructive detection of cowpea bruchid within cowpea seeds with a hand-held Raman spectrometer. *Anal. Chem.* **2019**, *91*, 1733–1737. [CrossRef] [PubMed]
25. Sanchez, L.; Pant, S.; Xing, Z.; Mandadi, K.; Kurouski, D. Rapid and noninvasive diagnostics of Huanglongbing and nutrient deficits on citrus trees with a handheld Raman spectrometer. *Anal. Bioanal. Chem.* **2019**, *411*, 3125–3133. [CrossRef]
26. Farber, C.; Mahnke, M.; Sanchez, L.; Kurouski, D. Advanced spectroscopic techniques for plant disease diagnostics. A review. *Trends Anal. Chem.* **2019**, *118*, 43–49. [CrossRef]
27. Hellerer, T.; Axäng, C.; Brackmann, C.; Hillertz, P.; Pilon, M.; Enejder, A. Monitoring of lipid storage in Caenorhabditis elegans using coherent anti-Stokes Raman scattering (CARS) microscopy. *Proc. Natl. Acad. Sci. USA* **2007**, *104*, 14658–14663. [CrossRef]
28. Maier, O.; Oberle, V.; Hoekstra, D. Fluorescent lipid probes: Some properties and applications (a review). *Chem. Phys. Lipids* **2002**, *116*, 3–18. [CrossRef]
29. Jones, R.R.; Hooper, D.C.; Zhang, L.; Wolverson, D.; Valev, V.K. Raman Techniques: Fundamentals and Frontiers. *Nanoscale Res. Lett.* **2019**, *14*, 231. [CrossRef]
30. Bergholt, M.S.; Serio, A.; Albro, M.B. Raman Spectroscopy: Guiding Light for the Extracellular Matrix. *Front. Bioeng. Biotechnol.* **2019**, *7*, 303. [CrossRef]
31. Novovic, D.; Dewes, R.C.; Aspinwall, D.K.; Voice, W.; Bowen, P. The effect of machined topography and integrity on fatigue life. *Mach. Tools Manuf.* **2004**, *44*, 125–134. [CrossRef]
32. Hickey, D.P.; Jones, K.S.; Elliman, R.G. Amorphization and graphitization of single crystal diamond–A transmission electron microscopy study. *Diam. Relat. Mater.* **2009**, *18*, 1353–1359. [CrossRef]
33. Yan, J.; Gai, X.; Harada, H. Subsurface damage of single crystalline silicon carbide in nanoindentation tests. *J. Nanosci. Nanotechnol.* **2010**, *10*, 7808–7811. [CrossRef]
34. Meng, B.B.; Zhang, Y.; Zhang, F.H. Material removal mechanism of 6H-SiC studied by nano-scratching with Berkovich indenter. *Appl. Phys. A* **2016**, *122*, 247. [CrossRef]
35. Hanninen, T.; Kontturi, E.; Vuorinen, T. Distribution of lignin and its coniferyl alcohol and coniferyl aldehyde groups in Picea abies and Pinus sylvestris as observed by Raman imaging. *Phytochemistry* **2011**, *72*, 1889–1895. [CrossRef]
36. Das, R.S.; Agrawal, Y.K. Raman spectroscopy: Recent advancements, techniques and applications. *Vib. Spectrosc.* **2011**, *57*, 163–176. [CrossRef]
37. Smekal, A. On the quantum theory of dispersion. *Naturwissenschaften* **1923**, *11*, 873–875. [CrossRef]
38. Raman, C.V.; Krishnan, K.S. A new type of secondary radiation. *Nature* **1928**, *121*, 501–502. [CrossRef]
39. Rostron, P. Raman Spectroscopy, a review. *Int. J. Eng. Tech. Res.* **2016**, *6*, 2454–4698.
40. Colomban, P.; Slodczyk, A. Raman Intensity: An Important Tool in the Study of Nanomaterials and Nanostructures. *Acta. Phys. Pol.* **2009**, *116*, 7–12. [CrossRef]
41. Nekvapil, F.; Brezestean, I.; Barchewitz, D.; Glamuzina, B.; Chiş, V.; Pinzaru, S.C. Citrus fruits freshness assessment using Raman spectroscopy. *Food Chem.* **2018**, *242*, 560–567. [CrossRef]
42. Gierlinger, N.; Schwanninger, M. The potential of Raman microscopy. *Spectroscopy* **2007**, *21*, 69–89. [CrossRef]
43. Haynes, C.L.; McFarland, A.D.; Duyne, R.P.V. Surface-enhanced Raman spectroscopy. *Anal. Chem.* **2005**, *77*, 338A. [CrossRef]
44. Chylińska, M.; Szymańska—Chargot, M.; Zdunek, A. Imaging of polysaccharides in the tomato cell wall with Raman microspectroscopy. *Plant Methods* **2014**, *10*, 14. [CrossRef]
45. Msaoubi, R.; Outeiro, J.C.; Chandrasekaran, H.; Dillon, O.W.J.; Jawahir, I.S. A Review of surface integrity in machining and its impact on functional performance and life of machined products. *Sustain. Manuf.* **2008**, *1*, 203–236. [CrossRef]
46. Zeise, I.; Heiner, Z.; Holz, S.; Joester, M.; Büttner, C.; Kneipp, J. Raman Imaging of Plant Cell Walls in Sections of Cucumis sativus. *Plants* **2018**, *7*, 7. [CrossRef]
47. Schulte, F.; Lingott, J.; Panne, U.; Kneipp, J. Chemical Characterization and Classification of Pollen. *Anal. Chem.* **2008**, *80*, 9551–9556. [CrossRef]
48. Schulte, F.; Panne, U.; Kneipp, J. Molecular changes during pollen germination can be monitored by Raman microspectroscopy. *J. Biophotonics* **2010**, *3*, 542–547. [CrossRef]
49. Schulz, H.; Baranska, M.; Baranski, R. Potential of NIR-FT-Raman spectroscopy in natural carotenoid analysis. *Biopolymers* **2005**, *77*, 212–221. [CrossRef]
50. Lopez-Casado, G.; Matas, A.J.; Dominguez, E.; Cuartero, J.; Heredia, A. Biomechanics of isolated tomato (Solanum lycopersicum L.) fruit cuticles: The role of the cutin matrix and polysaccharides. *J. Exp. Bot.* **2007**, *58*, 3875–3883. [CrossRef]
51. Roman, M.; Dobrowolski, J.C.; Baranska, M.; Baranski, R. Spectroscopic Studies on Bioactive Polyacetylenes and Other Plant Components in Wild Carrot Root. *J. Nat. Prod.* **2011**, *74*, 1757–1763. [CrossRef]

52. Atalla, R.H.; Agarwal, U.P. Raman Microprobe Evidence for Lignin Orientation in the Cell-Walls of Native Woody Tissue. *Science* **1985**, *227*, 636–638. [CrossRef] [PubMed]
53. Agarwal, U.P. Raman imaging to investigate ultrastructure and composition of plant cell walls: Distribution of lignin and cellulose in black spruce wood (*Picea mariana*). *Planta* **2006**, *224*, 1141–1153. [CrossRef] [PubMed]
54. Gierlinger, N.; Schwanninger, M. Chemical imaging of poplar wood cell walls by confocal Raman microscopy. *Plant Physiol.* **2006**, *140*, 1246–1254. [CrossRef] [PubMed]
55. Schmidt, M.; Schwartzberg, A.M.; Perera, P.N.; Weber-Bargioni, A.; Carroll, A.; Sarkar, P.; Bosneaga, E.; Urban, J.J.; Song, J.; Balakshin, M.Y.; et al. Label-free in situ imaging of lignification in the cell wall of low lignin transgenic Populus trichocarpa. *Planta* **2009**, *230*, 589–597. [CrossRef]
56. Gierlinger, N. Revealing changes in molecular composition of plant cell walls on the micron-level by Raman mapping and vertex component analysis (VCA). *Front. Plant Sci.* **2014**, *5*, 306. [CrossRef]
57. Ma, J.; Zhou, X.; Ma, J.; Ji, Z.; Zhang, X.; Xu, F. Raman Microspectroscopy Imaging Study on Topochemical Correlation Between Lignin and Hydroxycinnamic Acids in Miscanthus sinensis. *Microsc. Microanal.* **2014**, *20*, 956–963. [CrossRef]
58. Baranska, M.; Roman, M.; Dobrowolski, J.C.; Schulz, H.; Baranski, R. Recent Advances in Raman Analysis of Plants: Alkaloids, Carotenoids, and Polyacetylenes. *Curr. Anal. Chem.* **2013**, *9*, 108–127. [CrossRef]
59. Butler, H.J.; McAinsh, M.R.; Adams, S.; Martin, F.L. Application of vibrational spectroscopy techniques to non-destructively monitor plant health and development. *Anal. Methods* **2015**, *7*, 4059–4070. [CrossRef]
60. Heiner, Z.; Zeise, I.; Elbaum, R.; Kneipp, J. Insight into plant cell wall chemistry and structure by combination of multiphoton microscopy with Raman imaging. *J. Biophotonics* **2017**, *11*, e201700164. [CrossRef]
61. Ferraro, J.R. *Introductory Raman Spectroscopy*, 2nd ed.; Academic Press: San Diego, CA, USA, 2003.
62. Short, L.; Thoms, A.V.; Cao, B.; Sinyukov, A.M.; Joshi, A.; Scully, R.; Sanders, V.; Voronine, D.V. Facile residue analysis of recent and prehistoric cook stones using handheld Raman spectrometry. *J. Raman Spectrosc.* **2015**, *46*, 126–132. [CrossRef]
63. Krimmer, M.; Farber, C.; Kurouski, D. Rapid and noninvasive typing and assessment of nutrient content of maize kernels using a handheld raman spectrometer. *ACS Omega* **2019**, *4*, 16330–16335. [CrossRef]
64. Sanchez, L.; Pant, S.; Irey, M.S.; Mandadi, K.; Kurouski, D. Detection and identification of canker and blight on orange trees using a hand-held Raman spectrometer. *J. Raman Spectrosc.* **2019**, *50*, 1875–1880. [CrossRef]
65. Rong Sng, B.J.; Singh, G.P.; Van Vu, K.; Chua, N.-H.; Ram, R.J.; Jang, I.-C. Rapid metabolite response in leaf blade and petiole as a marker for shade avoidance syndrome. *Plant Methods* **2020**, *16*, 144. [CrossRef]
66. Gupta, S.; Huang, C.H.; Singh, G.P.; Park, B.S.; Chua, N.-H.; Ram, R.J. Portable Raman leaf-clip sensor for rapid detection of plant stress. *Sci. Rep.* **2020**, *10*, 20206. [CrossRef]
67. Lew, T.T.S.; Sarojam, R.; Jang, I.-C.; Park, B.S.; Naqvi, N.I.; Wong, M.H.; Singh, G.P.; Ram, R.J.; Shoseyov, O.; Saito, K.; et al. Species-independent analytical tools for next-generation agriculture. *Nat. Plants* **2020**, *6*, 1408–1417. [CrossRef]
68. Arendse, E.; Fawole, O.A.; Magwaza, L.S.; Opara, U.L. Non-destructive prediction of internal and external quality attributes of fruit with thick rind: A review. *J. Food Eng.* **2018**, *217*, 11–23. [CrossRef]
69. Saletnik, A.; Saletnik, B.; Puchalski, C. Overview of Popular Techniques of Raman Spectroscopy and Their Potential in the Study of Plant Tissues. *Molecules* **2021**, *26*, 1537. [CrossRef]
70. Li, J.L.; Sun, D.W.; Cheng, J.H. Recent Advances in Nondestructive Analytical Techniques for Determining the Total Soluble Solids in Fruits: A Review. *Compr. Rev. Food Sci. Food Saf.* **2016**, *15*, 897–911. [CrossRef]
71. Huang, C.H.; Singh, G.P.; Park, S.H.; Chua, N.H.; Ram, R.J.; Park, B.S. Early Diagnosis and Management of Nitrogen Deficiency in Plants Utilizing Raman Spectroscopy. *Front. Plant Sci.* **2020**, *11*, 663. [CrossRef]
72. Weng, S.; Hu, X.; Wang, J.; Tang, L.; Li, P.; Zheng, S.; Zheng, L.; Huang, L.; Xin, Z. Advanced Application of Raman Spectroscopy and Surface-Enhanced Raman Spectroscopy in Plant Disease Diagnostics: A Review. *J. Agric. Food Chem.* **2021**, *69*, 2950–2964. [CrossRef]
73. Cavaco, A.M.; Utkin, A.B.; Marques da Silva, J.; Guerra, R. Making Sense of Light: The Use of Optical Spectroscopy Techniques in Plant Sciences and Agriculture. *Appl. Sci.* **2022**, *12*, 997. [CrossRef]
74. Boyaci, I.H.; Temiz, H.T.; Geniş, H.E.; Soykut, E.A.; Yazgan, N.N.; Güven, B.; Uysal, R.S.; Bozkurt, A.G.; Ilaslan, K.; Torun, O.; et al. Dispersive and FT-Raman spectroscopic methods in food analysis. *RSC Adv.* **2015**, *5*, 56606–56624. [CrossRef]
75. Baranski, R.; Baranska, M.; Schulz, H. Changes in carotenoid content and distribution in living plant tissue can be observed and mapped in situ using NIR-FT-Raman spectroscopy. *Planta* **2005**, *222*, 448–457. [CrossRef]
76. Yang, D.; Ying, Y. Applications of Raman Spectroscopy in Agricultural Products and Food Analysis: A Review. *Appl. Spectrosc. Rev.* **2011**, *46*, 539–560. [CrossRef]
77. Qin, J.; Chao, K.; Kim, M.S. Chapter 14: Raman scattering for food quality and safety assessment. In *Light Scattering Technology for Food Property, Quality and Safety Assessment*; Lu, R., Ed.; Taylor & Francis: Boca Raton, FL, USA, 2016; pp. 387–428.
78. Skolik, P.; Morais, C.L.M.; Martin, F.L.; McAinsh, M.R. Determination of developmental and ripening stages of whole tomato fruit using portable infrared spectroscopy and Chemometrics. *BMC Plant Biol.* **2019**, *19*, 236. [CrossRef]
79. Chamuah, N.; Hazarika, A.; Hatiboruah, D.; Nath, P. SERS on paper: An extremely low cost technique to measure Raman signal. *J. Phys. D Appl. Phys.* **2017**, *50*, 485601. [CrossRef]
80. Piot, O.; Autran, J.C.; Manfait, M. Spatial Distribution of Protein and Phenolic Constituents in Wheat Grain as Probed by Cnfocal Raman Microscopy. *J. Cereal Sci.* **2000**, *32*, 57–71. [CrossRef]

81. Chylińska, M.; Szymańska-Chargot, M.; Deryło, K.; Tchórzewska, D.; Zdunek, A. Changing of biochemical parameters and cell wall polysaccharides distribution during physiological development of tomato fruit. *Plant Physiol. Biochem.* **2017**, *119*, 328–337. [CrossRef]
82. López-Sánchez, M.; Ayora-Cañada, M.J.; Molina-Díaz, A. Olive Fruit Growth and Ripening as Seen by Vibrational Spectroscopy. *J. Agric. Food Chem.* **2010**, *58*, 82–87. [CrossRef]
83. Szymańska-Chargot, M.; Chylińska, M.; Pieczywek, P.M.; Rösch, P.; Schmitt, M.; Popp, J.; Zdunek, A. Raman imaging of changes in the polysaccharides distribution in the cell wall during apple fruit development and senescence. *Planta* **2016**, *243*, 935–945. [CrossRef]
84. Qin, J.; Chao, K.; Kim, M.S. Nondestructive evaluation of internal maturity of tomatoes using spatially offset Raman spectroscopy. *Postharvest Biol. Technol.* **2012**, *71*, 21–31. [CrossRef]
85. Martin, D.; Gonzalvez, A.G.; Medina, R.M.; González Ureña, A. Modeling Tomato Ripening Based on Carotenoid Raman Spectroscopy: Experimental Versus Kinetic Model. *Appl. Spectrosc.* **2017**, *71*, 1310–1320. [CrossRef] [PubMed]
86. Legner, R.; Voigt, M.; Servatius, C.; Klein, J.; Hambitzer, A.; Jaeger, M. A Four-Level Maturity Index for Hot Peppers (Capsicum annum) Using Non-Invasive Automated Mobile Raman Spectroscopy for On-Site Testing. *Appl. Sci.* **2021**, *11*, 1614. [CrossRef]
87. Cabrales, L.; Abidi, N.; Manciu, F. Characterization of Developing Cotton Fibers by Confocal Raman Microscopy. *Fibers* **2014**, *2*, 285–294. [CrossRef]
88. Prats-Mateu, B.; Felhofer, M.; de Juan, A.; Gierlinger, N. Multivariate unmixing approaches on Raman images of plant cell walls: New insights or overinterpretation of results? *Plant Methods* **2018**, *14*, 52. [CrossRef]
89. Zhu, N.; Wu, D.; Chen, K. Label-free visualization of fruit lignification: Raman molecular imaging of loquat lignified cells. *Plant Methods* **2018**, *14*, 58. [CrossRef]
90. Nikbakht, A.M.; Hashjin, T.T.; Malekfar, R.; Gobadian, B. Nondestructive determination of tomato fruit quality parameters using Raman spectroscopy. *J. Agric. Sci. Technol.* **2011**, *13*, 517–526.
91. Morey, R.; Ermolenkov, A.; Payne, W.Z.; Scheuring, D.C.; Koym, J.W.; Vales, M.I.; Kurouski, D. Non-invasive identification of potato varieties and prediction of the origin of tuber cultivation using spatially offset Raman spectroscopy. *Anal. Bioanal. Chem.* **2020**, *412*, 4585–4594. [CrossRef]
92. Abreu, G.F.; Borem, F.M.; Oliveira, L.F.C.; Almeida, M.R.; Alves, A.P.C. Raman spectroscopy: A new strategy for monitoring the quality of green coffee beans during storage. *Food Chem.* **2019**, *287*, 241–248. [CrossRef]
93. Feng, X.; Zhang, Q.; Zhu, Z. Rapid classification of citrus fruits based on Raman spectroscopy and pattern recognition techniques. *Food Sci. Technol. Res.* **2013**, *19*, 1077–1084. [CrossRef]
94. Figueiredo, L.P.; Borem, F.M.; Almeida, M.R.; Oliveira, L.F.C.; Alves, A.P.C.; Santos, C.M.D.; Rios, P.A. Raman spectroscopy for the differentiation of Arabic coffee genotypes. *Food Chem.* **2019**, *288*, 262–267. [CrossRef]
95. Keidel, A.; von Stetten, D.; Rodrigues, C.; Maguas, C.; Hildebrandt, P. Discrimination of green arabica and Robusta coffee beans by Raman spectroscopy. *J. Agric. Food Chem.* **2010**, *58*, 11187–11192. [CrossRef]
96. Atalla, R.H.; Agarwal, U.P. Recording Raman spectra from plant cell walls. *J. Raman Spectrosc.* **1986**, *17*, 229–231. [CrossRef]
97. Pecinar, I.M. Raman Microscopy in Plant Science, Carotenoids Detection in Fruit Material. In *Application of Molecular Methods and Raman Microscopy/Spectroscopy in Agricultural Sciences and Food Technology*; Ubiquity Press: London, UK, 2019; pp. 177–186.
98. Agarwal, U.P. 1064 nm FT- Raman spectroscopy for investigations of plant cell walls and other biomass materials. *Front. Plant Sci.* **2014**, *5*, 1–12. [CrossRef]
99. Parker, S.F. A review of the theory of Fourier-transform Raman spectroscopy. Spectrochim. *Acta Part A Mol. Spectrosc.* **1994**, *50*, 1841–1856. [CrossRef]
100. Xie, Y.F.; Xu, L.; Wang, Y.Q.; Shao, J.D.; Wang, L.; Wang, H.Y.; Qian, H.; Yao, W.R. Label-free detection of the foodborne pathogens of Enterobacteriaceae by surface-enhanced Raman spectroscopy. *Anal. Methods.* **2013**, *5*, 946–952. [CrossRef]
101. Chan, J.W.; Esposito, A.P.; Talley, C.E.; Hollars, C.W.; Lane, S.M.; Huser, T. Reagentless identification of single bacterial spores in aqueous solution by confocal laser tweezers Raman spectroscopy. *Anal. Chem.* **2004**, *76*, 599–603. [CrossRef]
102. Cheng, J.X.; Volkmer, A.; Lewis, D.B.; Xie, X.S. Multiplex coherent anti-stokes Raman scattering microspectroscopy and study of lipid vesicles. *Phys. Chem. B* **2002**, *106*, 8493–8498. [CrossRef]
103. Cardona, M. *Light Scattering in Solid*; Springer: Berlin, Germany, 1975; p. 79.
104. Sugar, J.; Bour, P. Quantitative analysis of sugar composition in honey using 532-nm excitation Raman and Raman optical activity spectra. *J. Raman Spectrosc.* **2016**, *47*, 1298–1303. [CrossRef]
105. Gierlinger, N. New insights into plant cell walls by vibrational micro-spectroscopy. *Appl. Spectrosc. Rev.* **2017**, *53*, 517–551. [CrossRef]
106. Horbens, M.; Feldner, A.; Höfer, M.; Neinhuis, C. Ontogenetic tissue modification in Malus fruit peduncles: The role of sclereids. *Ann. Bot.* **2014**, *113*, 105–118. [CrossRef]
107. De Meester, B.; de Vries, L.; Özparpucu, M.; Gierlinger, N.; Corneillie, S.; Pallidis, A.; Goeminne, G.; Morreel, K.; De Bruyne, M.; De Rycke, R.; et al. Vessel-specific reintroduction of CINNAMOYL-COA REDUCTASE1 (CCR1) in dwarfed ccr1 mutants restores vessel and xylary fiber integrity and increases biomass. *Plant Physiol.* **2018**, *176*, 611–633. [CrossRef]
108. Jin, K.; Liu, X.; Wang, K.; Jiang, Z.; Tian, G.; Yang, S.; Shang, L.; Ma, J. Imaging the dynamic deposition of cell wall polymer in xylem and phloem in Populus × euramericana. *Planta* **2018**, *248*, 849–858. [CrossRef]

109. Dinant, S.; Wolff, N.; De Marco, F.; Vilaine, F.; Gissot, L.; Aubry, E.; Sandt, C.; Bellini, C.; Le Hir, R. Synchrotron FTIR and Raman spectroscopy provide unique spectral fingerprints for Arabidopsis floral stem vascular tissues. *J. Exp. Bot.* **2019**, *70*, 871–884. [CrossRef]
110. Borowska-Wykręt, D.; Dulski, M. Raman Spectroscopy in Nonwoody Plants. Methods in Molecular Biology (book series MIMB, 1992). In *Plant Cell Morphogenesis*; Humana: New York, NY, USA, 2019; pp. 83–107.
111. Gierlinger, N.; Keplinger, T.; Harrington, M. Imaging of plant cell walls by confocal Raman microscopy. *Nat. Protoc.* **2012**, *7*, 1694–1708. [CrossRef]
112. Prats Mateu, B.; Hauser, M.T.; Heredia, A.; Gierlinger, N. Waterproofing in Arabidopsis: Following Phenolics and Lipids In situ by Confocal Raman Microscopy. *Front. Chem.* **2016**, *4*, 10. [CrossRef]

MDPI
St. Alban-Anlage 66
4052 Basel
Switzerland
Tel. +41 61 683 77 34
Fax +41 61 302 89 18
www.mdpi.com

Molecules Editorial Office
E-mail: molecules@mdpi.com
www.mdpi.com/journal/molecules

www.ingramcontent.com/pod-product-compliance
Lightning Source LLC
Chambersburg PA
CBHW061849210926
51850CB00006B/1010
9783036582313